FUZZY PREFERENCE MODELLING AND MULTICRITERIA DECISION SUPPORT

THEORY AND DECISION LIBRARY

General Editors: W. Leinfellner (*Vienna*) and G. Eberlein (*Munich*)

Series A: Philosophy and Methodology of the Social Sciences

Series B: Mathematical and Statistical Methods

Series C: Game Theory, Mathematical Programming and Operations Research

Series D: System Theory, Knowledge Engineering and Problem Solving

SERIES D: SYSTEM THEORY, KNOWLEDGE ENGINEERING AND PROBLEM SOLVING

VOLUME 14

Scope: Design, study and development of structures, organizations and systems aimed at formal applications mainly in the social and human sciences but also relevant to the information sciences. Within these bounds three types of study are of particular interest. First, formal definition and development of fundamental theory and/or methodology, second, computational and/or algorithmic implementations and third, comprehensive empirical studies, observation or case studies. Although submissions of edited collections will appear occasionally, primarily monographs will be considered for publication in the series. To emphasize the changing nature of the fields of interest we refrain from giving a clear delineation and exhaustive list of topics. However, certainly included are: artificial intelligence (including machine learning, expert and knowledge based systems approaches), information systems (particularly decision support systems), approximate reasoning (including fuzzy approaches and reasoning under uncertainty), knowledge acquisition and representation, modeling, diagnosis, and control.

The titles published in this series are listed at the end of this volume.

FUZZY PREFERENCE MODELLING AND MULTICRITERIA DECISION SUPPORT

by

JÁNOS FODOR

Eötvös Loránd University,
Budapest, Hungary

and

MARC ROUBENS

University of Liège,
Liège, Belgium

SPRINGER-SCIENCE+BUSINESS MEDIA, B.V.

A C.I.P. Catalogue record for this book is available from the Library of Congress.

ISBN 978-90-481-4466-2 ISBN 978-94-017-1648-2 (eBook)
DOI 10.1007/978-94-017-1648-2

Printed on acid-free paper

Dedicated to Kinga and Andris

In memory of my mother
Irèn Kereczman-Roubens

Il y a diverses sortes de curiosités : l'une d'intérêt, qui nous porte à désirer d'apprendre ce qui nous peut être utile et l'autre d'orgueil, qui vient du désir de savoir ce que les autres ignorent.

La Rochefoucauld

Contents

Foreword

The encounter, in the late seventies, between the theory of triangular norms, issuing from stochastic geometry, especially the works of Menger, Schweizer and Sklar, on the one hand, and the theory of fuzzy sets due to Zadeh, on the other hand has been very fruitful. Triangular norms have proved to be ready-made mathematical models of fuzzy set intersections and have shed light on the algebraic foundations of fuzzy sets. One basic idea behind the study of triangular norms is to solve functional equations that stem from prescribed axioms describing algebraic properties such as associativity. Alternative operations such as means have been characterized in a similar way by Kolmogorov, for instance, and the methods for solving functional equations are now well established thanks to the efforts of Aczel, among others. One can say without overstatement that the introduction of triangular norms in fuzzy sets has strongly influenced further developments in fuzzy set theory, and has significantly contributed to its better acceptance in pure and applied mathematics circles.

The book by Fodor and Roubens systematically exploits the benefits of this encounter in the analysis of fuzzy relations. The authors apply functional equation methods to notions such as equivalence relations, and various kinds of orderings, for the purpose of preference modelling. Central to this book is the multivalued extension of the well-known result claiming that any relation expressing weak preference can be separated into three components respectively describing strict preference, indifference and incomparability. This result plays a basic role in the European school of decision analysis which has put forward the qualitative notion of preference relation as a basis for decision-support, as opposed to the American school which founds its methods on the utility function.

To some extent fuzzy relations offer a compromise between utility functions and preference relations. Fuzzy relations, as presented in this book, are numerical, but their power of expressivity is far beyond the one of utility functions since they allow for such phenomena as non-transitivity and incomparability. Especially a contribution of this book is to show that the problem of aggregating preference relations (first studied for voting purposes by eighteenth century French philosophers) and the one of aggregating utility functions as well as combining fuzzy sets membership functions can be done with the same tools. But interestingly the theory of multiattribute utility has put emphasis on trade-off operations such as averages, while fuzzy set theory has focused on multiple-valued counterparts of the logical connectives "and" and "or", and the theory of preference relations is especially known for proving impossibility theorems about aggregation schemes. This book puts everything together in a unified way, at the mathematical level.

This book might become a landmark in the mathematics of valued preference relations because it systematically adopts an axiomatic point of view on several basic problems of preference modelling, aggregation, and scoring. It elaborates on previous works dealing with similarity relations and fuzzy orderings by Zadeh, Orlowski, Ovchinnikov, Valverde, and others. However it also contains a lot of new technical results, unpublished elsewhere, on crucial issues such as the decomposition of valued preference structures, scale-invariance properties of numerical aggregation functions, the representation of fuzzy orderings and similarity, and so on.... The European school of decision analysis, as exemplified by the works of Bernard Roy, has traditionally been very skeptical about the practical relevance of utility-based approaches in real-world decision problems. However the latter have very sound mathematical foundations that relational methods were lacking, especially when valued preference had to be used. This book is a major step towards equipping the European school of decision analysis with sound mathematical underpinnings. The task is far from being finished: if you have your favorite properties for modelling and aggregating valued preferences, this book gives you the appropriate tool that will satisfy these properties. But only preliminary investigations are reported on another crucial problem, that is, how do you get the numbers. Future research might consider fuzzy relations from a measurement -theoretic point of view, a topic where utility approaches have also accumulated a lot of results. This leads to the question of the operational semantics of valued preference relations: here a crucial distinction should be made between intensity of preference between two perfectly described objects, and uncertainty about binary preference, due to the lack of precision in the description of objects. This question is briefly considered in chapter 7 but it opens new avenues of research. Especially, while much results exist on the links between valued preference and probability, very little work has been done pertaining to non-additive uncertainty theories, such as possibility theory, belief functions and the like, that may shed light on the semantics of some kinds of fuzzy preference relations, whose mathematical properties are described here.

Whatever developments are witnessed to-morrow in these directions, they will have to use the present book where a lot can be found about which procedures are mathematically meaningful and which are inconsistent, regardless of the considered semantics. This book is a new proof, if still needed, that the mathematics of fuzzy sets are precise, and that the acceptance of fuzzy sets by other research communities requires that fuzzy-set-based methods be developed in the setting of established alternative methodologies, with a view to unifying a field of investigation.

Didier Dubois Henri Prade
Toulouse, July 26, 1994

Introduction

Il y a diverses sortes de curiosités : l'une d'intérêt, qui nous porte à désirer d'apprendre ce qui nous peut être utile et l'autre d'orgueil, qui vient du désir de savoir ce que les autres ignorent.

La Rochefoucauld

The main purpose of this book is to present an axiomatical introduction to the concepts and procedures commonly used in multicriteria decision aid when the technique used is based on pairwise comparison of alternatives for every criterion (attribute).

The procedures are designed for structures of mixed qualitative-quantitative data. Each attribute can be represented by scales of different types : numeric, ordinal or linguistic, using the theory of fuzzy sets.

It is supposed that the importance parameters or weights linked to the criteria are independent of the scale of the attributes. Moreover the weights' elicitation is not described or discussed in this work.

The procedures for decision aid can be classified into two phases : aggregation and exploitation.

The aggregation process corresponds to the operation which transforms the marginal evaluations into a global outranking relation between every pair of alternatives a and b, giving the degree to which a is estimated to be not worse than b.

The outranking relation will not generally be transitive nor complete and it is interesting to define the degrees of indifference and incomparability of every pair of potential actions.

The exploitation process deals with the outranking relation in order to clarify the decision through a partial preordering reflecting some of the irreducible indifferences and incomparabilities. The main point consists in finding reasonable ways of dealing with the intransitivities without losing too much of the contents of the outranking relation.

Chapter 1 discusses basic concepts and models for logical connectives as AND, OR and NOT in the case where the valuation set is the unit-interval [0,1]. Representation of conjunctions, disjunctions, negations, implications and equivalences are emphasized.

Chapter 2 considers the valued binary relations. Traces, cut relations and transitive relations are defined and main properties like reflexivity, irreflexivity, symmetry, completeness, transitivity, semitransitivity, Ferrers and linearity are studied.

Chapter 3 deals with the problem of the axiomatical definition of strict preference, indifference and incomparability.

Chapter 4 is concerned with similarity relations and valued orders. These structures are linked to the Hasse diagrams and the partition trees.

Chapter 5 considers the operators $M(x_1,\ldots,x_m)$ which transform m elements $(x_1,\ldots,x_m)$ – say some marginal binary relations – into a unique value according to several criteria like monotonicity, consensus properties ($M(0,\ldots,0)=0,\ M(1,\ldots,1)=1$) and domain of variation (M should lie between complete conjunctiveness min and complete disjunctiveness max). These averaging operators present two important subclasses : the generalized weighted means and the generalized medians. Both correspond to fuzzy integrals.

Chapter 6 presents the axiomatical properties of some exploitation techniques called scoring procedures. The outranking relations are introduced in a scoring function which transforms the global relations into one function related to every alternative. It becomes obvious to order the alternatives according to these scores. Procedures like net flow scores and min scores and characterized.

Chapter 7 proposes different ways to build the marginal binary relations when the information given to the decision maker is expressed in terms of imprecise values (linguistic, nominal variables) of the attributes. Starting from these marginal valued binary relations some aggregation and exploitation processes are proposed with reference to the previous chapters.

Chapter 8 summarizes the most important results obtained in the previous chapters.

Parts of the contents of this book were developed with the help of grants from the scientific agreements between the French Community of Belgium and Hungary, from the Peregrinatio II Foundation (ELTE) and from the Hungarian Scientific Research Foundation. This assistance is gratefully acknowledged.

In preparing this book, we have benefited from the help of many people. We want especially to thank Didier Dubois, Michel Grabisch, Jean-Luc Marichal and Sergei Ovchinnikov.

For the task of typing the manuscript, we are indebted to the high qualification and incredible patience of Mrs. Nicole Dumont from the Institute of Mathematics at the University of Liège.

Finally, we must mention the invaluable help of our wives Joli and Michèle who gave up countless evenings and week-ends during the writing of this book.

Janos Fodor
Marc Roubens
July 1994
Solymár, Hungary
Waterloo, Belgium

Chapter 1

Fuzzy logical connectives

In order to be able to work with valued concepts (especially, relations in subsequent chapters) we provide a rather complete and up to date survey of connectives used in fuzzy set theory. This approach to the foundations is only one possible way (for others see Höhle and Stout (1991)).

What we offer is a sound mathematical base that, at the same time, can be useful for practicioners who apply the theory in different fields such as decision-making, approximate reasoning, fuzzy control and others.

All the introduced connectives are justified on an axiomatic basis, to avoid ad hoc constructions. A special emphasis is given on interrelations between different operator classes (such as conjunctions and implications, implications and negations, etc).

1.1 The need of fuzzy logics

Let Ω be a given set. In classical set theory intersection, union and complement of subsets of Ω are defined in a unique way. If $A, B \subseteq \Omega$ are crisp sets then

$$\begin{aligned} A \cap B &= \{a \mid a \in A \quad \textbf{AND} \quad a \in B\}, \\ A \cup B &= \{a \mid a \in A \quad \textbf{OR} \quad a \in B\}, \\ A^c &= \{a \mid \textbf{NOT} \quad a \in A\}. \end{aligned}$$

The uniqueness is due to the fact that **AND, OR** and **NOT** are two-valued logical operations. For example, if P_1 and P_2 are propositions being either **TRUE** or **FALSE** then P_1 **AND** P_2 is **TRUE** if and only if both P_1 and P_2 are **TRUE**.

We can use a *valuation* of any proposition P as follows:

$$\begin{aligned} v(P) = 1 \quad &\text{if and only if} \quad P \quad \text{is } \textbf{TRUE}, \\ v(P) = 0 \quad &\text{if and only if} \quad P \quad \text{is } \textbf{FALSE}. \end{aligned}$$

By this valuation, set-theoretic operations can be represented by

$$\begin{aligned} v(a \in A \cap B) &= v(a \in A) \wedge v(a \in B), \\ v(a \in A \cup B) &= v(a \in A) \vee v(a \in B), \\ v(a \in A^c) &= \neg v(a \in A), \end{aligned}$$

where

$$\begin{aligned} 0 \wedge 0 &= 0 \wedge 1 = 1 \wedge 0 = 0, \ 1 \wedge 1 = 1, \\ 0 \vee 1 &= 1 \vee 0 = 1 \vee 1 = 1, \ 0 \vee 0 = 0, \\ \neg 0 &= 1, \quad \neg 1 = 0. \end{aligned}$$

However, if for each element $a \in \Omega$ there is a grade of membership belonging to $A \subseteq \Omega$, that is, $v(a \in A)$ is a number from [0,1] rather than from $\{0, 1\}$, then the interpretation of logical connectives is not so obvious.

The aim of this chapter is to introduce appropriate extensions of logical connectives $\wedge, \vee$ and $\neg$ in the case when the valuation set is the unit interval [0,1]. This framework naturally leads us to extensions of set-theoretic operations when proposition '$a \in A$' has a grade of membership, i.e., when $v(a \in A)$ could be any number from [0,1]. This situation is dealt with by *fuzzy set theory* (Zadeh (1965)) where the basic idea is to use a function $\mu_A : \Omega \to [0, 1]$ (membership function) instead of the characteristic function $\chi_A : \Omega \to \{0, 1\}$. The valuation set [0,1] is choosen for convenience and for having applications in sight. Of course, it is possible to use other valuation sets, e.g. a lattice L instead of [0,1] (Goguen (1967)). In any case, we restrict our investigations to [0,1].

Suppose that one would like to extend operations $\wedge, \vee$ and $\neg$ from the set $\{0, 1\}$ to the closed unit interval [0,1] as τ, σ and ν, respectively. Then, of course, we must have

$$\wedge = \tau \Big|_{\{0,1\}}, \qquad \vee = \sigma \Big|_{\{0,1\}}, \qquad \neg = \nu \Big|_{\{0,1\}}, \tag{1.1}$$

that is, the restriction of each extended operation to $\{0, 1\}$ must coincide with the respective crisp operation. On the other hand, [0,1] must be closed under the extended operations, i.e.,

$$\tau(x, y), \sigma(x, y), \nu(x) \in [0, 1] \quad \forall x, y \in [0, 1]. \tag{1.2}$$

Through this chapter these two conditions are supposed to be satisfied.

Now we turn to the axiomatic extensions of fuzzy logical operations. In the rest of this chapter $P_1, P_2, \ldots$ denote propositions with truth values $v(P_1), v(P_2), \ldots$, respectively, with $v(P_i) \in [0, 1], \quad i = 1, 2, \ldots$.

1.2 Negations

Negation is a unary operation that identifies the truth value of **NOT** P for a proposition P. The following axioms seem to be acceptable as minimal requirements in the present setting.

N0. $v(\mathbf{NOT}P_1)$ depends only on $v(P_1)$.

N1. If $v(P_1) = 1$ then $v(\mathbf{NOT}P_1) = 0$.

N2. If $v(P_1) = 0$ then $v(\mathbf{NOT}P_1) = 1$.

N3. If $v(P_1) \leq v(P_2)$ then $v(\mathbf{NOT}P_1) \geq v(\mathbf{NOT}P_2)$.

Axiom N0 means that negation can be defined pointwisely. N1 and N2 require the coincidence with Boolean negation, see (1.1). By N3, the value of **NOT** P should not become greater when the value of P increases.

According to these axioms, there exists a function $n : [0,1] \to [0,1]$ such that

$$n(0) = 1, \quad n(1) = 0 \tag{1.3}$$

and

$$n \quad \text{is nonincreasing.} \tag{1.4}$$

Definition 1.1 A function $n : [0,1] \to [0,1]$ satisfying conditions (1.3) and (1.4) is called a *negation.*

Therefore, a negation is any nonincreasing function from $[0,1]$ to $[0,1]$ such that the classical properties $n(0) = 1$, $n(1) = 0$ are preserved. Both for theoretical and practical purposes, this class of functions should be tightened. To do so, one can modify N3 and choose additional axioms as follows:

N3'. If $v(P_1) < v(P_2)$ then $v(\mathbf{NOT}P_1) > v(\mathbf{NOT}P_2)$.

N4. $v(\mathbf{NOT}P_1)$ depends continuously on $v(P_1)$.

N5. $v(\mathbf{NOT}(\mathbf{NOT}P_1)) = v(P_1)$.

N3' yields that the value of **NOT** P should become smaller when the value of P increases. N4 excludes any chaotic reaction to a small change of the value of P. N5 means that negation is involutive.

Translating these axioms by the function n, we obtain the following conditions:

$$n \quad \text{is strictly decreasing,} \tag{1.5}$$

$$n \quad \text{is continuous,} \tag{1.6}$$

$$n(n(x)) = x \quad \text{for all} \quad x \in [0,1]. \tag{1.7}$$

Definition 1.2 A negation is called *strict* if it satisfies (1.5) and (1.6). A strict negation is *strong* if (1.7) also holds.

Since a strict negation n is a strictly increasing and continuous function, its inverse n^{-1} is also a strict negation, generally different from n. Obviously, we have $n^{-1} = n$ if and only if n is involutive: $n(n(x)) = x$ holds for all $x \in [0,1]$. This means that the graph of the function n is symmetric with respect to the line $\{(x,y) \mid x = y\}$.

Another important property of a strict negation n is that there exists a unique value $0 < \nu < 1$ such that $n(\nu) = \nu$. Then we also have $n^{-1}(\nu) = \nu$.

Before going on, let us see some examples for different types of negations. The first one is said to be the *intuitionistic* negation and defined by

$$n_i(x) = \begin{cases} 1 & \text{if } x = 0 \\ 0 & \text{if } x > 0 \end{cases},$$

see Yager (1980). By duality, we can define the *dual intiutionistic* negation by

$$n_{di}(x) = \begin{cases} 1 & \text{if } x < 1 \\ 0 & \text{if } x = 1 \end{cases},$$

see Ovchinnikov (1983). Obviously, these are not strict negations.

The standard negation is simply

$$N(x) = 1 - x.$$

This is strong.

An example of strict but not strong negation can be given by

$$n(x) = 1 - x^2.$$

Finally, a family of strong negations including the standard one is defined as follows (see Sugeno (1977) under the name λ-complement):

$$N_\lambda(x) = \frac{1 - x}{1 + \lambda x}, \quad \lambda > -1.$$

It is easy to see that for any negation n we have

$$n_i \leq n \leq n_{di}.$$

In general, n denotes a strict negation and N means a strong negation in the sequel.

1.2.1 Representation theorems for negations

The representation theorem of strong negations was obtained by Trillas (1979). We cite here that result in a slightly modified form which is more suitable in the sequel, see Ovchinnikov and Roubens (1991). First we need the definition of an *automorphism* of a real interval $[a, b] \subset \mathbb{R}$. This notion will be used extensively in the book, especially in the case of the unit interval $[0, 1]$.

Definition 1.3 A continuous, strictly increasing function $\varphi : [a, b] \to [a, b]$ with boundary conditions $\varphi(a) = a, \varphi(b) = b$ is called an *automorphism* of the interval $[a, b] \subset \mathbb{R}$.

Theorem 1.1 *A function $N : [0, 1] \to [0, 1]$ is a strong negation if and only if there exists an automorphism φ of the unit interval such that*

$$N(x) = N_\varphi(x) := \varphi^{-1}(1 - \varphi(x)). \tag{1.8}$$

In this case $N_\varphi(x)$ is called a *φ-transform* of the standard negation $1 - x$. Clearly, this representation is not unique. The following statement can easily be proved.

Proposition 1.1 *Suppose that N is a strong negation. Then we have*

$$N(x) = \varphi^{-1}(1 - \varphi(x)) = \psi^{-1}(1 - \psi(x))$$

with automorphisms φ, ψ of the unit interval if and only if there exists an automorphism η of the interval $[-\frac{1}{2}, \frac{1}{2}]$ such that

$$\psi(x) = \frac{1}{2} + \eta\left(\varphi(x) - \frac{1}{2}\right).$$

Proof. Suppose we have $\varphi^{-1}(1-\varphi(x)) = \psi^{-1}(1-\psi(x))$. Introducing $y = \varphi(x)$, this equality holds if and only if

$$\psi \circ \varphi^{-1}(1 - y) = 1 - \psi \circ \varphi^{-1}(y).$$

Define a function $\eta : [-\frac{1}{2}, \frac{1}{2}] \to [-\frac{1}{2}, \frac{1}{2}]$ by

$$\eta(z) = \psi \circ \varphi^{-1}\left(z + \frac{1}{2}\right) - \frac{1}{2}.$$

It is easily seen that thus defined η is an automorphism of $[-\frac{1}{2}, \frac{1}{2}]$. Moreover, we have

$$\psi \circ \varphi^{-1}(y) = \frac{1}{2} + \eta\left(y - \frac{1}{2}\right),$$

and finally

$$\psi(x) = \frac{1}{2} + \eta\left(\varphi(x) - \frac{1}{2}\right).$$

The converse statement is obviously true.

■

An extension of the representation result for strict negations was given by Fodor (1993b) as follows.

Theorem 1.2 *A function $n : [0, 1] \to [0, 1]$ is a strict negation if and only if there exists two automorphisms φ and ψ of the unit interval such that*

$$n(x) = \psi(1 - \varphi(x)). \tag{1.9}$$

1.2.2 A more general approach

For sake of completeness, it is worth mentioning that there is a more general approach to negations. In that case it is assumed that the negation depends also on the point $a \in \Omega$, i.e.,

$$v(\mathbf{NOT}a \in A) = n_a(v(a \in A)),$$

see Klement (1981a), Lowen (1978) and Ovchinnikov (1980, 1983). An interesting problem is to find conditions under which n_a is independent of $a \in \Omega$. One answer is given by using category theory, see Klement (1981a) and Lowen (1978). However, in the sequel we restrict ourselves to axiom N1, i.e., we assume that negations are independent of $a \in \Omega$.

1.2.3 Complement of a fuzzy subset

Suppose that Ω is a given (crisp) set. Then, a fuzzy subset A of Ω is defined by

$$\{(a, \mu_A(a)) \mid a \in \Omega\},$$

where μ_A is a function from Ω to the unit interval, called the *membership function* of A. For a particular element $a \in \Omega$, $\mu_A(a)$ is a degree to which a belongs to A. Clearly, the membership function is the extension of the classical *characteristic function* of a subset of Ω. In the sequel we use the same symbol (capital letter) to denote fuzzy subsets and membership functions.

By the outlined approach to negations, we can define the *complement* A^c of a fuzzy subset A of Ω as follows. Suppose that n is a negation. Then

$$A^c(a) = n(A(a)) \quad \text{for all } a \in \Omega. \tag{1.10}$$

1.3 Conjunctions

Conjunctions serve as a basis for defining intersections of fuzzy subsets of a given set. Taking into account classical properties, the following axioms are reasonable to be expected:

C0. $v(P_1\mathbf{AND}P_2)$ depends only on the values $v(P_1)$ and $v(P_2)$.

C1. If $v(P_1) = 1$ then $v(P_1\mathbf{AND}P_2) = v(P_2)$ for any proposition P_2.

C2. $v(P_1\mathbf{AND}P_2) = v(P_2\mathbf{AND}P_1)$.

C3. If $v(P_1) \leq v(P_2)$ then $v(P_1\mathbf{AND}P_3) \leq v(P_2\mathbf{AND}P_3)$ for any P_3.

C4. $v(P_1\mathbf{AND}(P_2\mathbf{AND}P_3)) = v((P_1\mathbf{AND}P_2)\mathbf{AND}P_3)$.

C0 means that a fuzzy conjunction is a pointwise operation: there exists a function $T : [0,1]^2 \to [0,1]$ such that

$$v(P_1\mathbf{AND}P_2) = T(v(P_1), v(P_2)).$$

C1 says that the value of the conjunction of any proposition and tautology is equal to the value of the proposition. C2 expresses that conjunction is independent of the order of the two propositions. C3 declares that the value of the conjunction cannot decrease when the value of one proposition increases. C4 asserts that one can extend conjunction for three argument in a unique way.

Using functional form of **AND**, the above axioms can be expressed as follows.

Axiom C1 is equivalent to condition

$$T(1, x) = x \quad \text{for all} \quad x \in [0,1]. \tag{1.11}$$

Axiom C2 implies that T is commutative, i.e.,

$$T(x, y) = T(y, x) \quad \text{for all} \quad x, y \in [0,1]. \tag{1.12}$$

By Axiom C3, T is nondecreasing in both arguments. That is,

$$T(x,y) \le T(u,v) \quad \text{for any} \quad 0 \le x \le u \le 1, 0 \le y \le v \le 1. \tag{1.13}$$

Finally, Axiom C4 means that T is associative, i.e.,

$$T(x, T(y,z)) = T(T(x,y), z) \quad \text{for all} \quad x, y, z \in [0,1]. \tag{1.14}$$

Definition 1.4 A function $T : [0,1]^2 \to [0,1]$ is a *triangular norm* (*t-norm* for short) if and only if it satisfies conditions (1.11)–(1.14).

Remark that (1.11) and (1.13) together imply

$$T(0,x) = 0 \quad \text{for all} \quad x \in [0,1].$$

Moreover, the extension of a t-norm for more than two arguments is unique since associativity is satisfied.

From an algebraic point of view, T is a semigroup operation on [0,1] with identity 1. It is worth noting that the class of t-norms was introduced in the theory of statistical (probabilistic) metric spaces as a tool for generalizing the classical triangular inequality by Menger in 1942 (see Menger (1979) and Schweizer and Sklar (1983)).

There are several examples of t-norms. We list here the most frequently used and important ones.

In his classical paper Zadeh (1965) proposed to use

$$\min(x,y)$$

as a conjunction. By (1.11) and (1.13) one can see immediately that min is the greatest t-norm. There are several axiomatic justifications of min as the only possible conjunction operation, see e.g. Alsina et al. (1983), Bellman and Giertz (1973), Bellman and Zadeh (1977), Fung and Fu (1975), Hamacher (1978), Klement (1981b).

Another example is given by the product

$$\Pi(x,y) = xy.$$

The t-norm introduced by Łukasiewicz in 1931 (see Łukasiewicz (1970)) will play a key role in several chapters of this book and is defined by

$$W(x,y) = \max\{x + y - 1, 0\}.$$

A new t-norm was found by Fodor (1993c), called *nilpotent minimum.* This is given by the following formula:

$$\min_0(x,y) = \begin{cases} \min(x,y) & \text{if } x + y > 1 \\ 0 & \text{otherwise} \end{cases}$$

Derivation of this t-norm is explained in details in Fodor (1993c) concerning contrapositive symmetry of fuzzy implications (see also Section 1.8).

Finally, the weakest t-norm can be defined by using condition (1.11) as follows:

$$Z(x,y) = \begin{cases} \min(x,y) & \text{if } \max(x,y) = 1 \\ 0 & \text{otherwise} \end{cases}$$

Thus, for any t-norm T we have

$$Z \leq T \leq \min .$$

Moreover, $W \leq \Pi$ and $W \leq \min_0$, but Π and $\min_0$ are not comparable in this sense. Indeed, we have

$$\Pi(0.5, 0.6) = 0.3 < 0.5 = \min_0(0.5, 0.6)$$

and

$$\Pi(0.2, 0.3) = 0.06 > 0 = \min_0(0.2, 0.3).$$

One of our main goal in this section is to obtain representation theorems for some t-norms. To do this, we need the following definitions.

Definition 1.5 A t-norm T is said to be

(a) *continuous* if T as a function is continuous on the unit interval;

(b) *Archimedean* if $T(x,x) < x$ for all $x \in (0,1)$.

The t-norm min is continuous but not Archimedean. Z is Archimedean but not continuous. Π and W are continuous and Archimedean. $\min_0$ is neither Archimedean nor continuous. However, it is left-continuous.

1.3.1 Representation of continuous Archimedean t-norms

We state here the representation theorem attributed very often to Ling (1965). In fact, her main theorem can be deduced from previously known results on topological semigroups, see Faucett (1955), Mostert and Shield (1957) and Paalman – De Miranda (1964). Nevertheless, the advantage of Ling's approach is twofold: treating two different cases in a unified manner and establishing elementary proofs.

Theorem 1.3 *A t-norm T is continuous and Archimedean if and only if there exists a strictly decreasing and continuous function $f : [0,1] \to [0,\infty]$ with $f(1) = 0$ such that*

$$T(x,y) = f^{(-1)}(f(x) + f(y)), \tag{1.15}$$

where $f^{(-1)}$ is the pseudoinverse of f defined by

$$f^{(-1)}(x) = \begin{cases} f^{-1}(x) & \text{if } x \leq f(0) \\ 0 & \text{otherwise} \end{cases}$$

Moreover, representation (1.15) is unique up to a positive multiplicative constant.

Definition 1.6 Let T be a continuous Archimedean t-norm. We say that T *is generated by* f if T has representation (1.15). In this case f is said to be an *additive generator* of T.

An additive generator of W is given by $f_W(x) = 1 - x$. Then $f_W^{(-1)}(x) = \max\{1 - x, 0\}$. Π is generated by $f_\Pi(x) = -\log(x)$ with $f_\Pi^{(-1)}(x) = f_\Pi^{-1}(x) = \exp(-x)$.

Notice that Ling (1965) proved also that the minimum cannot be represented in the form

$$\min(x, y) = g(f(x) + f(y)),$$

assuming either f, g to be continuous or strictly decreasing.

However, the weakest t-norm Z has a representation (see Ling (1965))

$$Z(x, y) = g(f(x) + f(y)),$$

for example, with the following functions:

$$f(x) = \begin{cases} 2 - x & \text{if } x < 1 \\ 0 & \text{if } x = 1 \end{cases},$$

$$g(x) = \begin{cases} 1 & \text{if } 0 \le x \le 1 \\ 2 - x & \text{if } 1 \le x \le 2 \\ 0 & \text{otherwise} \end{cases}.$$

Notice that f is decreasing but discontinuous while g is continuous and noninreasing.

Definition 1.7 A t-norm T has *zero divisors* if there exist $x, y \in (0, 1)$ such that $T(x, y) = 0$.

T is said to be *positive* if $x, y > 0$ imply $T(x, y) > 0$.

T is called *strict* if it is a strictly increasing function in each place on $(0, 1)^2$.

It is obvious that min and Π are positive t-norms, while W, $\min_0$ and Z have zero divisors. The only strict t-norm among the previous examples is Π. It is easy to see that a continuous Archimedean t-norm T is positive if and only if it is strict.

Proposition 1.2 *Let T be a continuous Archimedean t-norm with additive generator f. Then*

(a) *T has zero divisors if and only if $f(0) < +\infty$;*

(b) *T is strict if and only if $f(0) = \lim_{x \to 0} f(x) = +\infty$.*

Ling (1965) proved also the following theorem in an elementary and constructive way.

Theorem 1.4 *Every continuous Archimedean t-norm is the limit of a pointwise convergent sequence of continuous strict t-norms.*

1.3.2 Representation of continuous t-norms with zero divisors

Using the general representation theorem of continuous Archimedean t-norms we can give another form of representation for a class of continuous t-norms having zero divisors. More exactly, for continuous t-norms T such that the law of contradiction is valid (that is, $T(x, n(x)) = 0$ holds with a strict negation n for all $x \in [0,1]$). First we need the following lemma (see Ovchinnikov and Roubens (1992)).

Lemma 1.1 *If T is a continuous t-norm such that $T(x, n(x)) = 0$ holds for all $x \in [0,1]$ with a strict negation n then T is Archimedean.*

Proof. Suppose that T is not Archimedean. That is, there exists $x \in (0,1)$ such that $T(x,x) = x$. If $x \le n(x)$ then $x = T(x,x) \le T(x, n(x)) = 0$, a contradiction since $x \in (0,1)$. If $x > n(x)$ then, since T is a continuous function, there exists $y \le x$ such that $n(x) = T(x,y)$. Then we have $n(x) = T(x,y) = T(T(x,x),y) = T(x,T(x,y)) = T(x,n(x)) = 0$, again a contradiction since $x \in (0,1)$. Thus our proposition is proved.

■

The following theorem is established after Ovchinnikov and Roubens (1991).

Theorem 1.5 *A continuous t-norm T is such that $T(x, n(x)) = 0$ holds for all $x \in [0,1]$ with a strict negation n if and only if there exists an automorphism φ of the unit interval such that*

$$T(x,y) = \varphi^{-1}(\max\{\varphi(x) + \varphi(y) - 1, 0\}) \tag{1.16}$$

and

$$n(x) \le \varphi^{-1}(1 - \varphi(x)).$$

Proof. (Necessity) According to the previous proposition, T is Archimedean. Thus there exists a generator f of T such that (1.15) holds with $f(0) < +\infty$. Define

$$\varphi(x) = 1 - \frac{f(x)}{f(0)}. \tag{1.17}$$

Thus defined φ is an automorphism of the unit interval. From (1.17) we have

$$f(x) = f(0) - f(0)\varphi(x)$$

and

$$f^{-1}(x) = \varphi^{-1}\left(1 - \frac{x}{f(0)}\right).$$

Using (1.15) we can go on as

$$\begin{aligned} T(x,y) &= f^{(-1)}(f(x) + f(y)) \\ &= f^{-1}(\min\{f(x) + f(y), f(0)\}) \\ &= \varphi^{-1}\left(1 - \frac{\min\{f(0) - f(0)\varphi(x) + f(0) - f(0)\varphi(y), f(0)\}}{f(0)}\right) \\ &= \varphi^{-1}(\max\{\varphi(x) + \varphi(y) - 1, 0\}). \end{aligned}$$

On the other hand, $T(x, n(x)) = 0$ is equivalent to $\varphi(x) + \varphi(n(x)) \le 1$, whence we obtain the inequality $n(x) \le \varphi^{-1}(1 - \varphi(x))$.

Proof of sufficiency is immediate.

■

1.3.3 Representation of continuous strict t-norms

It is obvious that any strict t-norm is Archimedean since $T(x,x) < T(x,1) = x$ for any $x \in (0,1)$. The following form of representation can be found in Schweizer and Sklar (1983).

Theorem 1.6 *A continuous t-norm T is strict if and only if there exists an automorphism φ of the unit interval such that*

$$T(x,y) = \varphi^{-1}(\varphi(x)\varphi(y)). \tag{1.18}$$

Proof. (Necessity) If T is a continuous, strict t-norm then T admits representation (1.15) with $f(0) = +\infty$, i.e.,

$$T(x,y) = f^{-1}(f(x) + f(y)).$$

Define

$$\varphi(x) = \exp(-f(x)). \tag{1.19}$$

Thus defined φ is an automorphism of the unit interval. From (1.19) we obtain that

$$f(x) = -\log \varphi(x)$$

and

$$f^{-1}(x) = \varphi^{-1}(\exp(-x)).$$

Thus we can conclude that

$$\begin{aligned} T(x,y) &= f^{-1}(f(x) + f(y)) \\ &= \varphi^{-1}(\exp(\log \varphi(x) + \log \varphi(y))) \\ &= \varphi^{-1}(\varphi(x)\varphi(y)). \end{aligned}$$

Proof of sufficiency is obvious.

■

An automorphism φ such that representation (1.18) holds is called a *multiplicative generator* of T.

1.3.4 Classification of continuous t-norms

Suppose that $\{[a_m, b_m]\}$ is a countable family of non-overlapping, closed, proper subintervals of [0,1], denoted by $\mathcal{I}$. With each $[a_m, b_m] \in \mathcal{I}$ associate a continuous Archimedean t-norm T_m. Let T be a function defined on $[0,1]^2$ via

$$T(x,y) = \begin{cases} a_m + (b_m - a_m)T_m\left(\frac{x-a_m}{b_m-a_m}, \frac{y-a_m}{b_m-a_m}\right) & \text{if } (x,y) \in [a_m, b_m]^2 \\ \min(x,y) & \text{otherwise} \end{cases} \tag{1.20}$$

In this case T is called the *ordinal sum* of $\{([a_m, b_m], T_m)\}$ and each T_m is called a *summand.*

It is more or less known but not emphasized enough that for a continuous t-norm T there are three possibilities:

1. $T = \min$;

2. T is Archimedean;

3. there exist a family $\{([a_m, b_m], T_m)\}$ such that T is the ordinal sum of this family,

see Mostert and Shields (1957). A parametric family of t-norms based on ordinal sums is given by Fodor (1991c).

1.3.5 Intersection of fuzzy subsets

Now we are able to define the intersection of fuzzy subsets A, B of a given set Ω. Suppose that T is a t-norm. Then

$$(A \cap_T B)(a) = T(A(a), B(a)) \quad \text{for all } a \in \Omega. \tag{1.21}$$

The choice of a particular t-norm depends on the studied particular problem. There are some cases when we can decide on an axiomatic basis, at least up to an automorphism of the unit interval (see Chapter 3).

1.4 Disjunctions

Axioms for disjunctions are very similar to those emphasized for conjunctions. The only difference is in the boundary condition.

D0. $v(P_1 \mathbf{OR} P_2)$ depends only on the values $v(P_1)$ and $v(P_2)$.

D1. If $v(P_1) = 0$ then $v(P_1 \mathbf{OR} P_2) = v(P_2)$ for any proposition P_2.

D2. $v(P_1 \mathbf{OR} P_2) = v(P_2 \mathbf{OR} P_1)$.

D3. If $v(P_1) \leq v(P_2)$ then $v(P_1 \mathbf{OR} P_3) \leq v(P_2 \mathbf{OR} P_3)$ for any P_3.

D4. $v(P_1 \mathbf{OR} (P_2 \mathbf{OR} P_3)) = v((P_1 \mathbf{OR} P_2) \mathbf{OR} P_3)$.

Axiom D0 means that there exists a function $S : [0,1]^2 \to [0,1]$ such that

$$v(P_1 \mathbf{OR} P_2) = S(v(P_1), v(P_2)).$$

Axiom D1 is equivalent to condition

$$S(0, x) = x \quad \text{for all} \quad x \in [0,1]. \tag{1.22}$$

Axiom D2 implies that S is commutative, i.e.,

$$S(x, y) = S(y, x) \quad \text{for all} \quad x, y \in [0,1]. \tag{1.23}$$

By Axiom D3, S is nondecreasing in both arguments. That is,

$$S(x, y) \leq S(u, v) \quad \text{for any} \quad 0 \leq x \leq u \leq 1, 0 \leq y \leq v \leq 1. \tag{1.24}$$

Finally, Axiom D4 means that S is associative, i.e.,

$$S(x, S(y, z)) = S(S(x, y), z) \quad \text{for all} \quad x, y, z \in [0,1]. \tag{1.25}$$

Definition 1.8 A function $S : [0,1]^2 \to [0,1]$ is a *t-conorm* if and only if it satisfies conditions (1.22)–(1.25).

Remark that (1.22) and (1.24) together imply

$$S(1,x) = 1 \quad \text{for all} \quad x \in [0,1].$$

From an algebraic point of view, S is a semigroup operation on [0,1] with identity 0. On the other hand, it is easy to see that S is a t-conorm if and only if $T(x,y) = nS(nx,ny)$ is a t-norm. According to this association taking n as the standard negation, the following examples can be given:

$$\begin{array}{rcl}
\min(x,y) & \longleftrightarrow & \max(x,y), \\
\Pi(x,y) = xy & \longleftrightarrow & \Pi'(x,y) = x + y - xy, \\
W(x,y) = \max\{x+y-1, 0\} & \longleftrightarrow & W'(x,y) = \min\{x+y, 1\}, \\
\min_0(x,y) & \longleftrightarrow & \max_1(x,y) = \begin{cases} \max(x,y) & \text{if } x+y<1 \\ 1 & \text{otherwise} \end{cases} \\
Z(x,y) & \longleftrightarrow & Z'(x,y) = \begin{cases} \max(x,y) & \text{if } \min(x,y)=0 \\ 1 & \text{otherwise} \end{cases}
\end{array}$$

Obviously, for any t-conorm S we have

$$\max \le S \le Z'.$$

Moreover, $\Pi' \le W'$ and $\max_1 \le W'$ but Π' and $\max_1$ are not comparable.

Definition 1.9 A t-conorm S is said to be

(a) *continuous* if S as a function is continuous on the unit interval;

(b) *Archimedean* if $S(x,x) > x$ for all $x \in (0,1)$.

The t-conorm max is continuous but not Archimedean. Z' is Archimedean but not continuous. Π' and W' are continuous and Archimedean. $\max_1$ is neither Archimedean nor continuous. However, it is right-continuous.

1.4.1 Representation of continuous Archimedean t-conorms

We state here the representation theorem analogue to the one presented for t-norms. For proofs see e.g. Ling (1965).

Theorem 1.7 *A t-conorm S is continuous and Archimedean if and only if there exists a strictly increasing and continuous function $g : [0,1] \to [0,\infty]$ with $g(0) = 0$ such that*

$$S(x,y) = g^{(-1)}(g(x) + g(y)), \tag{1.26}$$

where $g^{(-1)}$ is the pseudoinverse of g defined by

$$g^{(-1)}(x) = \begin{cases} g^{-1}(x) & \text{if } x \le g(1) \\ 1 & \text{otherwise} \end{cases}$$

Moreover, representation (1.26) is unique up to a positive multiplicative constant.

Definition 1.10 Let S be a continuous Archimedean t-norm. We say that S *is generated by* g if S has representation (1.26). In this case g is said to be an *additive generator* of S.

An additive generator of W' is given by $g_{W'}(x) = x$. Then $g_{W'}^{(-1)}(x) = \min\{x, 1\}$. Π' is generated by $g_{\Pi'}(x) = -\log(1-x)$ with $g_{\Pi'}^{(-1)}(x) = g_{\Pi'}^{-1}(x) = 1 - \exp(-x)$.

Definition 1.11 A t-conorm S is *nilpotent* if there exist $x, y \in (0,1)$ such that $S(x,y) = 1$.
S is called *strict* if it is a strictly increasing function in each place on $(0,1)^2$.

It is obvious that W', $\max_1$ and Z' are nilpotent. The only strict t-conorm among the previous examples is Π'.

Proposition 1.3 *Let S be a continuous Archimedean t-norm with additive generator g. Then*

(a) *S is nilpotent if and only if $g(1) < +\infty$;*

(b) *S is strict if and only if $g(1) = \lim_{x\to 1} g(x) = +\infty$.*

1.4.2 Representation of continuous nilpotent t-conorms

Using the general representation theorem of continuous Archimedean t-conorms we can prove another representation of continuous nilpotent t-conorms having property $S(x, n(x)) = 1$ for all $x \in [0,1]$, with a strict negation n.

Lemma 1.2 *If S is a continuous t-conorm such that $S(x, n(x)) = 1$ holds for all $x \in [0,1]$ with a strict negation n then S is Archimedean.*

Proof. Suppose that S is not Archimedean. That is, there is $x \in (0,1)$ such that $S(x,x) = x$.
Suppose first $x \geq n(x)$. Then $x = S(x,x) \geq S(x, n(x)) = 1$, a contradiction. If $x < n(x)$ then, by continuity of S, there exists $y \geq x$ such that $n(x) = S(x,y)$. Therefore, we have $n(x) = S(x,y) = S(S(x,x),y) = S(x, S(x,y)) = S(x,n(x)) = 1$, a contradiction. Thus our proposition is proved.

■

The following theorem can be proved easily.

Theorem 1.8 *A continuous t-conorm S satisfies condition $S(x,n(x)) = 1$ for all $x \in [0,1]$ with a strict negation n if and only if there exists an automorphism φ of the unit interval such that*

$$S(x,y) = \varphi^{-1}(\min\{\varphi(x) + \varphi(y), 1\}) \tag{1.27}$$

and

$$n(x) \geq \varphi^{-1}(1 - \varphi(x)).$$

Proof. (Necessity) According to the previous lemma, S is Archimedean. Thus there exists a generator g of S such that (1.26) holds with $g(1) < +\infty$. Define

$$\varphi(x) = \frac{g(x)}{g(1)}. \tag{1.28}$$

Thus defined φ is an automorphism of the unit interval. From (1.28) we have

$$g(x) = g(1)\varphi(x)$$

and

$$g^{-1}(x) = \varphi^{-1}\left(\frac{x}{g(1)}\right).$$

Using (1.26) we can go on as

$$\begin{aligned} S(x,y) &= g^{(-1)}(g(x)+g(y)) \\ &= g^{-1}(\min\{g(x)+g(y), g(1)\}) \\ &= \varphi^{-1}\left(\frac{\min\{g(1)\varphi(x)+g(1)\varphi(y), g(1)\}}{g(1)}\right) \\ &= \varphi^{-1}(\min\{\varphi(x)+\varphi(y), 1\}). \end{aligned}$$

On the other hand,$S(x, n(x)) = 1$ is equivalent to $\varphi(x) + \varphi(n(x)) \geq 1$, whence we obtain the required inequality $n(x) \geq \varphi^{-1}(1 - \varphi(x))$.

Proof of sufficiency is immediate.

■

1.4.3 Representation of continuous strict t-conorms

It is obvious that any strict t-conorm is Archimedean since $S(x,x) > S(x,0) = x$ for any $x \in (0,1)$.

Theorem 1.9 *A continuous t-conorm S is strict if and only if there exists an automorphism φ of the unit interval such that*

$$S(x,y) = \varphi^{-1}(\varphi(x)+\varphi(y)-\varphi(x)\varphi(y)). \tag{1.29}$$

Proof. (Necessity) If S is a continuous, strict t-conorm then S admits representation (1.26) with $g(1) = +\infty$, i.e.,

$$S(x,y) = g^{-1}(g(x)+g(y)).$$

Define

$$\varphi(x) = 1 - \exp(-g(x)). \tag{1.30}$$

Thus defined φ is an automorphism of the unit interval. From (1.30) we obtain that

$$g(x) = -\log(1-\varphi(x))$$

and

$$g^{-1}(x) = \varphi^{-1}(1-\exp(-x)).$$

Thus we can conclude that

$$\begin{aligned} S(x,y) &= g^{-1}(g(x)+g(y)) \\ &= \varphi^{-1}(1-(1-\varphi(x))(1-\varphi(y))) \\ &= \varphi^{-1}(\varphi(x)+\varphi(y)-\varphi(x)\varphi(y)). \end{aligned}$$

Proof of sufficiency is obvious.

■

1.4.4 Classification of continuous t-conorms

Suppose that $\{[a_m, b_m]\}$ is a countable family of non-overlapping, closed, proper subintervals of [0,1], denoted by $\mathcal{I}$. With each $[a_m, b_m] \in \mathcal{I}$ associate a continuous Archimedean t-conorm S_m. Let S be a function defined on $[0,1]^2$ via

$$S(x,y) = \begin{cases} a_m + (b_m - a_m)S_m\left(\frac{x-a_m}{b_m-a_m}, \frac{y-a_m}{b_m-a_m}\right) & \text{if } (x,y) \in [a_m, b_m]^2 \\ \max(x,y) & \text{otherwise} \end{cases} \tag{1.31}$$

In this case S is called the *ordinal sum* of $\{([a_m, b_m], S_m)\}$ and each S_m is called a *summand.*

Mostert and Shield (1957) have proved that a continuous t-conorm S is

1. either the maximum;
2. or Archimedean;
3. or there exist a family $\{([a_m, b_m], S_m)\}$ such that S is the ordinal sum of this family.

1.4.5 Union of fuzzy subsets

Suppose that A, B are fuzzy subsets of a given set Ω and S is a t-conorm. Then the union of A and B (with respect to S) is defined by

$$(A \cup_S B)(a) = S(A(a), B(a)) \quad \text{for all } a \in \Omega. \tag{1.32}$$

As in the case of intersection, the choice of a particular t-conorm depends on the problem under investigation. Axiomatic considerations can help the user in this choice (see Chapter 3).

1.5 Laws for conjunctions and disjunctions

In this section we investigate the validity of classical laws such as idempotency, absorption, distributivity, the laws of contradiction and excluded middle in the fuzzy case. These kind of problems were studied by Weber (1983).

Assume that T is any t-norm, S is any t-conorm and n is a strict negation.

1.5.1 Idempotency

Idempotency for T and S means that

$$T(x,x) = x \quad \text{for all} \quad x \in [0,1], \tag{1.33}$$

$$S(x,x) = x \quad \text{for all} \quad x \in [0,1]. \tag{1.34}$$

Proposition 1.4 (a) *T is idempotent if and only if* $T = \min$.

(b) *S is idempotent if and only if* $S = \max$.

Proof. (a) Assume that $x \leq y$. Then $x = T(x,x) \leq T(x,y) \leq T(x,1) = x$ imply that $T(x,y) = x$ when $x \leq y$, i.e., by commutativity, if and only if $T(x,y) = \min(x,y)$.

(b) Similar to the case (a).

■

1.5.2 Absorption

There are two forms of the absorption law:

$$T(S(x,y),x) = x \quad \text{for all} \quad x \in [0,1], \tag{1.35}$$
$$S(T(x,y),x) = x \quad \text{for all} \quad x \in [0,1]. \tag{1.36}$$

Proposition 1.5 (a) *(1.35) holds if and only if* $T = \min$.

(b) *(1.36) is true if and only if* $S = \max$.

Proof. Let $y = 0$ in (1.35). This is idempotency for T. Thus, by the previous proposition, $T = \min$. On the other hand $S(x,y) \geq x$ whence $\min(S(x,y),x) = x$ holds for $x, y \in [0,1]$.
(b) Similarly to the case (a).

■

1.5.3 Distributivity

Laws of distributivity are defined as follows:

$$S(x,T(y,z)) = T(S(x,y),S(x,z)) \quad \text{for all} \quad x,y,z \in [0,1], \tag{1.37}$$
$$T(x,S(y,z)) = S(T(x,y),T(x,z)) \quad \text{for all} \quad x,y,z \in [0,1]. \tag{1.38}$$

Proposition 1.6 (a) *(1.37) holds if and only if* $T = \min$.

(b) *(1.38) is true if and only if* $S = \max$.

Proof. (a) Let $z = 0$ in (1.37). This implies that $x = T(S(x,y),x)$ and hence $T = \min$ by the previous proposition. On the other hand, if $T = \min$ then (1.37) is obviously true.
(b) See part (a).

■

1.5.4 The law of contradiction

Traditional law of contradiction expresses that 'nothing can be both A and not-A', see Rescher (1969) for more details. By translating this rule, we obtain the following form (n is a strict negation):

$$T(x,n(x)) = 0 \quad \text{for all } x \in [0,1]. \tag{1.39}$$

Examples of t-norms possessing this property are W, $\min_0$ and Z. Assuming also continuity, the following statement can be proved.

Proposition 1.7 *The law of contradiction (1.39) holds for a continuous t-norm* T *if and only if there exists an automorphism* φ *of the unit interval such that*

$$T(x,y) = \varphi^{-1}(\max\{\varphi(x) + \varphi(y) - 1, 0\})$$

and

$$n(x) \leq \varphi^{-1}(1 - \varphi(x)).$$

Proof. Assume that (1.39) holds. Then, T must have the above representation, by Theorem 1.5. Using this representation and equation (1.39) we obtain immediately the inequality for n.

The other part of the proof is obvious.

■

1.5.5 The law of the excluded middle

The law of the excluded middle means the principle that 'everything is either A or not-A', see Rescher (1969). In our context, this is expressed by

$$S(x, n(x)) = 1 \quad \text{for all } x \in [0,1], \tag{1.40}$$

where n is a strict negation.

Examples for t-conorms satisfying this condition are W', $\max_1$ and Z'. Supposing that S is continuous, we have the following result.

Proposition 1.8 *The law of the excluded middle (1.40) holds for a continuous t-conorm S if and only if there exists an automorphism φ of the unit interval such that*

$$S(x, y) = \varphi^{-1}(\min\{\varphi(x) + \varphi(y), 1\})$$

and

$$n(x) \geq \varphi^{-1}(1 - \varphi(x)).$$

Proof. The proof can be carried out as that of the previous proposition.

■

1.6 De Morgan triples

If T is a t-norm, S is a t-conorm and n_1, n_2 are negations (not necessarily strict or strong ones) then two types of the De Morgan law can be expressed:

$$n_1(S(x,y)) = T(n_1(x), n_1(y)), \tag{1.41}$$

$$n_2(T(x,y)) = S(n_2(x), n_2(y)). \tag{1.42}$$

Obviously, for strong negations these two laws are equivalent. It is also clear that if T and S are given then (1.41) is true with a strict negation $n_1 = n$ if and only if (1.42) holds with $n_2 = n^{-1}$. The proof of the following statement is easy, thus it is omitted (see Weber (1983)).

Proposition 1.9 *Let n be any strict negation.*
(a) For any t-norm T there exists a t-conorm S defined by

$$S(x, y) = n^{-1}(T(n(x), n(y))$$

which fulfils (1.41) with $n_1 = n$. *If* T *is continuous then* S *is continuous. In addition, if* T *is Archimedean with additive generator* f *then* S *is Archimedean with additive generator* $g = f \circ n$ *and* $g(1) = f(0)$.

(b) For any t-conorm S *there exists a t-norm* T *defined by*

$$T(x, y) = n^{-1}(S(n(x), n(y))$$

which satisfies (1.42) with $n_2 = n$. *If* S *is continuous then* T *is continuous. Moreover, if* S *is Archimedean with additive generator* g *then* T *is Archimedean with additive generator* $f = g \circ n$ *and* $f(0) = g(1)$.

In the next theorem we establish a simple condition under which there exists a strict negation n such that (1.41) holds with $n_1 = n$ and (1.42) is true with $n_2 = n^{-1}$ for a continuous, Archimedean t-norm T and for a continuous, Archimedean t-conorm S. This result extends Theorem 2.3 of Alsina et al. (1983) related to strong negations.

Theorem 1.10 *Assume that* T *is a continuous Archimedean t-norm with additive generator* f *and* S *is a continuous Archimedean t-conorm with additive generator* g. *Then there exists a strict negation* n *such that (1.41) holds with* $n_1 = n^{-1}$ *and (1.42) holds with* $n_2 = n$ *if and only if* $0 < g(1)/f(0) < \infty$, *where* ∞/∞ *is understood as 1.*

Proof. Let us define n by

$$n(x) = g^{(-1)}\left(\frac{g(1)}{f(0)} f(x)\right).$$

It is easy to check that under the conditions of this theorem n has the following properties:

- $0 \le \frac{g(1)}{f(0)} f(x) \le g(1)$ thus $g^{(-1)} = g^{-1}$.
- n is strictly decreasing since f is decreasing and g^{-1} is increasing.
- n is continuous since f and g^{-1} are continuous.
- (1.41) is satisfied with $n_1 = n^{-1}$.
- (1.42) is fulfilled with $n_2 = n$.

■

Definition 1.12 Assume that T is a t-norm, S is a t-conorm and n is a strict negation. We say that (T, S, n) is a *De Morgan triple* if and only if (1.41) is satisfied with $n_1 = n$.

A De Morgan triple is called *continuous* if T and S are continuous functions on the unit interval.

Two types of De Morgan triples will be important in the sequel. We define them now.

Definition 1.13 A De Morgan triple (T, S, N) is called a *strong* (or *Lukasiewicz-like*) De Morgan triple if and only if there exists an automorphism φ of the unit interval such that

$$\begin{aligned} T(x,y) &= \varphi^{-1}(\max\{\varphi(x)+\varphi(y)-1,0\}), \\ S(x,y) &= \varphi^{-1}(\min\{\varphi(x)+\varphi(y),1\}), \\ N(x) &= \varphi^{-1}(1-\varphi(x)). \end{aligned}$$

Definition 1.14 A De Morgan triple (T, S, N) is said to be a *strict* (or *product-like*) De Morgan triple if and only if there exists an automorphism φ of the unit interval such that

$$\begin{aligned} T(x,y) &= \varphi^{-1}(\varphi(x)\varphi(y)), \\ S(x,y) &= \varphi^{-1}(\varphi(x)+\varphi(y)-\varphi(x)\varphi(y)), \\ N(x) &= \varphi^{-1}(1-\varphi(x)). \end{aligned}$$

Closing this section we remark that De Morgan triples depending also on points in Ω have been studied by Klement (1981a) and Lowen (1978).

1.7 Parametric families of connectives

In this section we consider two parametric families of t-norms, t-conorms and strong negations. For other see e.g. Dombi (1982a).

1.7.1 The Frank family

Let $s > 0, s \neq 1$ be a real number. Define a parametric family of continuous Archimedean t-norms in the following way:

$$T^s(x,y) = \log_s\left(1 + \frac{(s^x-1)(s^y-1)}{s-1}\right).$$

We can extend this definition for $s = 0$, $s = 1$ and $s = \infty$ by taking limits. Thus

$$\begin{aligned} T^0(x,y) &= \lim_{s\to 0} T^s(x,y) = \min\{x,y\}, \\ T^1(x,y) &= \lim_{s\to 1} T^s(x,y) = xy, \\ T^\infty(x,y) &= \lim_{s\to\infty} T^s(x,y) = \max\{x+y-1,0\}. \end{aligned}$$

The family $\{T^s\}_{s\in[0,\infty]}$ is called the *Frank family* of t-norms. If φ is an automorphism of the unit interval then $T(x,y) = \varphi^{-1}(T^s(\varphi(x),\varphi(y)))$ is called a φ-transform of T^s.

The De Morgan law enables us to define the Frank family of t-conorms $\{S^s\}_{s\in[0,\infty]}$ by

$$S^s(x,y) = 1 - T^s(1-x, 1-y)$$

for any $s \in [0, \infty]$.

In Frank (1979) one can find the following interesting characterization of these parametrized families.

Theorem 1.11 *A continuous t-norm T and a continuous t-conorm S satisfy the functional equation*

$$T(x,y) + S(x,y) = x + y \tag{1.43}$$

if and only if

(a) there is a number $s \in [0,\infty]$ such that $T = T^s$ and $S = S^s$,

or

(b) T is representable as an ordinal sum of t-norms, each of which is a member of the family $\{T_s\}$, $0 < s \leq \infty$, and S is obtained from T via equation (1.43).

We remark that these families will play a central role in the investigation of valued preference structures.

1.7.2 The Hamacher family

Let us define three parametrized families of t-norms, t-conorms and strong negations, respectively, as follows.

$$\begin{aligned} T_\alpha(x,y) &= \frac{xy}{\alpha + (1-\alpha)(x+y-xy)}, \quad \alpha \geq 0, \\ S_\beta(x,y) &= \frac{x+y+(\beta-1)xy}{1+\beta xy}, \quad \beta \geq -1, \\ N_\gamma(x) &= \frac{1-x}{1+\gamma x}, \quad \gamma > -1. \end{aligned}$$

Hamacher (1978) proved the following characterization theorem.

Theorem 1.12 *(T,S,n) is a De Morgan triple such that*

$$\begin{aligned} T(x,y) = T(x,z) \quad &\Longrightarrow \quad y = z, \\ S(x,y) = S(x,z) \quad &\Longrightarrow \quad y = z, \\ \forall z \leq x \quad \exists y, y' \quad &\text{such that} \quad T(x,y) = z, S(z,y') = x \end{aligned}$$

and T and S are rational functions if and only if there are numbers $\alpha \geq 0, \beta \geq -1$ and $\gamma > -1$ such that $\alpha = \frac{1+\beta}{1+\gamma}$ and $T = T_\alpha, S = S_\beta$ and $n = N_\gamma$.

Remark that another characterization of the Hamacher family of t-norms with positive parameter has been obtained recently by Fodor and Keresztfalvi (1994) as solutions of a functional equation.

1.8 Implications

There exist several approaches to the definition of implications. We consider an implication as a connective. The following axioms try to capture the essentials of fuzzy implications, synthesizing diverse proposals that can be found in the literature.

I0. $v(P_1 \to P_2)$ depends only on the values $v(P_1)$ and $v(P_2)$.

I1. If $v(P_1) \leq v(P_3)$ then $v(P_1 \to P_2) \geq v(P_3 \to P_2)$ for any proposition P_2.

I2. If $v(P_2) \leq v(P_3)$ then $v(P_1 \to P_2) \leq v(P_1 \to P_3)$ for any P_1.

I3. If $v(P_1) = 0$ then $v(P_1 \to P) = 1$ for any P.

I4. If $v(P_1) = 1$ then $v(P \to P_1) = 1$ for any P.

I5. If $v(P_1) = 1$ and $v(P_2) = 0$ then $v(P_1 \to P_2) = 0$.

I0 sometimes is called the truth-functionality axiom, see Smets and Magrez (1987).

I1 and I2, two opposite kinds of monotonicity of the implication, are based on the idea that if the value of the antecedent decreases and/or the value of the consequent increases then the truth of the implication should not decrease. This is a natural condition since "the implication is essentially a measure of the fact that the consequent is more true than the antecedent", as it was stated by Smets and Magrez (1987).

I3 means that falsity implies anything while I4 declares that tautology is implied by anything. I5 claims that tautology cannot justify falsity.

More formally, our axioms declare the following conditions.

I0 asserts that there exists a function $I^{\to} : [0,1]^2 \to [0,1]$ such that

$$v(P_1 \to P_2) = I^{\to}(v(P_1), v(P_2)).$$

I1 and I2 means the following requirement for $I^{\to}$:

$$\text{If} \quad x \leq z \quad \text{then} \quad I^{\to}(x,y) \geq I^{\to}(z,y) \quad \forall y \in [0,1]. \tag{1.44}$$

$$\text{If} \quad y \leq t \quad \text{then} \quad I^{\to}(x,y) \leq I^{\to}(x,t) \quad \forall x \in [0,1]. \tag{1.45}$$

I3 expresses that

$$I^{\to}(0,x) = 1 \quad \forall x \in [0,1]. \tag{1.46}$$

I4 claims that

$$I^{\to}(x,1) = 1 \quad \forall x \in [0,1]. \tag{1.47}$$

I5 is simply a crisp condition which is not implied by the previous axioms:

$$I^{\to}(1,0) = 0. \tag{1.48}$$

Definition 1.15 An *implication* is a function $I^{\to} : [0,1]^2 \to [0,1]$ having properties (1.44)–(1.48).

One of the main motivations of Definition 1.15 is the following. If $I^{\to}$ is an implication and n is a (strong) negation then, although $I^{\to}(x,y) = I^{\to}(n(y), n(x))$ does not hold in general, we want to consider the function $I^{\to}(n(y), n(x))$ also as an implication. One can easily verify that this function satisfies the conditions of Definition 1.15 if $I^{\to}$ is an implication.

Now we recall further axioms, in terms of function $I^{\to}$. These properties are required in different papers and they could be important also in some applications (see e.g. Dubois and Prade (1984a, 1985, 1991b) and further references there).

I6. $I^{\rightarrow}(1, x) = x$.

I7. $I^{\rightarrow}(x, I^{\rightarrow}(y, z)) = I^{\rightarrow}(y, I^{\rightarrow}(x, z))$.

I8. $x \leq y$ if and only if $I^{\rightarrow}(x, y) = 1$.

I9. $I^{\rightarrow}(x, 0) = N(x)$ is a strong negation.

I10. $I^{\rightarrow}(x, y) \geq y$.

I11. $I^{\rightarrow}(x, x) = 1$.

I12. $I^{\rightarrow}(x, y) = I^{\rightarrow}(N(y), N(x))$ with a strong negation N.

I13. $I^{\rightarrow}$ is a continuous function.

I6 yields that the value of an implication when the antecedent is tautology, is equal to the value of the consequent.

I7 is called the exchange principle and is based on the following equivalence:

$$\text{"if } P_1 \text{ then (if } P_2 \text{ then } P_3\text{)"} \iff \text{"if } (P_1 \textbf{ AND } P_2) \text{ then } P_3\text{"}.$$

I8 expresses that implication defines an ordering.

I9 reflects that $P \rightarrow Q = \neg P$ if $v(Q) = \text{FALSE}$.

I10 is the numerical counterpart of $P \rightarrow (Q \rightarrow P)$.

I11 is called identity principle and it yields that $P \rightarrow P$ is always true.

I12, the contrapositive symmetry, expresses a relationship between modus ponens and modus tollens, see Smets and Magrez (1987). In general, this is a strong condition, see Fodor (1993c).

I13 prevents implication from reacting in a chaotic way to a small change of the truth value of either the antecedent or the consequent. This is also a fairly restrictive condition.

We would like to illustrate some interdependencies among these axioms now. At the same time, the following lemma plays an important role in the sequel. Some parts of it are based on an unpublished paper of Wu Wangming (1992).

Lemma 1.3 *Let* $I^{\rightarrow} : [0,1]^2 \rightarrow [0,1]$ *be any function which fulfils conditions I2, I7, I8. Then* $I^{\rightarrow}$ *satisfies also I1, I3, I4, I5, I6, I10 and I11.*

Proof. I3, I4 and I11 are obviously true by I8.

To prove I1, let $x \leq z$. Then $I^{\rightarrow}(z, I^{\rightarrow}(I^{\rightarrow}(z, y), y)) = I^{\rightarrow}(I^{\rightarrow}(z, y), I^{\rightarrow}(z, y)) = 1$ by I7, whence $z \leq I^{\rightarrow}(I^{\rightarrow}(z, y), y)$, using I8. Therefore,

$$1 = I^{\rightarrow}(x, z) \leq I^{\rightarrow}(x, I^{\rightarrow}(I^{\rightarrow}(z, y), y)) = I^{\rightarrow}(I^{\rightarrow}(z, y), I^{\rightarrow}(x, y)),$$

that is, $I^{\rightarrow}(x, y) \geq I^{\rightarrow}(z, y)$.

Turning to I6, $I^{\rightarrow}(x, I^{\rightarrow}(1, x)) = I^{\rightarrow}(1, I^{\rightarrow}(x, x)) = I^{\rightarrow}(1, 1) = 1$, thus $x \leq I^{\rightarrow}(1, x)$.

On the other hand, $I^{\rightarrow}(1, I^{\rightarrow}(I^{\rightarrow}(1, x), x)) = I^{\rightarrow}(I^{\rightarrow}(1, x), I^{\rightarrow}(1, x)) = 1$ implies $I^{\rightarrow}(I^{\rightarrow}(1, x), x) = 1$, that is, $I^{\rightarrow}(1, x) \leq x$. Therefore, we have $I^{\rightarrow}(1, x) = x$.

Now I5 is a consequence of I6.

Finally, $I^{\rightarrow}(y, I^{\rightarrow}(x, z)) = I^{\rightarrow}(x, I^{\rightarrow}(y, y)) = I^{\rightarrow}(x, 1) = 1$ and thus $y \leq I^{\rightarrow}(x, y)$ follows. ∎

1.8.1 Implications by t-norms, t-conorms and negations

It is clear that, in a consistent way, implications and conjunctions (or implications and disjunctions) cannot be studied independently of each other. Therefore, particular classes of implications should be introduced using t-norms, t-conorms and negations.

The two most important families of such implications is related either to the formalism of Boolean logic or to a residuation concept from intuitionistic logic. Thus we introduce the following definitions.

Definition 1.16 An *S-implication* associated with a t-conorm S and a strong negation N is defined by

$$I^{\rightarrow}_{S,N}(x,y) = S(N(x),y). \tag{1.49}$$

An *R-implication* associated with a t-norm T is defined by

$$I^{\rightarrow}_{T}(x,y) = \sup\{z|T(x,z) \leq y\}. \tag{1.50}$$

The idea behind the definition of $I^{\rightarrow}_{S,N}$ is obvious :

$$P \rightarrow Q = \neg P \vee Q.$$

One can justify expression for $I^{\rightarrow}_{T}$ by the following classical set-theoretic identity :

$$A^c \cup B = (A \setminus B)^c = \cup\{Z \mid A \cap Z \subseteq B\},$$

where $\setminus$ denotes set-difference operator.

It is easy to see that both $I^{\rightarrow}_{S,N}$ and $I^{\rightarrow}_{T}$ satisfy properties (1.44)–(1.48) for any t-norm T, t-conorm S and strong negation N, thus they are implications.

For the sake of completeness we mention a third type of implications used in quantum logic and called QL-implication:

$$I^{\rightarrow}_{T,S,N}(x,y) = S(N(x), T(x,y)).$$

In general, $I^{\rightarrow}_{T,S,N}$ violates property (1.44). Conditions under that (1.44) is satisfied can be found in Fodor (1993c).

Now we characterize S-implications (see also Trillas and Valverde (1981,1985)).

Theorem 1.13 *An implication is an S-implication with an appropriate t-conorm S and a strong negation N if and only if $I^{\rightarrow}$ satisfies I6,I7 and I12.*

Proof. Suppose first that $I^{\rightarrow}(x,y) = S(N(x),y)$ with a t-conorm and a strong negation N. Then I6 is satisfied since $I^{\rightarrow}(1,y) = S(0,y) = y$.

I7 holds because

$$I^{\rightarrow}(x, I^{\rightarrow}(y,z)) = S(N(x), S(N(y),z)) = S(N(y), S(N(x),z)) = I^{\rightarrow}(y, I^{\rightarrow}(x,z)),$$

by associativity and commutativity of S.

I12 is valid since $I^{\rightarrow}(x,y) = S(N(x),y) = S(y,N(x)) = I^{\rightarrow}(N(y),N(x))$, by commutativity of S and because N is involutive.

Turning to the converse direction, assume that $I^{\rightarrow}$ fulfils I6, I7 and I12.

By I12, let $N(x) = I^{\rightarrow}(x,0) = I^{\rightarrow}(1,N(x))$ and define $S(x,y) = I^{\rightarrow}(N(x),y)$. S is nondecreasing in both places by I1 and I2 ($I^{\rightarrow}$ is an implication).

S is commutative, by I12:

$$S(x,y) = I^{\rightarrow}(N(x),y) = I^{\rightarrow}(N(y),x) = S(y,x).$$

By I6, $S(0,y) = I^{\rightarrow}(1,y) = y$.

Finally, by I7 and I12, we have that

$$\begin{aligned} & S(x,S(y,z)) \\ = \; & I^{\rightarrow}(N(x),I^{\rightarrow}(N(y),z)) = I^{\rightarrow}(N(x),I^{\rightarrow}(N(z),y)) = I^{\rightarrow}(N(z),I^{\rightarrow}(N(x),y)) \\ = \; & I^{\rightarrow}(N(I^{\rightarrow}(N(x),y)),z) \\ = \; & S(S(x,y),z). \end{aligned}$$

Therefore, S is a t-conorm.

We obviously have that $I^{\rightarrow}(x,y) = S(N(x),y)$, by the definition of S.

■

Now we try to characterize implications which can be defined as R-implications based on left-continuous t-norms (see also Fodor (1991d, 1993b), Miyakoshi and Shimbo (1985)).

Remark. Left-continuous t-norms play a key role from several points of view. As Höhle (1988) pointed out, a rather general acceptable structure of the valuation set (instead of the unit interval) consists of an *integral cl-monoid* $(L,*,\leq)$ (see also Birkhoff (1948)). A triple $(L,*,\leq)$ is called a commutative cl-monoid if and only if

- $(L,\leq)$ is a complete lattice
- $(L,*)$ is a commutative monoid and we have

$$\alpha * (\bigvee A) = \bigvee\{\alpha * \beta | \beta \in A\} \;, \quad \forall A \subseteq L.$$

A cl-monoid $(L,*,\leq)$ is integral if and only if the universal upper bound is the unity with respect to $*$. It was stressed by Höhle (1988) that integral cl-monoids are common frameworks for complete Heyting algebras (see Birkhoff (1948)) and for left-continuous t-norms on the unit interval. In addition, they form a basic structure in which abstract ideal theory takes place. For more details we refer to Höhle (1988).

We would like to emphasize that, as far as we know, the nilpotent minimum is the only t-norm (up to an automorphism) which is left-continuous and not continuous on the unit interval.

It is worth noting that a t-norm T is left-continuous if and only if the following condition is satisfied :

$$T(x,z) \leq y \Longleftrightarrow I_T^{\rightarrow}(x,y) \geq z. \tag{1.51}$$

Here is the result concerning R-implications based on left-continuous t-norms.

Theorem 1.14 *A function $I^{\rightarrow} : [0,1]^2 \rightarrow [0,1]$ is an R-implication based on a left-continuous t-norm if and only if $I^{\rightarrow}$ satisfies conditions I2, I7, I8 and $I^{\rightarrow}(x,.)$ is right-continuous for any fixed $x \in [0,1]$.*

Proof. Suppose first that

$$I^{\rightarrow}(x,y) = I_T^{\rightarrow}(x,y) = \sup\{z|T(x,z) \leq y\},$$

where $T(x,.)$ is a left-continuous function on the unit interval for any fixed $x \in [0,1]$.

Then I2 holds trivially. Left-continuity of T implies right-continuity of $I^{\rightarrow}$ with respect to its second argument.

$T(1,x) = x$ and left-continuity of T imply I8. Finally, I7 is valid since, again by left-continuity and associativity of T, $I_T^{\rightarrow}(x, I_T^{\rightarrow}(y,z)) = I_T^{\rightarrow}(T(x,y),z)$ is true.

Indeed, we have by (1.51) that

$$\begin{aligned} I_T^{\rightarrow}(x, I_T^{\rightarrow}(y,z)) &= \sup\{t \mid T(x,t) \leq I_T^{\rightarrow}(y,z)\} \\ &= \sup\{t \mid T(y,T(x,t)) \leq z\} \\ &= \sup\{t \mid T(T(x,y),t) \leq z\} \\ &= I_T^{\rightarrow}(T(x,y),z). \end{aligned}$$

Assume now that $I^{\rightarrow}$ satisfies the required conditions. Then, by the previous lemma, $I^{\rightarrow}$ fulfils also I1, I3, I4, I5 and I6. Define a binary operation $T_{I^{\rightarrow}}$ by

$$T_{I^{\rightarrow}}(x,y) = \inf\{t|I^{\rightarrow}(x,t) \geq y\}.$$

We prove that $T_{I^{\rightarrow}}$ is a t-norm and $I^{\rightarrow}(x,y) = \sup\{z|T_{I^{\rightarrow}}(x,z) \leq y\}$. For sake of simplicity, we write simply T insted of $T_{I^{\rightarrow}}$ during the proof.

$T(1,y) = y$ since

$$T(1,y) = \inf\{t|I^{\rightarrow}(1,t) \geq y\} = \inf\{t|t \geq y\} = y,$$

by I6.

Commutativity:

By I7, I8 and $I^{\rightarrow}(x,t) \geq y$ we have

$$1 = I^{\rightarrow}(y, I^{\rightarrow}(x,t)) = I^{\rightarrow}(x, I^{\rightarrow}(y,t)),$$

that is, $I^{\rightarrow}(x,t) \geq y$ if and only if $I^{\rightarrow}(y,t) \geq x$. This implies that T is commutative.

Associativity:

By commutativity, $T(T(x,y),z) = T(z,T(x,y))$. That is,

$$I^{\rightarrow}(T(x,y),t) \geq z \quad \Longleftrightarrow \quad I^{\rightarrow}(z,t) \geq T(x,y).$$

Therefore, it is sufficient to prove that

$$I^{\rightarrow}(z,t) \geq T(x,y) \quad \Longleftrightarrow \quad I^{\rightarrow}(x,t) \geq T(z,y).$$

Since $I^{\rightarrow}(x,.)$ is right-continuous, we have from definition of T that

$$I^{\rightarrow}(x,T(x,y)) \geq y$$

and

$$T(x, I^{\rightarrow}(x,y)) \leq y$$

since $I^{\rightarrow}(x,y) \leq I^{\rightarrow}(x,y)$.

Therefore, $I^{\rightarrow}(z,t) \geq T(x,y)$ implies $I^{\rightarrow}(x, I^{\rightarrow}(z,t)) \geq I^{\rightarrow}(x, T(x,y) \geq y$. On the other hand,

$$T(z,y) \leq T(z, I^{\rightarrow}(x,(I^{\rightarrow}(z,t))) = T(z, I^{\rightarrow}(z, I^{\rightarrow}(x,t))) \leq I^{\rightarrow}(x,t).$$

Thus, we have proved that $I^{\rightarrow}(z,t) \geq T(x,y)$ implies $I^{\rightarrow}(x,t) \geq T(z,y)$. The opposite direction can be proved similarly.

Thus, T is a left-continuous t-norm.

To complete the proof, we need that

$$I^{\rightarrow}(x,y) = I^{\rightarrow}_{T_{I^{\rightarrow}}}(x,y) = \sup\{z|T_{I^{\rightarrow}}(x,z) \leq y\}.$$

But we have that

$$1 = I^{\rightarrow}_{T_{I^{\rightarrow}}}(T_{I^{\rightarrow}}(x,z),y) = I^{\rightarrow}_{T_{I^{\rightarrow}}}(z, I^{\rightarrow}_{T_{I^{\rightarrow}}}(x,y)),$$

as we established in the first part of the present proof. That is, $T_{I^{\rightarrow}}(x,z) \leq y$ if and only if $I^{\rightarrow}_{T_{I^{\rightarrow}}}(x,y) \geq z$. It was also proved in the previous part that $T_{I^{\rightarrow}}(x, I^{\rightarrow}(x,y)) \leq y$, whence

$$I^{\rightarrow}_{T_{I^{\rightarrow}}}(x,y) \geq I^{\rightarrow}(x,y).$$

On the other hand, $I^{\rightarrow}(x, T_{I^{\rightarrow}}(x,z)) \geq z$. Let $z = I^{\rightarrow}_{T_{I^{\rightarrow}}}(x,y)$. Then we have

$$I^{\rightarrow}_{T_{I^{\rightarrow}}}(x,y) \leq I^{\rightarrow}(x, T_{I^{\rightarrow}}(x, I^{\rightarrow}_{T_{I^{\rightarrow}}}(x,y))) \leq I^{\rightarrow}(x,y),$$

that is, $I^{\rightarrow}(x,y) = \sup\{z|T_{I^{\rightarrow}}(x,z) \leq y\}$.

■

Using (1.51), it is easy to prove the following proposition which we will need in Chapter 2 (see also Valverde (1985)).

Proposition 1.10 *If T fulfils (1.51) then the following inequality holds for all $x,y,z \in [0,1]$:*

$$T(I^{\rightarrow}_{T}(x,y), I^{\rightarrow}_{T}(y,z)) \leq I^{\rightarrow}_{T}(x,z). \tag{1.52}$$

Proof. By (1.51), inequality (1.52) is true if and only if

$$T\left[x,\ T(I^{\rightarrow}_{T}(x,y), I^{\rightarrow}_{T}(y,z))\right] \leq z.$$

Since $I^{\rightarrow}_{T}(x,y) \leq I^{\rightarrow}_{T}(x,y)$, (1.51) implies that

$$T(x, I^{\rightarrow}_{T}(x,y)) \leq y$$

for all $x,y \in [0,1]$. Therefore, using also associativity of T, we have the following chain of inequalities :

$$\begin{aligned} T\left[x, T(I^{\rightarrow}_{T}(x,y), I^{\rightarrow}_{T}(y,z))\right] &= T\left[T(x, I^{\rightarrow}_{T}(x,y)), I^{\rightarrow}_{T}(y,z)\right] \\ &\leq T(y, I^{\rightarrow}_{T}(y,z)) \leq z. \end{aligned}$$

■

Considering the weakest t-norm Z, it is easily verified that

$$I^{\rightarrow}_{Z}(x,y) = \begin{cases} y & \text{if } x = 1 \\ 1 & \text{otherwise} \end{cases}.$$

If $z < y < x = 1$, then

$$\begin{aligned} Z(I^{\rightarrow}_{Z}(x,y), I^{\rightarrow}_{Z}(y,z)) &= Z(y, I^{\rightarrow}_{Z}(y,z) \\ &= Z(y,1) = y \\ &> z = I^{\rightarrow}_{Z}(x,z). \end{aligned}$$

Therefore, (1.51) does not hold for all $x,y,z \in [0,1]$ for Z.

1.8.2 Negations defined by implications

As we mentioned before, several types of negations can be introduced in fuzzy logics. In the classical (two-valued) case a connection between implications and negations is expressed by

$$v(P \to Q) = v(\mathbf{NOT}P) \quad \text{if} \quad v(Q) = \mathbf{FALSE}. \tag{1.53}$$

What can we state regarding fuzzy implications and negations? Now (1.53) corresponds to

$$I^{\to}(x, 0) = n(x) \quad \forall x \in [0, 1],$$

where $I^{\to}$ is a fuzzy implication. The next result was established by Wu Wangmin (1992).

Proposition 1.11 *Suppose that $I^{\to} : [0,1]^2 \to [0,1]$ is a function satisfying I2, I7, I8 and define $n(x) = I^{\to}(x, 0)$. Then*

(a) n is a negation

(b) $x \le n(n(x))$ for all $x \in [0, 1]$

(c) $n(n(n(x))) = n(x)$ for all $x \in [0, 1]$.

Proof. (a) $n(0) = I^{\to}(0, 0) = 1$, by I8. $n(1) = I^{\to}(1, 0) = 0$, by I5. $n(x)$ is nonincreasing by Lemma 1.3.

(b) $I^{\to}(x, n(n(x))) = I^{\to}(x, I^{\to}(I^{\to}(x, 0), 0)) = I^{\to}(I^{\to}(x, 0), I^{\to}(x, 0)) = 1$ implies, by I8, that $x \le n(n(x))$.

(c) Since n is nonincreasing and (b) is true, we have $n(x) \ge n(n(n(x)))$. On the other hand,

$$\begin{aligned} I^{\to}(n(x), n(n(n(x)))) &= I^{\to}(I^{\to}(x,0), I^{\to}(I^{\to}(I^{\to}(x,0),0),0)) \\ &= I^{\to}(I^{\to}(I^{\to}(x,0),0), I^{\to}(I^{\to}(x,0),0)) = 1, \end{aligned}$$

whence $n(x) \le n(n(n(x)))$.

■

Corollary 1.1 *Suppose that the conditions of the previous proposition are satisfied. If, in addition, $n(x) = I^{\to}(x, 0)$ is continuous then it is involutive.*

Proof. If n is continuous then for any $x \in [0, 1]$ there exists $y \in [0, 1]$ such that $n(y) = x$. Therefore, $n(n(x)) = n(n(n(y))) = n(y) = x$.

■

This corollary implies also that, under the above conditions, $n(x) = I^{\to}(x, 0)$ cannot be strict: it is either discontinuous or a strong negation.

Corollary 1.2 *Under the conditions of the previous proposition, if $n(x) = I^{\to}(x, 0)$ is continuous then $I^{\to}$ fulfils I12 with $n(x) = I(x, 0)$.*

Proof. By the previous Corollary, in our case $n(n(x)) = x$. Thus

$$\begin{aligned} I^{\rightarrow}(x,y) &= I^{\rightarrow}(x,n(n(y))) = I^{\rightarrow}(x,I^{\rightarrow}(I^{\rightarrow}(y,0),0)) = I^{\rightarrow}(I^{\rightarrow}(y,0),I^{\rightarrow}(x,0)) \\ &= I^{\rightarrow}(n(y),n(x)). \end{aligned}$$

■

For positive t-norms like min or Π, the negation obtained via R-implication is not continuous at all as we claim now.

Proposition 1.12 *Suppose T is a positive t-norm. Then $I_T^{\rightarrow}(x,0) = n_i(x)$.*

Proof. Obviously,

$$I_T^{\rightarrow}(x,0) = \begin{cases} 1 & \text{if } x = 0 \\ 0 & \text{if } x > 0 \end{cases},$$

which is the intuitionistic negation n_i.

■

Continuity of the implication is sufficient but not necessary to obtain strong negation via residuation. As an example, consider

$$I_{\text{mino},\varphi}^{\rightarrow}(x,y) = \begin{cases} 1 & \text{if } x \leq y \\ \max(N(x),y) & \text{otherwise} \end{cases},$$

where $N(x) = \varphi^{-1}(1-\varphi(x))$ is a strong negation. Then $I_{\text{mino},\varphi}^{\rightarrow}$ is not continuous but $I_{\text{mino},\varphi}^{\rightarrow}(x,0) = N(x)$ is a strong negation.

When $I_T^{\rightarrow}$ is continuous then we can represent $I_T^{\rightarrow}$ as a φ-transform of the Łukasiewicz implication. This was proved by Smets and Magrez (1987). They required more conditions than it is necessary. To obtain axioms of Smets and Magrez (1987), see Lemma 1.3.

Theorem 1.15 *A function $I^{\rightarrow} : [0,1]^2 \rightarrow [0,1]$ is such that I2, I7, I8 and I13 are satisfied if and only if there exists an automorphism φ of the unit interval such that*

$$I^{\rightarrow}(x,y) = \varphi^{-1}(\min\{1-\varphi(x)+\varphi(y),1\}). \tag{1.54}$$

Proof. Consider first an implication having form (1.54). Then $I^{\rightarrow}$ obviously satisfies conditions listed above.

Suppose now that $I^{\rightarrow}$ is any function which fulfils I2, I7, I8 and I13. Then, by Corollary 1.1, $n(x) = I^{\rightarrow}(x,0)$ is a strong negation. Define a function T by

$$T(x,y) = \inf\{z | I^{\rightarrow}(x,z) \geq y\}.$$

By Theorem 1.14, thus defined function is a t-norm. It is continuous since $I^{\rightarrow}$ is continuous. In addition,

$$T(x,n(x)) = \inf\{z | I^{\rightarrow}(x,z) \geq n(x)\} = 0.$$

Thus, by Theorem 1.5 , there exists an automorphism φ of the unit interval such that

$$T(x,y) = \varphi^{-1}(\max\{\varphi(x)+\varphi(y)-1,0\}).$$

Then

$$I_T^{\rightarrow}(x,y) = \varphi^{-1}(\min\{1-\varphi(x)+\varphi(y),1\}),$$

and, by Theorem 1.14, $I^{\rightarrow} = I_T^{\rightarrow}$.

■

Of course, one can obtain representation of R-implications defined by any continuous Archimedean t-norm as follows.

Theorem 1.16 *Assume that* T *is a continuous Archimedean t-norm with additive generator* f*. Then*

$$I_T^{\rightarrow}(x,y) = f^{-1}(\max\{f(y) - f(x), 0\}). \tag{1.55}$$

Proof. We have

$$\begin{aligned} I_T^{\rightarrow}(x,y) &= \sup\{z|T(x,z) \le y\} = \sup\{z|f^{-1}(\min\{f(x)+f(z), f(0)\}) \le y\} \\ &= \sup\{z|\min\{f(x)+f(z), f(0)\} \ge f(y)\} = \sup\{z|f(x)+f(z) \ge f(y)\} \\ &= \sup\{z|f(z) \ge f(y) - f(x)\} = f^{-1}(\max\{f(y)-f(x), 0\}), \end{aligned}$$

since f is continuous and strictly decreasing function.

■

The following simple statement will be needed in Chapter 4.

Lemma 1.4 *If* T *is a continuous Archimedean t-norm and* $z > 0$ *then*

$$T(x,y) = z \quad \textit{implies} \quad y = I_T^{\rightarrow}(x,z).$$

Proof. By representation

$$T(x,y) = f^{-1}(\min\{f(x)+f(y), f(0)\}),$$

$T(x,y) = z > 0$ is equivalnet to $f(x) + f(y) = f(z)$. We have, by (1.55), that

$$I_T^{\rightarrow}(x,z) = f^{-1}(\max\{f(z) - f(x), 0\}) = y.$$

■

Remarks. 1) Generally speakig, $T(x,y) = 0$ does not imply $y = I_T^{\rightarrow}(x,0)$. For example, if $T(x,y) = \max\{x+y-1, 0\}$ then $T(0.2, 0.3) = 0$ but $0.3 \neq I_T^{\rightarrow}(0.2, 0) = 0.8$.

2) In general, the statement of the previous lemma is not true if T is not Archimedean. Consider the t-norm defined by

$$T(x,y) = \begin{cases} 0.5 + 0.5\max\{2x+2y-3, 0\} & \text{if } x \ge 0.5, y \ge 0.5 \\ \min(x,y) & \textit{otherwise} \end{cases}.$$

Thus defined function is a continuous t-norm (see Section 1.3.4 on ordinal sums). We have $T(0.75, 0.6) = 0.5$ but $I_T^{\rightarrow}(0.75, 0.5) = 0.75$.

Closing this subsection, in Table 1.1 we list the most common implications which are based on t-norms, t-conorms and the strong negation $1 - x$ (for more details see e.g. Dubois and Prade (1991b), Fodor (1993c), Goguen (1969)). Remark that the implication itself in the last row of Table 1.1 was defined by several authors in a heuristic way. However, the fact that it is an R-, S- and QL-implication based on appropriate t-norms and t-conorms was pointed out by Fodor (1993c).

Form	Type	Name	Properties
$\max(1-x,y)$	S with $S=\max$ QL with $S=W', T=W$	Kleene–Dienes	1-7, 10, 12, 13
$1-x+xy$	S with $S=\Pi'$	Rechenbach	1–7, 10, 12, 13
$\min\{1-x+y,1\}$	S with $S=W'$ R with $T=W$ QL with $S=W'$, $T=\min$	Łukasiewicz	1-13
$\begin{cases} 1 & \text{if } x\le y \\ y & \text{otherwise}\end{cases}$	R with $T=\min$	Gödel	1–8,10,11
$\min\{y/x,1\}$	R with $T=\Pi$	Goguen	1-8,10,11
$\max(1-x,\min(x,y))$	QL with $S=\max, T=\min$	Zadeh	2, 3, 4,5, 13
$\begin{cases} 1 & \text{if } x\le y \\ \max(1-x,y) & \text{if } x>y\end{cases}$	R with $T=\min_0$ S with $S=\max_1$ QL with $T=\min$, $S=\max_1$	Fodor	1–12

Table 1.1: Fuzzy implications

1.9 Other operations on the unit interval

1.9.1 Coimplications

In this section we define operations which are dual to implications. Since t-conorms are linked to t-norms like these new operations do to implications, we will call them *coimplications* (see also Dubois and Prade (1984c)). The reason to introduce them will be clear in Chapter 5.

We start with the following axioms.

I0'. $v(P_2 \setminus P_1)$ depends only on the values $v(P_1)$ and $v(P_2)$.

I1'. If $v(P_1) \le v(P_3)$ then $v(P_2 \setminus P_1) \ge v(P_2 \setminus P_3)$.

I2' If $v(P_2) \le v(P_3)$ then $v(P_2 \setminus P_1) \le v(P_3 \setminus P_1)$.

I3'. If $v(P_1) = 0$ then $v(P_1 \setminus P) = 0$ for any P.

I4'. If $v(P_1) = 1$ then $v(P \setminus P_1) = 0$ for any P.

I5'. If $v(P_1) = 0$ and $v(P_2) = 1$ then $v(P_2 \setminus P_1) = 1$.

In fact, these operations support set-theoretic difference.

By axiom I0', there exists a function $I_c^{\rightarrow} : [0,1]^2 \to [0,1]$ such that

$$v(P_2 \setminus P_1) = I_c^{\rightarrow}(v(P_1), v(P_2)).$$

Moreover, axioms I1'–I5' imply the following conditions for $I_c^{\rightarrow}$:

$$\text{If } x \le z \text{ then } I_c^{\rightarrow}(x,y) \ge I_c^{\rightarrow}(z,y)\ \forall y \in [0,1]. \tag{1.56}$$

$$\text{If } y \le t \text{ then } I_c^{\rightarrow}(x,y) \le I_c^{\rightarrow}(x,t)\ \forall x \in [0,1]. \tag{1.57}$$

$$I_c^{\rightarrow}(x,0) = 0 \quad \forall x \in [0,1]. \tag{1.58}$$
$$I_c^{\rightarrow}(1,y) = 0 \quad \forall y \in [0,1]. \tag{1.59}$$
$$I_c^{\rightarrow}(0,1) = 1. \tag{1.60}$$

Definition 1.17 A function $I_c^{\rightarrow} : [0,1]^2 \rightarrow [0,1]$ is called *coimplication* if it satisfies conditions (1.56) – (1.60).

The simple connection between implications and coimplications is given by the following proposition.

Proposition 1.13 *A function $I_c^{\rightarrow} : [0,1]^2 \rightarrow [0,1]$ is a coimplication if and only if the function*

$$I^{\rightarrow}(x,y) = N(I_c^{\rightarrow}(N(x),N(y))) \tag{1.61}$$

is an implication for any strong negation N.

Proof. Let N be a strong negation, $I_c^{\rightarrow}$ be any function from $[0,1]^2$ to $[0,1]$ and define $I^{\rightarrow}$ by (1.61).

Then it is obvious that $I_c^{\rightarrow}$ fulfils conditions (1.56)–(1.60) if and only if $I^{\rightarrow}$ satisfies (1.44)–(1.48), respectively.

■

Therefore, there is no need to deal with coimplications in details since results on implications can be applied by the duality expressed by (1.61).

For example, coimplications corresponding to R- and S-implications are defined as

$$I_c^{\rightarrow}(x,y) = \inf\{z \mid S(x,z) \geq y\} \tag{1.62}$$

and

$$I_c^{\rightarrow}(x,y) = T(N(x),y) \tag{1.63}$$

Notice that these two classes have been investigated by Dubois and Prade (1984c). To see that both are connectives which support set-difference, consider the following classical identity (X is a given set, $A, B \subseteq X$):

$$B \setminus A = B \cap A^c = \cap\{Z \mid A \cup Z \supseteq B\}.$$

Some particular coimplications are listed below.

Form ($I_c^{\rightarrow}$)	Name
$\min(1-x,y)$	Kleene-Dienes
$y - xy$	Rechenbach
$\max(y-x,0)$	Lukasiewicz
$\begin{cases} 0 & \text{if } x \geq y \\ y & \text{if } x < y \end{cases}$	Gödel
$\max\left(\frac{y-x}{1-x}, 0\right)$	Goguen

Further properties of coimplications can be required as follows.

I6'. $I_c^{\rightarrow}(0, x) = x$.

I7'. $I_c^{\rightarrow}(x, I_c^{\rightarrow}(y, z)) = I_c^{\rightarrow}(y, I_c^{\rightarrow}(x, z))$.

I8'. $x \geq y$ iff $I_c^{\rightarrow}(x, y) = 0$.

I9'. $I_c^{\rightarrow}(x, 1) = N(x)$ is a strong negation.

I10'. $I_c^{\rightarrow}(x, y) \leq y$.

I11'. $I_c^{\rightarrow}(x, x) = 0$.

I12'. $I_c^{\rightarrow}(x, y) = I_c^{\rightarrow}(N(y), N(x))$ with a strong negation N.

I13'. $I_c^{\rightarrow}$ is continuous.

1.9.2 Equivalences

In the spirit of the previous sections, we give the definition of an *equivalence* as a binary operation on the unit interval. Our axioms are given as follows.

E0. $v(P_1 \leftrightarrow P_2)$ depends only on the values $v(P_1)$ and $v(P_2)$.

E1. $v(P_1 \leftrightarrow P_2) = v(P_2 \leftrightarrow P_1)$.

E2. $v(P_1 \leftrightarrow P_2) = 0$ if and only if $v(P_1) = 0$ and $v(P_2) = 1$.

E3. $v(P \leftrightarrow P) = 1$ for all proposition P.

E4. $v(P_0) \leq v(P_1) \leq v(P_2)$ implies $v(P_0 \leftrightarrow P_2) \leq v(P_1 \leftrightarrow P_2)$.

By axiom E0, there exists a function $E : [0,1]^2 \to [0,1]$ such that

$$v(P_1 \leftrightarrow P_2) = E(v(P_1), v(P_2)).$$

Furthermore, E1 – E4 imply the following conditions for E :

$$\begin{aligned} E(x,y) &= E(y,x) \quad \forall x,y \in [0,1], & (1.64)\\ E(0,1) &= E(1,0) = 0 & (1.65)\\ E(x,x) &= 1 \quad \forall x \in [0,1] & (1.66) \end{aligned}$$

$$x \leq x' \leq y' \leq y \Rightarrow E(x,y) \leq E(x',y'). \tag{1.67}$$

Definition 1.18 A function $E : [0,1]^2 \to [0,1]$ is called *equivalence* if it satisfies conditions (1.64) – (1.67).

The following proposition shows a close link between equivalences and implications, as it is in the classical case.

Proposition 1.14 *A function $E : [0,1]^2 \to [0,1]$ is an equivalence if and only if there exists an implication $I^{\rightarrow}$ having also property* I11 *such that*

$$E(x,y) = \min\{I^{\rightarrow}(x,y), I^{\rightarrow}(y,x)\}. \tag{1.68}$$

Proof. Suppose first that $I^{\rightarrow}$ is an implication, i.e., conditions (1.44)–(1.48) are satisfied by $I^{\rightarrow}$. Moreover, I11 is also true for $I^{\rightarrow}$. Then, obviously, E defined by (1.68) is an equivalence.

To prove the converse, suppose that E is an equivalence and define a binary operation $I^{\rightarrow}$ by

$$I^{\rightarrow}(x,y) = \begin{cases} 1 & \text{if} \quad x \leq y \\ E(x,y) & \text{if} \quad x > y. \end{cases} \tag{1.69}$$

We prove that thus defined $I^{\rightarrow}$ satisfies I1 – I5 and I11.

I1 : $x \leq z$ implies $I^{\rightarrow}(x,y) \geq I^{\rightarrow}(z,y)$.

This is obviously true when $x \leq y$ (then $I^{\rightarrow}(x,y) = 1$, by (1.69)).

Thus, suppose that $x > y$. Then applying E4 and E1 to $y = y < x \leq z$, we obtain $E(z,y) \leq E(x,y)$. But $I^{\rightarrow}(z,y) = E(z,y)$ and $I^{\rightarrow}(x,y) = E(x,y)$ since $x > z$ and $x > y$.

I2 : $y \leq t$ implies $I^{\rightarrow}(x,y) \leq I^{\rightarrow}(x,t)$.

The proof is similar to that of I1, we have to apply E4 and E1 to $y \leq t < x = x$.

I3 : $I^{\rightarrow}(0,x) = 1$ is obvious.
I4 : $I^{\rightarrow}(x,1) = 1$ is trivial.
I5 : $I^{\rightarrow}(1,0) = 0$ is true by E2.
I11 : $I^{\rightarrow}(x,x) = 1$ by (1.69).

Thus, we proved that $I^{\rightarrow}$ defined by (1.69) is an implication which fulfils I11.
We have to prove that (1.68) holds. This is obvious when $x = y$.
If $x < y$ then

$$\min(I^{\rightarrow}(x,y), I^{\rightarrow}(y,x)) = I^{\rightarrow}(y,x) = E(y,x) = E(x,y).$$

If $x > y$ then

$$\min(I^{\rightarrow}(x,y), I^{\rightarrow}(y,x)) = I^{\rightarrow}(x,y) = E(x,y).$$

Thus our proposition is proved.

■

Corollary 1.3 *A function $E : [0,1]^2 \rightarrow [0,1]$ is an equivalence if and only if there exists an implication $I^{\rightarrow}$ having also property* I11 *such that*

$$E(x,y) = I^{\rightarrow}(\max(x,y), \min(x,y)). \tag{1.70}$$

Proof. By the previous proposition, (1.69) holds. Then we have

$$E(x,y) = \min(I^{\rightarrow}(x,y), I^{\rightarrow}(y,x)) = I^{\rightarrow}(\max(x,y), \min(x,y)),$$

using I1, I2.

■

Some particular equivalences are listed below.

Form	Name
$1 - \|x - y\|$	Lukasiewicz
$\begin{cases} 1 & \text{if } x = y \\ \min(x,y) & \text{if } x \neq y \end{cases}$	Gödel
$\dfrac{\min(x,y)}{\max(x,y)}$	Goguen

Notice the appearance of R-implications in the particular equivalences. In fact, if one would like to use S-implications then I11 implies

$$S(N(x), x) = 1 \quad \forall x \in [0,1],$$

that is, the law of excluded middle must be satisfied by S and N. For a continuous t-conorm S this condition implies the existence of an automorphism φ of the unit interval such that S is the φ-transform of the Lukasiewicz t-conorm. In this case, using $N(x) = \varphi^{-1}(1 - \varphi(x))$, both R- and S-implications coincide. That is, there is no chance to use an S-implication different from the corresponding R-implication in formula (1.68) or (1.70).

In further investigations of equivalences, there may be considered several other conditions to be satisfied. However, this will be done in Chapter 4 when we study similarity relations in details.

1.9.3 Symmetric sums

Symmetric sums have been introduced and investigated by Silvert (1979). He writes in the last paragraph of the summary: "Perhaps the most interesting feature of the symmetric sum is that it is truly an operator on fuzzy sets and cannot be applied in ordinary set theory, since the symmetric sum of two ordinary sets is a fuzzy set (except for the sum of a set with itself."

Definition 1.19 A function $s : [0,1]^2 \to [0,1]$ is called *symmetric sum* if it satisfies the following conditions:

S1. $s(0,0) = 0,\ s(1,1) = 1$

S2. $s(x,y) = s(y,x)$

S3. s is continuous and nondecreasing in both places

S4. $1 - s(x,y) = s(1-x, 1-y)$.

Condition S4 expresses self-duality of s (compare with De Morgan laws). This condition can be extended by using any strong negation instead of $1 - x$, see Dombi (1982b).

Silvert has proved the following result.

Theorem 1.17 *Any symmetric sum has the following form*

$$s(x,y) = \frac{g(x,y)}{g(x,y)+g(1-x,1-y)},$$

where g is any continuous, nondecreasing and positive function from $[0,1]^2$ to $[0,1]$ with $g(0,0)=0$.

Remark that this representation is not unique: $\lambda g(x,y)$ and $s(x,y)$ generate also $s(x,y)$. The representation implies that

$$s(x, 1-x) = 1/2 \quad \text{for} \quad x \in (0,1).$$

If $g(0,x)=0$ for all $x \in [0,1)$ then $g(0,1)$ is not defined. Otherwise $s(0,1)=1/2$.

Obviously, g can be choosen from t-norms and t-conorms (among others). For example, if $g(x,y)=x+y$ then $s(x,y)=\frac{x+y}{2}$.

Silvert has proved also that the only symmetric sum which is both idempotent and associative on $(0,1)$ is $\text{med}_{1/2}$, where for any $\alpha \in [0,1]$ define

$$\text{med}_\alpha(x,y) = \begin{cases} x & \text{if } \min(\alpha,y) \le x \le \max(\alpha,y) \\ y & \text{if } \min(\alpha,x) \le y \le \max(\alpha,x) \\ \alpha & \text{if } \min(x,y) \le \alpha \le \max(x,y) \end{cases}.$$

Remark that $\text{med}_\alpha(x,y) = \text{med}(x,y,\alpha)$.

When $1-x$ is substituted by a strong negation $N_\varphi(x) = \varphi^{-1}(1-\varphi(x))$ then one can see easily that s is self-dual with respect to N_φ if and only if the function

$$s^\varphi(x,y) = \varphi(s(\varphi^{-1}(x), \varphi^{-1}(y)))$$

is a symmetric sum in the original sense. Using this function and the representation theorem of symmetric sums, s has the following form:

$$s(x,y) = \varphi^{-1}\left(\frac{g(\varphi(x),\varphi(y))}{g(\varphi(x),\varphi(y)) + g(1-\varphi(x), 1-\varphi(y))}\right),$$

where g is any generator function of symmetric sums. In this case $s(x, N_\varphi(x)) = \nu$, where $N_\varphi(\nu) = \nu = \varphi^{-1}(1/2)$.

We mention that Dombi (1982b) has investigated strictly increasing associative symmetric sums on $(0,1)$. Properties and representation of such operations can be found in his paper.

Chapter 2

Valued binary relations

Binary relations play a central role in various fields of mathematics. Especially, equivalence relations and different kinds of ordering relations are employed in basic mathematical models. Typical areas are decision making and measurement theory. In addition, applications of binary relations appear naturally in social sciences.

More realistic theoretical description of relations by using numbers to describe links between two elements of a certain universe of discourse goes back at least to the fifties, see Menger (1951). Then, so-called probabilistic relations have been studied and applied for decision making, mathematical psychology, etc. (see e.g. Fishburn (1973) and Roberts (1976)).

Fuzzy logics provide a natural framework for extending the concept of crisp binary relations, without restricting ourselves to probabilistic structures, by assigning to each ordered pair of elements in the universe of discourse a number from the unit interval — the degree to which the elements in question are in relation, or in other words, the strength of the link between any two elements. This idea was already used in the first paper on fuzzy sets by Zadeh (1965).

Since this book is entirely based on valued binary relations, our aim is to give full details of them in this chapter. To reach this goal, we need some ideas from the theory of crisp (two-valued) binary relations as guide for the valued case.

2.1 Basic notions of crisp binary relations

Assume that A is a given set. A *binary relation* R on A is a subset of the Cartesian product $A \times A$ (very often we write A^2). That is, R is a set of ordered pairs (a, b) such that a and b are in A: $R \subseteq A^2$. If the ordered pair (a, b) belongs to R then both the notations $(a, b) \in R$ or aRb are used indifferently.

The *complement* R^c, the *inverse* R^{-1} and the *dual* R^d are respectively defined as follows:

$$\begin{aligned} (a,b) \in R^c &\iff (a,b) \notin R, \\ (a,b) \in R^{-1} &\iff (b,a) \in R, \\ (a,b) \in R^d &\iff (b,a) \notin R. \end{aligned}$$

It is obvious from this definition that $R^d = (R^{-1})^c = (R^c)^{-1}$.

Since R, R^c, R^{-1}, R^d are subsets of A^2, we can use set-theoretic notations as union, intersection, etc. Let R and Q be two relations on A. R is *contained* in Q ($R \subseteq Q$)

if aRb implies aQb for $a, b \in A$. We can also define the *union* $R \cup Q$, the *intersection* $R \cap Q$ and the *composition* $R \circ Q$ of two binary relations R and Q in the following way:

$$\begin{array}{lcl} a(R \cup Q)b & \iff & aRb \text{ or } aQb, \\ a(R \cap Q)b & \iff & aRb \text{ and } aQb, \\ a(R \circ Q)b & \iff & \text{there exists } c \in A \ : \ aRc \text{ and } cQb. \end{array}$$

Very often $R \circ R$ is denoted by R^2.

2.1.1 Valuation, matrix and graph representation of binary relations

Suppose that R is a binary relation on the set A. One can naturally and easily associate a number from the set $\{0, 1\}$, called *valuation* and denoted by $R(a, b)$, for any ordered pair (a, b) as follows:

$$R(a,b) = \begin{cases} 1 & \text{if } aRb \\ 0 & \text{if } aR^cb \end{cases}$$

When A is a finite set, we can associate also a matrix M^R to the relation R, taking the entry in line a and column b to be $R(a, b)$.

Example 2.1 Suppose that $A = \{a, b, c, d\}$ and

$$R = \{(a, a), (a, c), (b, c), (c, b), (c, d), (d, a), (d, d)\}.$$

Then the matrix M^R is given by

M^R	a	b	c	d
a	1	0	1	0
b	0	0	1	0
c	0	1	0	1
d	1	0	0	1

Every binary relation R on the finite set A can be represented by a *directed graph* (A, R), where A is the set of nodes (vertices) and R is the set of arcs (edges). There exists an arc from a to b if and only if aRb holds. When aRa, one undirected loop is used instead of two directed loops. For example, relation defined in 2.1 is represented by the following directed graph.

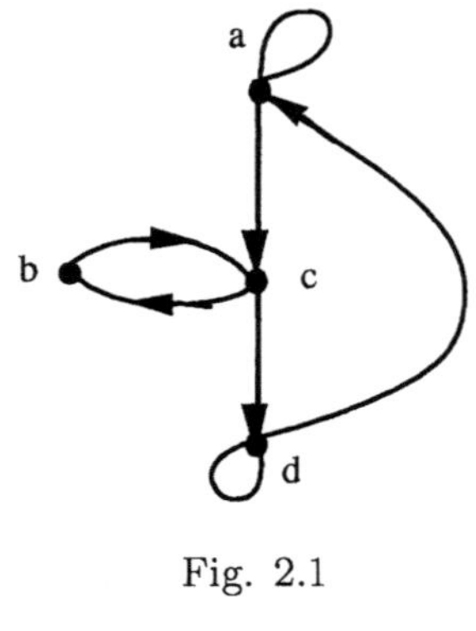

Fig. 2.1

2.1.2 Basic properties of binary relations

Definitions of basic properties are given in Table 2.1. Most of them should be well-known. This terminology corresponds to that of Roubens and Vincke (1985). We include qualification 'for all $a, b, c \in A$' among the definitions below.

A binary relation R on A is

PROPERTY		DEFINITION
reflexive	$\Longleftrightarrow$	aRa
irreflexive	$\Longleftrightarrow$	aR^ca
symmetric	$\Longleftrightarrow$	aRb implies bRa
antisymmetric	$\Longleftrightarrow$	aRb, bRa together imply $a = b$
asymmetric	$\Longleftrightarrow$	aRb implies bR^ca
complete	$\Longleftrightarrow$	aRb or bRa for $a \neq b$
strongly complete	$\Longleftrightarrow$	aRb or bRa
transitive	$\Longleftrightarrow$	aRb, bRc imply aRc
negatively transitive	$\Longleftrightarrow$	aR^cb, bR^cc imply aR^cc
Ferrers relation	$\Longleftrightarrow$	aRb, cRd imply aRd or cRb
semitransitive	$\Longleftrightarrow$	aRb, bRc imply aRd or dRc

Table 2.1: Definition of basic properties of crisp relations

Here we explain only Ferrers property and semitransitivity in details, using graph representation. Ferrers property means that if both arcs (a, b) and (c, d) exist in directed graph (A, R) then either arc (a, d) or arc (c, b) should also exist. This is illustrated on Figure 2.2 below. Semitransitivity indicates that if both arcs (a, b) and (b, c) belongs to directed graph (A, R) then either arc (a, d) or arc (d, c) should also belong to (A, R), see Figure 2.3.

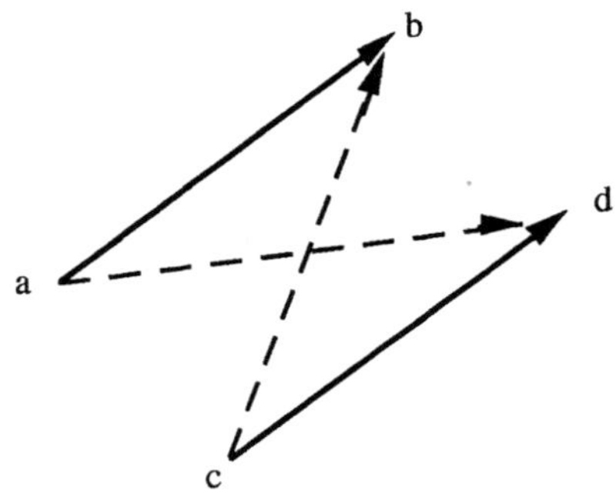

Fig. 2.2 Ferrers property

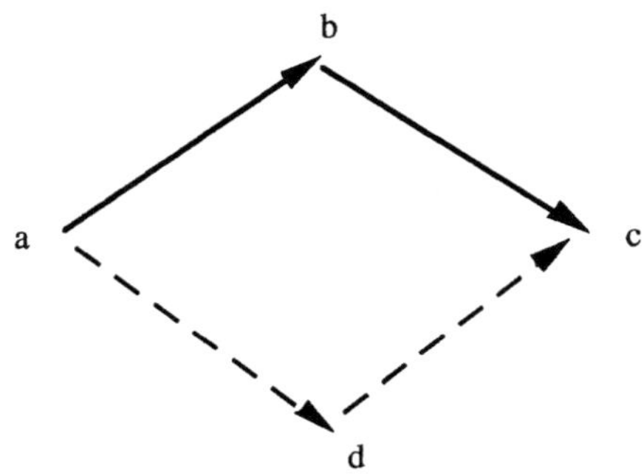

Fig. 2.3 Semitransitivity

Proposition 2.1 *The following statements hold for a binary relation R on A:*

(a) If R is asymmetric then R is irreflexive.

(b) If R is irreflexive and transitive then R is asymmetric.

(c) R is negatively transitive if and only if $(a,b) \notin R$, $(b,c) \notin R$ imply $(a,c) \notin R$.

(d) If R is irreflexive and either semitransitive or Ferrers relation then R is transitive.

Proof. Immediate. ■

All the basic properties defined in Table 2.1 can be expressed in terms of linear inequalities using evaluations $R(a,b)$ as follows.

A binary relation R on A is
- reflexive if $R(a,a) = 1$ for all $a \in A$;
- irreflexive if $R(a,a) = 0$ for all $a \in A$;
- symmetric if $R(a,b) = R(b,a)$ for all $a,b \in A$;
- antisymmetric if $R(a,b) + R(b,a) \leq 1$ for all $a \neq b$;
- asymmetric if $R(a,b) + R(b,a) \leq 1$ for all $a,b \in A$;
- complete if $R(a,b) + R(b,a) \geq 1$ for all $a \neq b \in A$;

– strongly complete if $R(a,b) + R(b,a) \geq 1$ for all $a, b \in A$;
– transitive if $R(a,c) \geq R(a,b) + R(b,c) - 1$ for all $a, b, c \in A$;
– negatively transitive if $R(a,c) \leq R(a,b) + R(b,c)$ for all $a, b, c \in A$;
– Ferrers relation if $R(a,d) + R(c,b) \geq R(a,b) + R(c,d) - 1$ for all $a, b, c, d \in A$;
– semitransitive if $R(a,d) + R(d,c) \geq R(a,b) + R(b,c) - 1$ for all $a, b, c, d \in A$.

These expressions may recall t-norms and t-conorms in Chapter 1 to readers' mind. This will be clear in the next section when we investigate valued binary relations in detail.

It is obvious that properties of R and those of its dual R^d correspond to each other as it is given in Table 2.2.

R	R^d
reflexive	irreflexive
irreflexive	reflexive
symmetric	symmetric
antisymmetric	complete
asymmetric	strongly complete
complete	antisymmetric
strongly complete	asymmetric
transitive	negatively transitive
negatively transitive	transitive
Ferrers relation	Ferrers relation
semitransitive	semitransitive

Table 2.2: Properties of R and R^d

Closing this section, it is worth mentioning that particular classes (equivalence and different kinds of orders) of binary relations are usually introduced in the following way.

A binary relation R on A is called
– *equivalence* relation if R is reflexive, symmetric and transitive;
– *tournament* if R is asymmetric and complete;
– *partial preoder* (or a quasiorder) if R is reflexive and transitive;

– *partial order* if R is antisymmetric and transitive;
– *strict partial order* if R is asymmetric and transitive;
– *total preorder* (or a linear quasiorder) if R is strongly complete and transitive;
– *total order* (or a linear order) if R is a complete partial order (that is, R is antisymmetric, complete and transitive);
– *strict total order* if R is a complet strict partial order (that is, R is asymmetric, complete and transitive);
– *weak order* if R is asymmetric and negatively transitive;
– *interval order* if R is complete and Ferrers;
– *semiorder* if R is a semitransitive interval order (that is, R is complete, Ferrers and semitransitive).

We will analyze corresponding valued classes together with some other important valued preference structures in Chapter 4.

2.2 Valued binary relations

Assume that A is a given set and (T, S, n) is a De Morgan triple modeling **AND**, **OR** and **NOT**, respectively. Valued binary relations are introduced naturally in the following way.

Definition 2.1 A *valued binary relation* R on the set A is a function $R : A \times A \to [0, 1]$.

That is, R is such that for any $a, b \in A$ the value $R(a, b)$ is understood as a degree (truth value) to which the interaction between a and b holds. For example, if A is a set of alternatives and R is a *valued preference relation* on A then $R(a, b)$ is the truth value of the statement 'a is not worse than b'.

Using the terminology of fuzzy sets, $R : A \times A \to [0, 1]$ is a fuzzy subset of the set $A \times A$, thus R might be called a fuzzy binary relation. Therefore, we can take the intersection or the union of valued binary relations as it was introduced for fuzzy subsets in Chapter 1.

The *complement* R^c, the *inverse* R^{-1} and the *dual* R^d of a given valued relation R are defined as follows $(a, b \in A)$:

$$\begin{aligned} R^c(a,b) &= n(R(a,b)), \\ R^{-1}(a,b) &= R(b,a), \\ R^d(a,b) &= n(R(b,a)). \end{aligned}$$

Notice that $R^d = (R^{-1})^c = (R^c)^{-1}$.

Suppose that R_1, R_2 are valued binary relations on A. Their composition can be defined in a natural way, taking into account the corresponding notion for crisp binary relations.

Definition 2.2 The *T-composition* $(R_1 \circ_T R_2)$ of R_1 and R_2 is a valued binary relation defined by

$$(R_1 \circ_T R_2)(a,b) = \sup_{c \in A} T(R_1(a,c), R_2(c,b)). \tag{2.1}$$

Indeed, if Q_1 and Q_2 are crisp binary relations on A then $a(Q_1 \circ Q_2)b$ if and only if **THERE EXISTS** an element $c \in A$ such that aQ_1c **AND** cQ_2b. This corresponds to (2.1) in the valued case.

Definition 2.3 For any valued binary relations R_1, R_2 on A we say that R_1 is *contained in* R_2 and denote by $R_1 \subseteq R_2$ if and only if for all $a, b \in A$ we have inequality $R_1(a,b) \leq R_2(a,b)$. R_1 and R_2 are said to be *equal* if and only if $R_1(a,b) = R_2(a,b)$ for all $a, b \in A$.

Due to associativity and nondecreasingness of T, one can prove easily the following statement.

Proposition 2.2 *Suppose that R_1, R_2 and R_3 are valued binary relations on A. Then*

(a) $R_1 \circ_T (R_2 \circ_T R_3) = (R_1 \circ_T R_2) \circ_T R_3$,

(b) $R_1 \subseteq R_2$ *implies* $R_1 \circ_T R_3 \subseteq R_2 \circ_T R_3$ *and* $R_3 \circ_T R_1 \subseteq R_3 \circ_T R_2$.

Proof. (a) For any $a, b \in A$ we have the following chain of equalities:

$$\begin{aligned}
((R_1 \circ_T R_2) \circ_T R_3)(a,b) &= \sup_{c \in A} T((R_1 \circ_T R_2)(a,c), R_3(c,b)) \\
&= \sup_{c \in A} T(\sup_{d \in A}(T(R_1(a,d), R_2(d,c)), R_3(c,b)) \\
&= \sup_{c \in A} \sup_{d \in A} T(T(R_1(a,d), R_2(d,c)), R_3(c,b)) \\
&= \sup_{c \in A} \sup_{d \in A} T(R_1(a,d), T(R_2(d,c), R_3(c,b))) \\
&= \sup_{d \in A} T(R_1(a,d), \sup_{c \in A} T(R_2(d,c), R_3(c,b))) \\
&= \sup_{d \in A} T(R_1(a,d), R_2 \circ_T R_3(d,b)) \\
&= (R_1 \circ_T (R_2 \circ_T R_3))(a,b).
\end{aligned}$$

(b) can be proved similarly, using nondecreasingness of T. ■

2.3 Traces of valued binary relations

The main aim of this section is to define the right and left trace of a valued binary relation (see Fodor (1992a)). Their role will be emphasized later in characterizing basic properties of valued binary relations.

The history of those binary relations what we call now traces goes back at least to the fifties. In the crisp case the term 'trace' was explicitly used first by Luce (1956). In case of probabilistic relations, it is also Luce (1958) who introduced this notion with its name (without name, this notion can also be found in Fishburn (1973), Roberts (1971) and Miller (1980)).

Right and left traces of crisp and valued binary relations have also been defined and exploited in Doignon et al. (1986). Supposing that A is a given set and R is a crisp binary relation on A, two binary relations R^ℓ and R^r can be associated with R as follows:

$$aR^\ell b \iff cRa \text{ implies } cRb \; \forall c \in A, \tag{2.2}$$
$$aR^r b \iff bRc \text{ implies } aRc \; \forall c \in A, \tag{2.3}$$

for all $a, b \in A$. R^ℓ and R^r are called the left and right trace of R, respectively. Observe that R^ℓ is the greatest relation U on A such that

$$R \circ U \subseteq R, \tag{2.4}$$

while R^r is the greatest relation V on A such that

$$V \circ R \subseteq R. \tag{2.5}$$

Indeed, if we assume that (2.4) is satisfied by a binary relation U and there exists $a, b \in A$ such that aUb and not $aR^\ell b$, then, by definition (2.2) of R^ℓ, there exists $c \in A$ such that cRa and not cRb. But cRa, aUb imply cRb; this is a contradiction. The other case can be verified similarly.

If R is a valued binary relation on A then Doignon et al. (1986) defined two crisp binary relations on A associated with R by

$$aR^\ell b \iff R(c,a) \leq R(c,b) \; \forall c \in A, \tag{2.6}$$
$$aR^r b \iff R(a,c) \geq R(b,c) \; \forall c \in A, \tag{2.7}$$

calling R^ℓ and R^r the left and right trace of R, respectively.

These definitions were revised by Fodor (1992a). We follow that line now. To define traces, we set out from (2.2) and (2.3). Suppose that (T, S, n) is a De Morgan triple with T being left-continuous and denote the R-implication defined from T by $I_T^{\rightarrow}$, as usual.

Definition 2.4 Let R be a valued binary relation on A. The *left trace* of R is defined by

$$R^\ell(a,b) = \inf_{c \in A} I_T^{\rightarrow}(R(c,a), R(c,b)), \tag{2.8}$$

and the *right trace* of R is determined by

$$R^r(a,b) = \inf_{c \in A} I_T^{\rightarrow}(R(b,c), R(a,c)). \tag{2.9}$$

Main properties of the traces are summarized in the next proposition.

Proposition 2.3 *For any valued binary relation R on A the following statements are true:*

(a) If R is a crisp relation then Definition 2.4 gives back formulae (2.2) and (2.3).
(b) R^ℓ is the greatest valued binary relation U on A such that $R \circ U \subseteq R$.
(c) R^r is the greatest valued binary relation V on A such that $V \circ R \subseteq R$.
(d) $R = R \circ R^\ell = R^r \circ R$.

Proof. (a) is obviously true.

(b) $R \circ U \subseteq R$ yields

$$\sup_{c \in A} T(R(a,c), U(c,b)) \leq R(a,b),$$

or equivalently,

$$T(R(a,c), U(c,b)) \leq R(a,b) \;\forall a,b,c \in A.$$

By left-continuity of T, $T(x,y) \leq z$ if and only if $I_T^{\rightarrow}(x,z) \geq y$, see Chapter 1. Applying this for the last inequality, it is true if and only if

$$U(c,b) \leq I_T^{\rightarrow}(R(a,c), R(a,b)) \;\forall a,b,c \in A,$$

that is, if and only if $U(c,b) \leq R^{\ell}(c,b)$.

(c) can be proved as statement (b).

(d) We have

$$R \circ R^{\ell}(a,b) = \sup_{c \in A} T(R(a,c), R^{\ell}(c,b)) \geq T(R(a,b), R^{\ell}(b,b)) = R(a,b)$$

since R^{ℓ} is reflexive. On the other hand, $R \circ R^{\ell} \subseteq R$ holds by (b).

Equality $R = R^{r} \circ R$ can be proved similarly.

■

The definition of traces is closely related to *composite fuzzy relational equations*:

$$R_1 = R_2 \circ_T R_3,$$

where T is a left-continuous t-norm and in the general case R_1, R_2 and R_3 are fuzzy binary relations on some finite sets $A \times B$, $A \times C$ and $C \times B$, respectively (see Miyakoshi and Shimbo (1985), Di Nola et al. (1989)). Then two problems can be formulated:

(a) determine R_2 if R_1 and R_3 are given,

(b) determine R_3 if R_1 and R_2 are given.

Considering the particular case when $B = C = A$ and $R_1 = R_2 = R$, the left trace $R_3 = R^{\ell}$ is the greatest solution of the relational equation

$$R = R \circ_T R_3,$$

while choosing $R_1 = R_2 = R$, the right trace $R_2 = R^{r}$ is the greatest solution of

$$R = R_2 \circ_T R.$$

For other issues on fuzzy relational equations we refer to the book of Di Nola et al. (1989).

2.4 Cut relations

A usual way to study valued binary relations is to generalize definitions and results known for crisp relations using our 'dictionary' of two-valued and fuzzy logic. This will be done in the next section.

Another way is to consider *cut relations* associated with the valued relation: for a given $\lambda \in [0,1]$ the relation R_λ is defined as the set of ordered pairs with values not less than λ:

$$R_\lambda = \{(a,b) \in A^2 \mid R(a,b) \geq \lambda\}.$$

These λ-cuts R_λ form a chain (a nested family) of relations.

It is intuitively clear that these two approaches are equivalent when we use connectives 'min' and 'max' for modelling fuzzy conjunction and disjunction, respectively. However, this is not the case for other models (e.g. t-norms and t-conorms), as we can see by the following proposition below which is a slightly modified version of a result proved by Klement (1981b). Moreover, it is also obvious that studying these chains of λ-cuts is the same as studying valued binary relations. We give a more complete and rigorous explanation after the next statement, based on the paper of Doignon et al. (1986).

Proposition 2.4 *Suppose that for all valued binary relations R and Q on A and for all $\lambda \in [0,1]$ we have*

$$\begin{aligned} (R \cap Q)_\lambda &= R_\lambda \cap Q_\lambda, \\ (R \cup Q)_\lambda &= R_\lambda \cup Q_\lambda. \end{aligned}$$

Then $(R \cap Q)(a,b) = \min(R(a,b), Q(a,b))$ and $(R \cup Q)(a,b) = \max(R(a,b), Q(a,b))$.

Definition 2.5 A *tower of relations* (or simply a tower) is a mapping F from $[0,1]$ to the subsets of A^2 satisfying the following conditions:

AT1. F is nonincreasing: $\lambda \leq \lambda'$ implies $F(\lambda) \supseteq F(\lambda')$

AT2. for every $(a,b) \in A^2$ there exists $\max\{\lambda | (a,b) \in F(\lambda)\}$.

It is clear that F satisfies also the following conditions:

AT3. for every $Q \subseteq A^2$ there exists $\max\{\lambda | Q \subseteq F(\lambda)\}$

AT4. $F^{-1}(A^2) \neq \emptyset$.

Definition 2.6 A *chain of relations* (or simply a chain) is a mapping C from $\Sigma = \{0 \leq \lambda_1 < \lambda_2 < \ldots < \lambda_m \leq 1\}$ (any finite subset of $[0,1]$) to the subsets of A^2 satisfying the following three conditions:

AC0. $\emptyset \subset C(\lambda_m)$

AC1. $\lambda_i < \lambda_{i+1}$ implies $C(\lambda_i) \supset C(\lambda_{i+1})$

AC2. for every $(a,b) \in A^2$ there exists $\max\{\lambda \in \Sigma | (a,b) \in C(\lambda)\}$.

It is clear that C satisfies also the following two conditions:

AC3. for every $Q \subseteq A^2$ there exists $\max\{\lambda \in \Sigma | Q \subseteq C(\lambda)\}$

AC4. $C(\lambda_1) = A^2$.

The image $C(\lambda_i)$ is a crisp binary relation on A and we denote it by Q_{λ_i}. Thus, the chain C can also be described by the following chain $\mathcal{C}$ of relations:

$$\emptyset \subset Q_{\lambda_m} \subset Q_{\lambda_{m-1}} \subset \ldots \subset Q_{\lambda_1} = A^2.$$

Given any object of one of the three types defined above, we can associate with this one objects of the two other types via the following constructions:

If R is a given valued binary relation then define

- a tower F_R by $F_R(\lambda) = \{(a,b) \in A^2 | R(a,b) \geq \lambda\}, \quad \lambda \in [0,1]$;
- a chain C_R by $C_R(\lambda_i) = \{(a,b) \in A^2 | R(a,b) \geq \lambda_i\}, \quad \lambda_i \in \Sigma$.

Then $F_R(\lambda)$ (or $C_R(\lambda)$) is called the λ-*cut* of the valued binary relation R.

If F is a given tower then define

- a valued binary relarion R_F on A by $R_F(a,b) = \max\{\lambda | (a,b) \in F(\lambda)\}, \quad (a,b) \in A^2$;
- a chain C_F by $C_F(\lambda_i) = F(\lambda_i), \quad \lambda_i \in \Sigma$, where

$$\Sigma = \{\lambda \in [0,1) | \forall\ \lambda < \lambda',\ F(\lambda') \subset F(\lambda)\} \ \cup\ \{\max\{\lambda | F(\lambda) \supset \emptyset\}\}.$$

If C is a given chain then define

- a valued binary relation R_C by $R_C(a,b) = \max\{\lambda | (a,b) \in C(\lambda)\}, \quad (a,b) \in A^2$;
- a tower F_C by

$$F_C(\lambda) = \begin{cases} A^2 & \text{for every } \lambda \in [0, \lambda_1] \\ C(\lambda_i) & \text{for every } \lambda \in (\lambda_{i-1}, \lambda_i],\ i = 2,3,\ldots,m \\ \emptyset & \text{for every } \lambda \in (\lambda_m, 1] \end{cases}$$

Then one can easily prove the following result.

Proposition 2.5 *Both the set of all towers of relations on A and the set of all chains of relations on A are bijective images of the set of all valued binary relations on A.*

Proof. The composite mappings R_{F_R}, F_{R_F}, etc, are the identity mappings whence the statement follows. ■

Although we will not use all the above notions and results extensively in the book, the explained connections can help the reader in theoretically sound applications.

2.5 Basic properties of valued binary relations

In this section we list the basic properties of valued binary relations, supposing that a De Morgan triple (T, S, N) is given, so that N is a strong negation and T is a left-continuous t-norm. That is, the following condition is satisfied for all $x, y, z \in [0, 1]$ (see Section 1.8) :

$$T(x, z) \leq y \quad \Leftrightarrow \quad I_T^{\rightarrow}(x, y) \geq z. \tag{2.10}$$

All definitions are based on translations of crisp properties using our dictionary. If $T = \min$ and/or $S = \max$, we simply write e.g. complete, transitive instead of max-complete, min-transitive, etc.

2.5.1 Reflexivity, irreflexivity and symmetry

Properties mentioned in the title are defined so that they are independent of the De Morgan triple (T, S, N).

Definition 2.7 A valued binary relation R on A is called *reflexive* if

$$R(a, a) = 1 \quad \text{for all } a \in A.$$

This definition have been proposed by Zadeh (1971). Other possible extensions of reflexivity for the valued case are known as well. For instance, Yeh (1973) called a valued binary relation R on A

- ε-*reflexive* if $R(a, a) \geq \varepsilon \; \forall a \in A$ ($\varepsilon \in (0, 1]$);
- *weakly reflexive* if $R(a, a) \geq R(a, b) \; \forall \; a, b \in$A.

Each of these notions gives back classical reflexivity when we apply them for crisp binary relations. Remark that ε-reflexivity appears naturally in Section 2.5.2 as a property which is implied by T-asymmetry when T has zero divisors. Moreover, ε-reflexivity with $\varepsilon = 1$ is just usual reflexivity. Clearly, if R is reflexive then it is ε-reflexive for all $\varepsilon \leq 1$ and is also weakly reflexive.

Concerning traces, they are always reflexive relations.

Proposition 2.6 *For any valued binary relation R on A, its traces R^r and R^ℓ are reflexive valued binary relations on A.*

Proof.

$$R^\ell(a, a) = \inf_{c \in A} I_T^{\rightarrow}(R(c, a), R(c, a)) = \inf_{c \in A} 1 = 1$$

since $I_T^{\rightarrow}(x, x) = 1$ for all $x \in [0, 1]$.

The same holds for R^r. ■

Using traces of R, the following statement can be proved easily.

Proposition 2.7 *The following three statements are equivalent :*
(a) *R is reflexive;*
(b) *$R^\ell \subseteq R$;*
(c) *$R^r \subseteq R$.*

Proof. (a) $\Longrightarrow$ (b) :

$$R^{\ell}(a,b) = \inf_{c\in A} I_T^{\rightarrow}(R(c,a), R(c,b)) \leq I_T^{\rightarrow}(R(a,a), R(a,b)) = R(a,b).$$

(b) $\Longrightarrow$ (a) : If $R^{\ell}(a,b) \leq R(a,b)$ then $1 = R^{\ell}(a,a) \leq R(a,a) \leq 1$ implies $R(a,a) = 1$.
(a) $\Longleftrightarrow$ (c) can be proved analogously.

■

The following simple proposition supports Definition 2.7.

Proposition 2.8 *If a valued binary relation R on A is reflexive then crisp relations R_λ are reflexive for all $\lambda \in (0,1]$.*

Proof. Immediate. ■

This statement is not true in general for ε-reflexive or weakly reflexive valued binary relations.

Irreflexive (sometimes called antireflexive) valued binary relations are introduced now.

Definition 2.8 A valued binary relation R on A is *irreflexive* if

$$R(a,a) = 0 \quad \text{for every } a \in A.$$

We left to the reader the definition of both ε-irreflexive and weakly irreflexive valued relations.

Characterizations of irreflexivity are given in terms of traces.

Proposition 2.9 *All the following statements are equivalent:*
(a) *R is irreflexive;*
(b) $(R^d)^{\ell} \subseteq R^d$;
(c) $(R^d)^r \subseteq R^d$;
(d) $(R^c)^{\ell} \subseteq R^c$;
(e) $(R^c)^r \subseteq R^c$;

Proof. R is irreflexive if and only if R^d is reflexive and if and only if R^c is reflexive. Then applying Proposition 2.7 for R^d and R^c, we obtain immediately the statement.

■

Cut relations behave as it is expected.

Proposition 2.10 *If a valued binary relation R on A is irreflexive then crisp relations R_λ are irreflexive for all $\lambda \in (0,1]$.*

Proof. Obvious. ■

Finally, symmetry of valued binary relations is defined in the most natural way.

Definition 2.9 A valued binary relation R on A is called *symmetric* if

$$R(a,b) = R(b,a) \quad \text{for all } a,b \in A.$$

Clearly, for any valued binary relation R on A, the following valued binary relation

$$I(a,b) = i(R(a,b), R(b,a))$$

is symmetric for any symmetric mapping $i : [0,1]^2 \to [0,1]$. This type of valued relations will be investigated in an axiomatic framework in Chapter 3 as indifference relation associated with a given (weak) preference R.

2.5.2 Antisymmetry and asymmetry

It is obvious that antisymmetry and asymmetry should depend on the particular model of conjunction. Indeed, classical antisymmetry is clearly equivalent to

$$a \neq b \text{ implies } (a,b) \notin R \cap R^{-1}.$$

while asymmetry holds if and only if

$$(a,b) \notin R \cap R^{-1}.$$

Therefore, we give the following definition.

Definition 2.10 A valued binary relation R on A is *T-antisymmetric* if

$$a \neq b \quad \text{implies} \quad T(R(a,b), R(b,a)) = 0.$$

R is *T-asymmetric* if

$$T(R(a,b), R(b,a)) = 0 \quad \text{for every } a,b \in A.$$

If T is a continuous Archimedean t-norm then we have the following equivalent form of T-antisymmetry and T-asymmetry.

Lemma 2.1 *Suppose that T is a continuous Archimedean t-norm with additive generator f. Then the following inequality*

$$f(R(a,b)) + f(R(b,a)) \geq f(0) \tag{2.11}$$

is equivalent to T-antisymmetry if (2.11) holds for $a \neq b$ and equivalent to T-asymmetry if (2.11) holds for all $a,b \in A$.

Proof. By Definition 2.10, R is T-asymmetric if and only if

$$f^{(-1)}(f(R(a,b)) + f(R(b,a))) = 0$$

for all $a,b \in A$. Applying f to both sides of this equation, we have

$$\min\{f(R(a,b)) + f(R(b,a)), f(0)\} = f(0),$$

by definition of $f^{(-1)}$. This last equation holds if and only if

$$f(R(a,b)) + f(R(b,a)) \geq f(0).$$

This proves the statement for T-asymmetry. T-antisymmetry can be handled in the same way.

■

Note that Zadeh (1971) called a relation R *perfectly antisymmetric* when Definition 2.10 holds with $T = \min$. Exactly the same property can be revealed when T is a positive t-norm (e.g. the product Π).

Proposition 2.11 *Suppose that T is a positive t-norm. Then a valued binary relation R is T-antisymmetric (T-asymmetric) if and only if R is antisymmetric (asymmetric).*

Proof. $T(R(a,b), R(b,a)) = 0$ if and only if at least one of $R(a,b)$ and $R(b,a)$ is zero. ■

In some cases, antisymmetry is a very restrictive condition (see e.g. Chapter 3). For example, relations when $0 < R(a,b) < 1$ and $0 < R(b,a)$ is small (close to zero) are excluded.

Concerning cut relations, one can prove easily the following statement.

Proposition 2.12 *If a valued binary relation R on A is antisymmetric (asymmetric) then its cut relations are antisymmetric (asymmetric) crisp relations for all $\lambda \in (0,1]$.*

Proof. Easy. ■

Clearly, if a valued binary relation R on A is T-antisymmetric (T-asymmetric) for a certain t-norm T, then R is T'-antisymmetric (T'-asymmetric) for any t-norm T' such that $T' \leq T$. Therefore, an antisymmetric relation is T-antisymmetric for any t-norm T.

Remarkable new definitions of antisymmetry (asymmetry) (that is, different from the case min) are obtained by using t-norms with zero divisors. In these cases, roughly speaking, values $R(a,b)$ and $R(b,a)$ cannot be too high at the same time. For example, considering the Lukasiewicz t-norm $T(x,y) = \max\{x+y-1, 0\}$ (or the nilpotent minimum), T-antisymmetry (T-asymmetry) indicates the following condition:

$$R(a,b) + R(b,a) \leq 1.$$

Finally, connections between irreflexivity and T-asymmetry are given in the following statement.

Proposition 2.13 *For each valued binary relation R on A,*

(a) T-asymmetry implies irreflexivity if and only if T is a positive t-norm;

(b) T-asymmetry implies ε-irreflexivity with $\varepsilon < 1$ if and only if T has zero divisors and $\varepsilon \leq \sup\{x | T(x,x) = 0\}$.

Proof. (a) $T(R(a,a), R(a,a)) = 0$ implies $R(a,a) = 0$ if and only if $[T(x,x) = 0 \iff x = 0]$ holds, i.e., if and only if T is positive.

(b) $T(R(a,a), R(a,a)) = 0 \Leftrightarrow R(a,a) \leq \sup\{x \mid T(x,x) = 0\}$.

■

2.5.3 Completeness and strong completeness

In this subsection we define and investigate possible extensions of crisp complete and strongly complete relations for the valued case.

Definition 2.11 A valued binary relation R on A is *S-complete* if

$$a \neq b \quad \text{implies} \quad S(R(a,b), R(b,a)) = 1.$$

R is *strongly S-complete* if

$$S(R(a,b), R(b,a)) = 1 \quad \text{for all } a, b \in A.$$

Recalling that $S(x,y) = N(T(N(x), N(y)))$ in (T, S, N), S-completeness and T-antisymmetry (strong S-completeness and T-asymmetry) are dual properties. That is, the following statement is valid.

Proposition 2.14 *A valued binary relation R on A is S-complete (strongly S-complete) if and only if its dual R^d is T-antisymmetric (T-asymmetric) on A.*

Proof. The statement follows from the fact that

$$S(x,y) = 1 \quad \text{if and only if} \quad T(N(x), N(y)) = 0.$$

■

When S is a continuous Archimedean t-conorm then the following equivalent forms of S-completeness (strong S-completeness) can be proved.

Lemma 2.2 *Suppose that S is a continuous Archimedean t-conorm with additive generator g. Then the following inequality*

$$g(R(a,b)) + g(R(b,a)) \leq g(1)$$

is equivalent to S-completeness when it holds for $a \neq b$ and to strong S-completeness if it holds for all $a, b \in A$.

Proof. One can prove this statement similarly to Lemma 2.1.

■

Using duality and Proposition 2.14, it is easy to prove the following result.

Proposition 2.15 *Assume that T is a positive t-norm in the De Morgan triple (T, S, N). Then a valued binary relation R on A is S-complete (strongly S-complete) if and only if R is complete (strongly complete) on A.*

Proof. Since $S(x,y) = N(T(N(x), N(y)))$, $S(R(a,b), R(b,a)) = 1$ if and only if either $R(a,b) = 1$ or $R(b,a) = 1$, or both. That is, if and only if $\max(R(a,b), R(b,a)) = 1$. ■

Cut relations behave in this case as it is expected.

Proposition 2.16 *If a valued binary relation R on A is complete (strongly complete) then its cut relations R_λ are complete (strongly complete) crisp binary relations for all $\lambda \in (0, 1]$.*

Proof. Easy. ■

Proposition 2.17 *If R is S-complete (strongly S-complete) on A then it is S'-complete (strongly S'-complete) on A for any t-conorm S' such that $S' \geq S$.*

Proof. Obvious. ■

As in the case of antisymmetry or asymmetry, S-completeness (strong S-completeness) differs from the case of 'max' if and only if S is a nilpotent t-conorm. In this case one may have $R(a,b) < 1, R(b,a) < 1$ if (at least) one of these two values is high enough. For instance, considering the Lukasiewicz t-conorm $S(x,y) = \min\{x+y, 1\}$, or the nilpotent maximum, S-completeness (strong S-completeness) means that

$$R(a,b) + R(b,a) \geq 1.$$

There is some connection between reflexivity and strong completeness, similarly to the crisp case. However, different kinds of reflexivity are obtained for positive t-norms and nilpotent t-conorms, respectively.

Proposition 2.18 *For each valued binary relation R on A,*

(a) strong S-completeness implies reflexivity if and only if T is a positive t-norm;

(b) strong S-completeness implies ε-reflexivity with $\varepsilon < 1$ if and only if S is a nilpotent t-conorm and $\varepsilon \geq \inf\{x | S(x,x) = 1\}$.

Proof. Statements follow from the properties of positive t-norms and nilpotent t-conorms, respectively. ■

2.5.4 Transitivity

Transitivity is certainly one of the most important properties concerning either equivalences or different types of orders. The idea behind transitivity is that "the strength of the link between two elements must be greater than or equal to the strength of any indirect chain (i.e., involving other elements)", see Dubois and Prade (1980). This is expressed in the following definition (see also Zadeh (1971)).

Definition 2.12 A valued binary relation R on A is *T-transitive* if

$$T(R(a,c), R(c,b)) \leq R(a,b) \tag{2.12}$$

for all $a, b, c \in A$.

Both left and right traces of a valued binary relation are T-transitive relations as we prove now.

Proposition 2.19 *For any valued binary relation R on A, its traces R^ℓ and R^r are T-transitive valued binary relations on A.*

Proof. We prove the statement only for R^ℓ, the other case can be handled in the same way. R^ℓ is T transitive if and only if, for all $a, b, d \in A$, we have

$$T(R^\ell(a,d), R^\ell(d,b)) \leq R^\ell(a,b).$$

But

$$\begin{aligned} T(R^\ell(a,d), R^\ell(d,b)) &= T\left[\inf_{c\in A} I_T^{\rightarrow}(R(c,a),R(c,d)), \inf_{c\in A} I_T^{\rightarrow}(R(c,d),R(c,b))\right] \\ &\leq \inf_{c\in A} T\left[I_T^{\rightarrow}(R(c,a),R(c,d)), I_T^{\rightarrow}(R(c,d),R(c,b))\right] \\ &\leq \inf_{c\in A} I_T^{\rightarrow}(R(c,a),R(c,b)) = R^\ell(a,b). \end{aligned}$$

Here we used Proposition 1.10.

■

Characterization of T-transitivity with the help of traces is given in the next proposition.

Proposition 2.20 *The next three statements are equivalent :*
(a) *R is T-transitive;*
(b) *$R \subseteq R^\ell$;*
(c) *$R \subseteq R^r$.*

Proof. R is T-transitive if and only if

$$T(R(a,c), R(c,b)) \leq R(a,b) \quad \forall a,b,c \in A.$$

T is left-continuous, thus this inequality is equivalent to either

$$I_T^{\rightarrow}(R(a,c), R(a,b)) \geq R(c,b),$$

or

$$I_T^{\rightarrow}(R(c,b), R(a,b)) \geq R(a,c),$$

that is, either $R^\ell \supseteq R$ or $R^r \supseteq R$.

■

2.5.5 Representation of transitive relations

Using the previous proposition, general representation theorems of T-transitive relations can be established.

Remark that T-transitive and reflexive relations (in other words : T-preorders) have been characterized by Valverde (1985), using some important ideas of Ovchinnikov (1984). Our result has relevance in the theory of orders (see Chapter 4) where reflexivity does not play any role. Instead, a kind of asymmetry or antisymmetry is used with transitivity to obtain an order. Therefore, in that context, the representation theorem of Valverde (1985) is not applicable.

Theorem 2.1 *Let R be a valued binary relation on A. Then R is T-transtive if and only if there exist two families $\{f_\gamma\}_{\gamma\in\Gamma}$, $\{g_\gamma\}_{\gamma\in\Gamma}$ of functions from A to $[0,1]$ such that $f_\gamma(a) \geq g_\gamma(a)$ for all $a \in A$, $\gamma \in \Gamma$ and*

$$R(a,b) = \inf_{\gamma\in\Gamma} I_T^{\rightarrow}(f_\gamma(a), g_\gamma(b)). \tag{2.13}$$

Proof. Suppose first that R is given by (2.13). Then we have

$$\begin{aligned} T(R(a,b),R(b,c)) &= T\left[\inf_{\gamma\in\Gamma} I_T^{\rightarrow}(f_\gamma(a),g_\gamma(b)), \inf_{\gamma\in\Gamma} I_T^{\rightarrow}(f_\gamma(b),g_\gamma(c))\right] \\ &\leq \inf_{\gamma\in\Gamma} T\left[I_T^{\rightarrow}(f_\gamma(a),g_\gamma(b)), I_T^{\rightarrow}(f_\gamma(b),g_\gamma(c))\right] \\ &\leq \inf_{\gamma\in\Gamma} T\left[I_T^{\rightarrow}(f_\gamma(a),f_\gamma(b)), I_T^{\rightarrow}(f_\gamma(b),g_\gamma(c))\right] \\ &\leq \inf_{\gamma\in\Gamma} I_T^{\rightarrow}(f_\gamma(a),g_\gamma(c)) = R(a,c). \end{aligned}$$

Here we used Proposition 1.10 and the fact that $I_T^{\rightarrow}$ is nondecreasing with respect to its second argument.

To prove the converse, suppose that R is T-transitive and define

$$\Gamma = A, \quad f_c(a) = R^r(c,a), \quad g_c(a) = R(c,a).$$

By the previous proposition, $R \subseteq R^r$, i.e. $g_c(a) \leq f_c(a)$ since R is T-transitive.

Then, we have

$$\inf_c I_T^{\rightarrow}(R^r(c,a),R(c,b)) \leq I_T^{\rightarrow}(R^r(a,a),R(a,b)) = R(a,b).$$

Knowing that $R^r \circ R = R$ (see Proposition 2.3), it follows that

$$T(R^r(c,a),R(a,b)) \leq R(c,b)$$

for all $a,b,c \in A$. This inequality is equivalent to

$$I_T^{\rightarrow}(R^r(c,a),R(c,b)) \geq R(a,b),$$

by left-continuity of T, which implies

$$\inf_c I_T^{\rightarrow}(R^r(c,a),R(c,b)) \geq R(a,b).$$

That is, finally we have

$$R(a,b) = \inf_{c\in A} I_T^{\rightarrow}(R^r(c,a),R(c,b)).$$

■

Another form of the representation theorem is given now.

Theorem 2.2 *Let R be a valued binary relation on A. Then R is T-transitive on A if and only if there exist two families $\{h_\gamma\}_{\gamma\in\Gamma}$, $\{k_\gamma\}_{\gamma\in\Gamma}$ of functions from A to $[0,1]$ such that $k_\gamma(a) \leq h_\gamma(a)$ for all $a \in A$, $\gamma \in \Gamma$ and*

$$R(a,b) = \inf_{\gamma\in\Gamma} I_T^{\rightarrow}(h_\gamma(b),k_\gamma(a)). \tag{2.14}$$

Proof. If R is given by (2.14), then T-transitivity can be proved as in the previous theorem.

Supposing that R is T-transitive, let $\Gamma = A$, $h_c(a) = R^\ell(a,c)$, $k_c(a) = R(a,c)$.

One can go on the proof as in the previous theorem.

■

2.5.6 Transitive closure

Using T-composition of valued binary relations, T-transitivity is equivalent to the condition $R \circ_T R \subseteq R$. Notice that if R is reflexive, then R is T-transitive if and only if $R \circ_T R = R$. Indeed,

$$R(a,b) \geq \sup_{c \in A} T(R(a,c), R(c,b)) \geq T(R(a,a), R(a,b)) = R(a,b).$$

Moreover, if $R^m = R \circ_T R^{m-1}$ with $m > 1$ and $R^1 = R$ then $R^m \subseteq R$ for $m \geq 1$ when R is T-transitive. It is easy to prove that R^m is nondecreasing in m when R is reflexive. The relation $\hat{R}$ defined by

$$\hat{R}(a,b) = \sup_m R^m(a,b)$$

is called the *T-transitive closure* of R.

Proposition 2.21 *(a) For any valued binary relation R on A, the T-transitive closure $\hat{R}$ is a T-transitive valued binary relation.*

(b) A valued binary relation R on A is T-transitive if and only if $R = \hat{R}$.

Proof. (a) By the definition of R^m, for any $a, b, c \in A$ we have that

$$R^{m+k}(a,b) \geq T(R^m(a,c), R^k(c,b)).$$

T-transitivity of $\hat{R}$ follows immediately by taking limits $m, k \to +\infty$ on both sides.

(b) Clearly, $R = \hat{R}$ implies that R is T-transitive. Suppose now that R is T-transitive. That is, $R^2 = R \circ_T R \subseteq R$ and thus $R^m \subseteq R$ for any $m \geq 1$. Hence,

$$\hat{R}(a,b) = \sup_m R^m(a,b) = R(a,b)$$

for all $a, b \in A$.

■

Another characterization of T-transitive closures can be given as follows (see Chakraborty and Das (1983)).

Proposition 2.22 *For any valued binary relation R on A, its T-transitive closure $\hat{R}$ is the intersection of all T-transitive valued relations containing R.*

Proof. Let R_k be any T-transitive valued binary relation containing R. Then $R \subseteq R_k$ implies

$$\min_{k \in K} R_k \subseteq \hat{R},$$

where $K = \{k \mid R_k \supseteq R$ and R_k is T-transitive$\}$, since $\hat{R}$ is a T-transitive valued relation containing R.

Conversely, $R \subseteq R_k$ implies $\hat{R} \subseteq \hat{R}_k$, by Proposition 2.2. On the other hand, R_k is T-transitive if and only if $R_k = \hat{R}_k$, by Proposition 2.21. Therefore, $\hat{R} \subseteq R_k$ and thus

$$\hat{R} \subseteq \min_{k \in K} R_k.$$

■

Using this Proposition, it is easy to prove the following statement.

Proposition 2.23 *The nearest T-transitive relation to any valued binary relation R that contains it is given by the T-transitive closure of R.*

Proof.

$$\begin{aligned}
\min_{k\in K} d(R_k, R) &= \min_k \sum_{a,b} |R_k(a,b) - R(a,b)| \\
&= \min_k \sum_{a,b} (R_k(a,b) - R(a,b)) \\
&= \sum_{a,b} \left[\min_k R_k(a,b) - R(a,b)\right] \\
&= \sum_{a,b} (\hat{R}(a,b) - R(a,b)) \\
&= \sum_{a,b} |\hat{R}(a,b) - R(a,b)| \\
&= d(\hat{R}, R).
\end{aligned}$$

■

Notice that $\hat{R}$ is minimal T-transitive relation containing R in the sense that if Q is a T-transitive relation such that $R \subseteq Q \subseteq \hat{R}$ then $Q = \hat{R}$. This follows from the previous proposition.

When A is finite with m elements then $\hat{R} = R^k$ for some $k < m$ since taking more than m elements, there must exist cycles whence the strength of the chains cannot be increased.

Remark that different methods of computing transitive closures are presented by Larsen and Yager (1990) and Li (1990).

2.5.7 Maximal transitive relations

In this section we present a procedure to obtain a maximal T-transitive relation $\check{R}$ contained in a given valued binary relation R. $\check{R}$ is maximal in the sense that if Q is a T-transitive valued binary relation on A such that $\check{R} \subseteq Q \subseteq R$ then $Q = \check{R}$. However, $\check{R}$ is not the greatest one. In general, there exist several maximal T-transitive relations contained in R, as we will see later. For more details see Fodor and Roubens (1994b).

It is easy to construct a T-transitive relation contained in a valued binary relation R. In fact, we have the following statement.

Proposition 2.24 *Any valued binary relation on R is the union of transitive valued binary relations on A.*

Proof. Suppose that R is a given relation on A. For such $c \in A$, we define a transitive relation by

$$R_c(a,b) = \begin{cases} R(c,b) & \text{if } a = c \\ 0 & \text{otherwise} \end{cases}$$

Clearly, R_c is T-transitive on A and $R = \sup_{c\in A} R_c$.

■

Using traces R^ℓ and R^r of R, nontrivial T-transitive relations contained in R can be obtained as follows. Suppose that T is a left-continuous t-norm.

Theorem 2.3 *For any valued binary relation R on A, both relations*

$$\min(R, R^\ell) \text{ and } \min(R, R^r)$$

are T-transitive on A and are contained in R.

Proof. It is obvious that $\min(R, R^\ell) \subseteq R$ and $\min(R, R^r) \subseteq R$.

To prove that they are T-transitive on R, we use the representation theorem of T-transitive valued binary relations (Theorem 2.1). In fact, the following equality is true for any valued binary relation R and for all $a, b \in A$:

$$\min(R(a,b), R^r(a,b)) = \inf_{c\in A} I_T^{\rightarrow}(\max(R(b,c), R^\ell(b,c)), R(a,c)). \tag{2.15}$$

Let us prove this formula first. Since $I_T^{\rightarrow}$ is nonincreasing in its first place, we have that

$$\begin{aligned}\inf_{c\in A} I_T^{\rightarrow}(\max(R(b,c), R^\ell(b,c)), R(a,c)) &\le \inf_{c\in A} I_T^{\rightarrow}(R(b,c), R(a,c)) \\ &= R^r(a,b),\end{aligned}$$

and

$$\begin{aligned}\inf_{c\in A} I_T^{\rightarrow}(\max(R(b,c), R^\ell(b,c)), R(a,c)) &\le I_T^{\rightarrow}(\max(R(b,b), R^\ell(b,b)), R(a,b)) \\ &= R(a,b).\end{aligned}$$

Turning to the converse,

$$\begin{aligned}&\inf_{c\in A} I_T^{\rightarrow}(\max(R(b,c), R^\ell(b,c)), R(a,c)) \\ &\quad= \inf_{c\in A} \min\{I_T^{\rightarrow}(R(b,c), R(a,c)), I_T^{\rightarrow}(R^\ell(b,c), R(a,c))\} \\ &\quad\ge \min\{\inf_{c\in A} I_T^{\rightarrow}(R(b,c), R(a,c)), \inf_{c\in A} I_T^{\rightarrow}(R^\ell(b,c), R(a,c))\} \\ &\quad= \min\{R^r(a,b), R(a,b)\}.\end{aligned}$$

Here we used that

$$R(a,b) \ge \inf_{c\in A} I_T^{\rightarrow}(R^\ell(b,c), R(a,c)) \ge R(a,b),$$

and this is true since $R \circ_T R^\ell = R$.

Therefore, we proved (2.15). T-transitivity of $\min(R, R^r)$ follows immediately by defining

$$\Gamma = A, \quad h_c(a) = \max(R(a,c), R^\ell(a,c)), \quad k_c(a) = R(a,c)$$

and by using Theorem 2.2.

On can prove the statement for $\min(R, R^\ell)$ in a similar way.

■

Generally speaking, $\min(R, R^\ell)$ and $\min(R, R^r)$ are not comparable in the sense of $\subseteq$, although both are contained in R and both are T-transitive relations on A. Consider the following example.

Example 2.2 Let $T(x,y) = W(x,y) = \max(x+y-1, 0)$ be the Łukasiewicz t-norm, $A = \{a,b,c\}$ and define R on A by

$$R = \begin{bmatrix} 0.5 & 0.8 & 0.3 \\ 0.4 & 0.5 & 0.7 \\ 0.6 & 0.9 & 0.5 \end{bmatrix}.$$

Clearly, R is not W-transitive since

$$R(a,c) = 0.3 < 0.5 = \max(R(a,b) + R(b,c) - 1, 0).$$

Traces of R are easily obtained as

$$R^\ell = \begin{bmatrix} 1 & 1 & 0.8 \\ 0.7 & 1 & 0.5 \\ 0.7 & 0.8 & 1 \end{bmatrix}, \quad R^r = \begin{bmatrix} 1 & 0.6 & 0.8 \\ 0.7 & 1 & 0.6 \\ 1 & 0.8 & 1 \end{bmatrix}.$$

Finally,

$$\min(R, R^\ell) = \begin{bmatrix} 0.5 & 0.8 & 0.3 \\ 0.4 & 0.5 & 0.5 \\ 0.6 & 0.8 & 0.5 \end{bmatrix}, \quad \min(R, R^r) = \begin{bmatrix} 0.5 & 0.6 & 0.3 \\ 0.4 & 0.5 & 0.6 \\ 0.6 & 0.8 & 0.5 \end{bmatrix},$$

and

$$\begin{aligned} \min(R, R^\ell)(a,b) = 0.8 &> 0.6 = \min(R, R^r)(a,b), \\ \min(R, R^\ell)(b,c) = 0.5 &< 0.6 = \min(R, R^r)(b,c), \end{aligned}$$

that is, the two relations $\min(R, R^\ell)$ and $\min(R, R^r)$ are not comparable.

In addition, neither $\min(R, R^\ell)$ nor $\min(R, R^r)$ are maximal. Indeed, define two relations Q_1 and Q_2 by

$$\begin{aligned} Q_1(c,b) = 0.9 \quad &\text{and} \quad Q_1 = \min(R, R^\ell) \text{ otherwise}, \\ Q_2(b,c) = 0.7 \quad &\text{and} \quad Q_2 = \min(R, R^r) \text{ otherwise}. \end{aligned}$$

Then W-transitivity of Q_1 and Q_2 can be justified easily and

$$\min(R, R^\ell) \subset Q_1 \subset R\,, \quad \min(R, R^r) \subset Q_2 \subset R.$$

■

Now, we turn to the construction of a maximal T-transitive relation contained in a given valued binary relation R when A is supposed to be finite and T is left-continuous. The construction is carried out in a recursive way and extends a result of Defays (1978) concerning the case when R is reflexive and $T = \min$.

Suppose that $A = \{a_1, a_2, \ldots, a_n\}$. We will define $\check{R}(a_1, a)$ first for all $a \in A$, then $\check{R}(a_2, a)$ for all $a \in A$, ..., finally $\check{R}(a_n, a)$ for all $a \in A$. Therefore, let

$$\check{R}(a_1, a) = R(a_1, a) \quad \forall a \in A.$$

Assuming that $\check{R}(a_j, a)$ is defined, let

$$\check{R}(a_{j+1}, a_k) = \begin{cases} \min\{R(a_{j+1}, a_k), U(a_{j+1}, a_k), V(a_{j+1}, a_k)\} \text{ if } k \le j \\ \min\{R(a_{j+1}, a_k), V(a_{j+1}, a_k)\} \text{ if } k > j \end{cases}, \tag{2.16}$$

where

$$U(a_{j+1}, a_k) = \min_{c \in A} I_T^{\rightarrow}(\check{R}(a_k, c), R(a_{j+1}, c)), \tag{2.17}$$

$$V(a_{j+1}, a_k) = \min_{i \le j} I_T^{\rightarrow}(\check{R}(a_i, a_{j+1}), \check{R}(a_i, a_k)). \tag{2.18}$$

Here function $U(a_{j+1}, a_k)$ is responsible for keeping $\check{R}(a_{j+1}, a_k)$ under the level which does not hurt T-transitivity in advance. Indeed, (2.17) is equivalent to

$$T(U(a_{j+1}, a_k), \check{R}(a_k, c)) \leq R(a_{j+1}, c) \;\; \forall c \in A. \tag{2.19}$$

If (2.19) is not satisfied then, since $\check{R} \subseteq R$, $\check{R}$ cannot be T-transitive.

The role of $V(a_{j+1}, a_k)$ is also to control T-transitivity, but for the already defined values of $\check{R}$. This is clear since (2.18) is equivalent to

$$T(\check{R}(a_i, a_{j+1}), V(a_{j+1}, a_k)) \leq \check{R}(a_i, a_k) \tag{2.20}$$

for all $i \leq j$.

Notice that if T is continuous, (2.17) implies that there exists $c \in A$ such that (2.19) is valid with equality. Similarly, it follows from (2.18) that there exists an index $i \leq j$ such that (2.20) holds with equality when T is continuous.

We present the main steps of the proof that the relation $\check{R}$ defined by the previous recursive procedure is T-transitive on A. For more details see Fodor and Roubens (1994b).

Theorem 2.4 *For any valued binary relation R on A, the relation $\check{R}$ defined by (2.16) is T transitive and $\check{R} \subseteq R$.*

Proof. For all $j \leq n$, we define a valued binary relation Q_j as follows :

$$Q_j(a_k, a) = \begin{cases} \check{R}(a_k, a) & \text{for } k \leq j \text{ and for all } a \in A \\ 0 & \text{for } k > j \end{cases}$$

We prove, by mathematical induction, that Q_j is T-transitive for $j = 1, 2, \ldots, n$.

For $j = 1$, the statement is obvious.

Suppose it holds for j, we prove for $j + 1$ that

$$T(Q_{j+1}(a, c), Q_{j+1}(c, b)) \leq Q_{j+1}(a, b) \tag{2.21}$$

is satisfied for all $a, b, c \in A$.

Four cases are considered.

Case (i) : If $a, c \in \{a_1, \ldots, a_j\}$ then, by the induction hypothesis, (2.21) is fulfilled.

Case (ii) : If $a = c = a_{j+1}$ then (2.21) is trivially satisfied since $T(x, y) \leq y$ for all $x, y \in [0, 1]$.

Case (iii) : If $a = a_\ell$ with $\ell \leq j$ and $c = a_{j+1}$ then (2.21) yields the following inequality:

$$T(\check{R}(a_\ell, a_{j+1}), \check{R}(a_{j+1}, b)) \leq \check{R}(a_\ell, b).$$

This inequality follows from the definition of $V(a_{j+1}, a_k)$ and by using Proposition 1.10, since we have $\check{R}(a_{j+1}, b) \leq V(a_{j+1}, b)$, by the construction.

Case (iv) : If $c = a_\ell$ with $\ell \leq j$ and $a = a_{j+1}$ then we have to prove that

$$T(\check{R}(a_{j+1}, a_\ell), \check{R}(a_\ell, b) \leq \check{R}(a_{j+1}, b). \tag{2.22}$$

If $\check{R}(a_{j+1}, b) = R(a_{j+1}, b)$ then we obtain this inequality by the definition of $U(a_{j+1}, a_k)$ and by Proposition 1.10.

Suppose that $\check{R}(a_{j+1}, b) < R(a_{j+1}, b)$. If $b = a_r$ with $r \leq j$ then, by (2.16), either there exists $c \in A$ such that

$$\check{R}(a_{j+1}, a_r) = I_T^{\rightarrow}(\check{R}(a_r, c), R(a_{j+1}, c)), \tag{2.23}$$

or there exists an index $i \leq j$ such that

$$\check{R}(a_{j+1}, a_r) = I_T^{\rightarrow}(\check{R}(a_i, a_{j+1}), \check{R}(a_i, a_r)). \tag{2.24}$$

If (2.23) is true then (2.22) follows from the induction hypothesis $(\ell, r \leq j)$ and from $\check{R}(a_{j+1}, a_\ell) \leq U(a_{j+1}, a_\ell)$.

Suppose now that (2.24) is satisfied. Then we have to prove that

$$T(\check{R}(a_{j+1}, a_\ell), \check{R}(a_\ell, a_r)) \leq I_T^{\rightarrow}(\check{R}(a_i, a_{j+1}), \check{R}(a_i, a_r)).$$

This inequality is implied by Case (iii) and the induction hypothesis.

Finally, suppose that $b = a_r$ with $r > j$. Then we must have (2.24) for some $i \leq j$, by the definition of $\check{R}$. Thus, we can use the previous arguments to obtain (2.22).

Therefore, we proved that Q_{j+1} is T-transitive if Q_j is T-transitive. This implies that $\check{R} = Q_n$ is T-transitive.

■

Now we prove that $\check{R}$ is maximal T-transitive relation contained in R.

Theorem 2.5 *For any valued binary relation R on A, $\check{R}$ is maximal T-transitive relation contained in R.*

Proof. Suppose Q is a T-transitive relation on A such that

$$\check{R} \subseteq Q \subseteq R.$$

We have to prove that $Q = \check{R}$.

Clearly, $Q(a_1, a) = \check{R}(a_1, a) = R(a_1, a)$ for all $a \in A$.

Supposing that for each $r \leq j$ and all $a \in A$ we have

$$Q(a_r, a) = \check{R}(a_r, a),$$

we prove $Q(a_{j+1}, a) = \check{R}(a_{j+1}, a)$ for all $a \in A$.

Case (i) : $\check{R}(a_{j+1}, a) = R(a_{j+1}, a)$. Then we trivially have $Q(a_{j+1}, a) = \check{R}(a_{j+1}, a)$.

Case (ii) : there exists $c \in A$ such that(2.23) holds.
If $Q(a_{j+1}, a_r) > \check{R}(a_{j+1}, a_r)$ then

$$\begin{aligned} R(a_{j+1}, c) &< T(Q(a_{j+1}, a_r), \check{R}(a_r, c)) \\ &\leq T(Q(a_{j+1}, a_r), Q(a_r, c)), \end{aligned}$$

therefore Q cannot be transitive since $R(a_{j+1}, c) \geq Q(a_{j+1}, c)$.

Case (iii) : there exists $i \leq j$ such that (2.24) is satisfied. Then $\check{R}(a_{j+1}, a_r) < Q(a_{j+1}, a_r)$ implies

$$\begin{aligned} \check{R}(a_i, a_r) &< T(\check{R}(a_i, a_{j+1}), Q(a_{j+1}, a_r)) \\ &\leq T(Q(a_i, a_{j+1}), Q(a_{j+1}, a_r)), \end{aligned}$$

therefore, Q cannot be T-transitive since, by the induction hypothesis, we have $Q(a_i, a_r) = \check{R}(a_i, a_r)$.

This proves our theorem.

■

2.5.8 Miscellaneous results on transitivity

For continuous Archimedean t-norms, the following equivalent form of T-transitivity can be formulated, see Ovchinnikov (1992b).

Lemma 2.3 *Suppose that T is a continuous Archimedean t-norm with additive generator f. Then T-transitivity condition (2.12) is equivalent to*

$$f(R(a,c)) + f(R(c,b)) \geq f(R(a,b)) \tag{2.25}$$

for all $a, b, c \in A$.

Proof. Assume that (2.12) is satisfied. Then we have

$$f^{(-1)}(f(R(a,c)) + f(R(c,b))) \leq R(a,b).$$

Applying f to both sides, we obtain

$$\min\{f(R(a,c)) + f(R(c,b)), f(0)\} \geq f(R(a,b)),$$

since $f^{(-1)}(x) = f^{-1}(\min\{x, f(0)\})$, by the definition of $f^{(-1)}$. This last inequality implies (2.25).

Suppose now that (2.25) is fulfilled. Since $f(0) \geq f(R(a,b))$, we have

$$\min\{f(R(a,c)) + f(R(c,b)), f(0)\} \geq f(R(a,b)).$$

Applying f^{-1} to both sides, we obtain that

$$f^{(-1)}(f(R(a,c)) + f(R(c,b))) \leq R(a,b),$$

which implies (2.12).

■

It follows by inequality (2.12) immediately that

$$R(a,c) = 1 \quad \text{implies} \quad R(c,b) \leq R(a,b) \quad \text{and} \quad R(b,a) \leq R(b,c). \tag{2.26}$$

Ovchinnikov (1986) proved that, in a sense, the converse is also true when A is a finite set. It is easy to extend his result for the infinite case if R satisfies an additional condition.

Proposition 2.25 *For any valued binary relation R on a set A for which*

$$\sup\{R(a,b) | R(a,b) < 1\} = \alpha < 1$$

and condition (2.26) is satisfied, there exists an automorphism φ such that R is T-transitive, where $T(x,y) = \varphi^{-1}(W(\varphi(x), \varphi(y)))$ is the φ-transform of the Lukasiewicz t-norm.

Proof. For $\alpha < 1$ defined above, there exists an automorphism φ such that $\varphi(\alpha) = 1/2$. If $R(a,c) = 1$ or $R(c,b) = 1$ then inequality (2.12) holds obviously. Otherwise,

$$T(R(a,c), R(c,b)) \leq T(\alpha, \alpha) = 0 \leq R(a,b).$$

■

Following ideas of Ovchinnikov (1986) further, if T is a positive t-norm then inequality (2.12) indicates that

$$R(a,c) > 0 \text{ and } R(c,b) > 0 \quad \text{imply} \quad R(a,b) > 0 \tag{2.27}$$

for all $a, b, c \in A$.

Then the following proposition can be proved.

Proposition 2.26 *Suppose that conditions of Theorem 2.25 are satisfied and*

$$\inf\{R(a,b) | R(a,b) > 0\} = \beta > 0.$$

If (2.27) holds then there exists an automorphism φ of the unit interval such that R is T-transitive, where $T(x,y) = \varphi^{-1}(\Pi(\varphi(x), \varphi(y)))$ is the φ-transform of the product t-norm.

Proof. For α, β defined above, there exists an automorphism φ such that $\varphi(\alpha) = 1/2$ and $\varphi(\beta) = 1/4$ if $\beta < \alpha$. Define $\varphi(\alpha) = \varphi(\beta) = 1/2$ otherwise. Then

$$T(R(a,c), R(c,b)) \leq R(a,b)$$

follows from (2.26), if $R(a,c) = 1$ or $R(c,b) = 1$; follows from (2.27) if $R(a,c) = 0$, and is also true otherwise, since

$$T(R(a,c), R(c,b)) = \varphi^{-1}(\Pi(\varphi(R(a,c)), \varphi(R(c,b)))) \leq \beta \leq R(a,b).$$

■

Proposition 2.27 *Suppose that R is T-transitive for a given T. Then R is T'-transitive for any t-norm T' such that $T' \leq T$. In particular, every transitive valued binary relation is T-transitive for all t-norm T.*

Proof. $T'(R(a,c), R(c,b)) \leq T(R(a,c), R(c,b)) \leq R(a,b)$. ■

Proposition 2.28 *If R_1, R_2 are T-transitive valued relations then $R = T(R_1, R_2)$ is a T-transitive valued binary relation.*

Proof. T-transitivity of R_1 and R_2 is equivalent to the following inequalities ($a, b, c \in A$):

$$\begin{aligned} R_1(a,b) &\geq T[R_1(a,c), R_1(c,b)], \\ R_2(a,b) &\geq T[R_2(a,c), R_2(c,b)]. \end{aligned}$$

Then we have

$$\begin{aligned} R(a,b) &= T[R_1(a,b), R_2(a,b)] \\ &\geq T[T[R_1(a,c), R_1(c,b)], T[R_2(a,c), R_2(c,b)]] \\ &= T[T[R_1(a,c), R_2(a,c)], T[R_2(c,b), R_2(c,b)]] \\ &= T[R(a,c), R(c,b)]. \end{aligned}$$

■

Proposition 2.29 *If R is transitive on A then each λ-cut of R is a transitive relation for $\lambda \in (0, 1]$.*

Proof. Trivial. ■

Suppose that A is a finite set. In this case R can be represented by a matrix M^R, as we mentioned earlier. Hashimoto (1983a) proved that for a transitive valued binary relation R, there exists an ordering of elements $A = \{a_1, a_2, \ldots, a_m\}$ such that $R(a_i, a_j) \geq R(a_j, a_i)$ for $i \leq j$.

Kolodziejczyk (1987) extended this result for so-called *s-transitive* valued binary relations. R is called s-transitive if the relation (see Ovchinnikov (1981))

$$P(a,b) = \begin{cases} R(a,b) & \text{if } R(a,b) > R(b,a) \\ 0 & \text{if } R(a,b) \leq R(b,a) \end{cases}$$

satisfies condition (2.27). It is easy to see that transitive valued binary relations form a proper subclass of s-transitive valued binary relations. It was proved by Kolodziejczyk (1987) that for any s-transitive valued binary relation on a finite A, there exists an ordering of elements $A = \{a_1, \ldots, a_m\}$ such that $R(a_i, a_j) \geq R(a_j, a_i)$ for $i \leq j$.

Closing this subsection, we prove that irreflexivity and transitivity together imply asymmetry, as in the classical case.

Proposition 2.30 *If R is irreflexive and T-transitive then it is T-asymmetric.*

Proof. $T(R(a,b), R(b,a)) \leq R(a,a) = 0$ holds for all $a, b \in A$. ■

2.5.9 Negative transitivity

Negative S-transitivity is the dual concept of T-transitivity and vice versa. Therefore, only some main points are explained in details. The others can be obtained by corresponding results on T-transitivity.

Definition 2.13 A valued binary relation on A is *negatively S-transitive* if

$$R(a,b) \leq S(R(a,c), R(c,b))$$

for all $a, b, c \in A$.

Proposition 2.31 *A valued binary relation R on A is negatively S-transitive if and only if its dual R^d is T-transitive.*

Proof. $R(a,b) \leq S(R(a,c), R(c,b))$ if and only if $N(R(a,b)) \geq T(N(R(a,c)), N(R(c,b)))$, if and only if $R^d(b,a) \geq T(R^d(b,c), R^d(c,a))$. ■

One can prove a representation theorem for negatively S-transitive relations, using the R-coimplication

$$I_c^{\rightarrow}(x,y) = \inf\{z \mid S(x,z) \geq y\}$$

defined in Section 1.9. As it was mentioned,

$$I_c^{\rightarrow}(x,y) = N(I_T^{\rightarrow}(N(x), N(y)))$$

when $T(x,y) = N(S(N(x), N(y))$ and n is a strong negation. Therefore, using the representation of T-transitive relations and the previous proposition, the following results can be established.

Theorem 2.6 *A valued binary relation R on A is negatively S-transitive if and only if there exist two families $\{f_\gamma\}_{\gamma\in\Gamma}$, $\{g_\gamma\}_{\gamma\in\Gamma}$ of functions from A to $[0,1]$ such that $g_\gamma(a) \leq f_\gamma(a)$ for all $a \in A$, $\gamma \in \Gamma$ and we have*

$$R(a,b) = \sup_{\gamma\in\Gamma}\{I_c^{\rightarrow}(g_\gamma(b), f_\gamma(a))\} \tag{2.28}$$

for all $a, b \in A$.

Theorem 2.7 *A valued binary relation R on A is negatively S-transtive if and only if there exist two families $\{h_\gamma\}_{\gamma\in\Gamma}$, $\{k_\gamma\}_{\gamma\in\Gamma}$ of functions from A to $[0,1]$ such that $k_\gamma(a) \leq h_\gamma(a)$ for all $a \in A$, $\gamma \in \Gamma$ and we have*

$$R(a,b) = \sup_{\gamma\in\Gamma}\{I_c^{\rightarrow}(k_\gamma(a), h_\gamma(b))\} \tag{2.29}$$

for all $a, b \in A$.

We can express negative S-transitivity in the following equivalent way when S is a continuous Archimedean t-conorm.

Lemma 2.4 *Suppose that S is a continuous Archimedean t-norm with additive generator g. Then a valued binary relation R is negatively S-transitive on A if and only if*

$$g(R(a,c)) + g(R(c,b)) \leq g(R(a,b))$$

holds for all $a, b, c \in A$.

Proof. One can prove this statement the same way as Lemma 2.3.

■

Proposition 2.32 *Suppose that R is negatively S-transitive for a given S. Then R is negatively S'-transitive for any t-conorm S' such that $S' \geq S$. In particular, a negatively transitive relation is negatively S'-transitive for any t-conorm S'.*

Proof. $R(a,b) \leq S(R(a,c), R(c,b)) \leq S'(R(a,c), R(c,b))\ \forall a, b, c \in A$.

■

For crisp binary relations, strong completeness and transitivity together imply negative transitivity of the relation. Fortunately, we can prove similar statement for valued binary relations when the De Morgan triple (T, S, N) is such that T is either the minimum (see Ovchinnikov (1991)) or T is continuous and Archimedean, whence for all continuous t-norms T, by using ordinal sums.

Theorem 2.8 *Suppose that in the De Morgan triple (T, S, N), T is a continuous t-norm. If R is strongly S-complete and T-transitive on A then R is negatively S-transitive on A.*

Proof. We need to prove that

$$S(R(a,c), R(c,b)) \geq R(a,b) \tag{2.30}$$

when

$$S(R(a,b), R(b,a)) = 1, \ \forall a, b \in A$$

and

$$T(R(a,c), R(c,b)) \leq R(a,b).$$

Suppose first that T is positive. Then, by Proposition 2.15, R is strongly S-complete if and only if

$$\max(R(a,b), R(b,a)) = 1$$

for all $a, b \in A$.

Inequality (2.30) trivially holds when $R(a,c) = 1$ or $R(c,b) = 1$.

Thus, assume that $R(a,c) < 1$, $R(c,b) < 1$. Then, by strong completeness of R, $R(c,a) = 1$ and $R(b,c) = 1$. Now we have

$$\begin{aligned} S(R(a,c), R(c,b)) &\geq S(T(R(a,b), R(b,c)), T(R(c,a), R(a,b))) \\ &= S(R(a,b), R(a,b)) \\ &\geq R(a,b). \end{aligned}$$

That is, R is negatively S-transitive.

Suppose now that T is continuous Archimedean having zero divisors. Then, by Theorem 1.5, we have

$$T(x,y) = \varphi^{-1}(\max\{\varphi(x) + \varphi(y) - 1, 0\})$$

with an automorphism φ of the unit interval.

Therefore, we have to prove that

$$\varphi(R(a,c)) + \varphi(R(c,b)) \geq \varphi(R(a,b)).$$

In this case strong S-completeness is equivalent to

$$\varphi(R(a,b)) + \varphi(R(b,a)) \geq 1.$$

Therefore,

$$\begin{aligned} \varphi(R(a,c)) + \varphi(R(c,b)) &\geq [1 - \varphi(R(c,a))] + \varphi(R(c,a)) + \varphi(R(a,b)) - 1 \\ &= \varphi(R(a,b)). \end{aligned}$$

To prove the statement for any continuous t-norm T, consider the ordinal sum representations of T and S, respectively.

Since the statement has already been proved for the min and for all continuous Archimedean t-norms, it also follows for ordinal sums, using similar argumentations.

Thus we proved that R is negatively S-transitive in all cases.

■

Corollary 2.1 *Suppose that T is a continuous t-norm. If R is strongly S-complete and T-transitive then R^d is T-transitive on A.*

Proof. By Theorem 2.8, R is negatively S-transitive on A, which is equivalent to T-transitivity of R^d since $S(x,y) = N(T(N(x), N(y)))$.

■

2.5.10 Semitransitivity

According to the classical case we establish the following definition.

Definition 2.14 A valued binary relation on A is *T-S-semitransitive* if for every $a, b, c, d \in A$ we have

$$T(R(a,d), R(d,b)) \leq S(R(a,c), R(c,b)).$$

Proposition 2.33 *If R is reflexive and T-S-semitransitive then R is negatively S-transitive. If R is irreflexive and T-S-semitransitive then R is T-transitive.*

Proof. If R is reflexive and T-S-semitransitive then

$$S(R(a,c), R(c,b)) \geq T(R(a,a), R(a,b)) = R(a,b).$$

If R is irreflexive and T-S-semitransitive then

$$T(R(a,d), R(d,b)) \leq S(R(a,a), R(a,b)) = R(a,b).$$

■

Proposition 2.34 *If R is T-transitive and negatively S-transitive then R is T-S-semitransitive.*

Proof. In this case

$$T(R(a,d), R(d,b)) \leq R(a,b) \leq S(R(a,c), R(c,b)).$$

■

Proposition 2.35 *If R is semitransitive then it is T-S-semitransitive for any De Morgan triple (T, S, N).*

Proof. This is obvious since $T \leq \min$ and $S \geq \max$ for any t-norm T and for any t-conorm S.

■

Proposition 2.36 *Suppose that T is a continuous t-norm in the De Morgan triple (T, S, N). If R is T-asymmetric and negatively S-transitive on A then R is T-S-semitransitive on A.*

Proof. It sufficies to prove, by Proposition 2.34, that T-asymmetry and negative S-transitivity of R together imply that R is T-transitive on A.

Consider R^d, the dual of R. By our conditions, R^d is strongly S-complete and T-transitive. Therefore, Theorem 2.8 yields that R^d is negatively S-transitive, which is equivalent to the T-transitivity of R.

■

2.5.11 Ferrers property

Definition 2.15 A valued binary relation R on A is *T-S-Ferrers relation* if for every $a, b, c, d \in A$ we have

$$T(R(a,b), R(c,d)) \le S(R(a,d), R(c,b)).$$

Proposition 2.37 *If a valued binary relation R is a Ferrers relation on A then it is T-S-Ferrers for any De Morgan triple (T,S,N).*

Proof. The statement follows from the fact that $T \le \min$ and $S \ge \max$. ■

Proposition 2.38 *If R is reflexive and T-S-Ferrers then R is negatively S-transitive. If R is irreflexive and T-S-Ferrers then it is T-transitive on A.*

Proof. We have

$$S(R(a,d), R(c,a)) \ge T(R(a,a), R(c,d)) = R(c,d)$$

and

$$T(R(a,b), R(b,d)) \le S(R(a,d), R(b,b)) = R(a,d).$$

■

Proposition 2.39 *Suppose that T is a continuous t-norm. If R is T-asymmetric and negatively S-transitive then R is T-S-Ferrers relation on A.*

Proof. We have to prove that

$$T(R(a,b), R(c,d)) \le S(R(a,d), R(c,b))$$

holds for all $a, b, c, d \in A$. This is trivially true when $R(a,b) = 0$ or $R(c,d) = 0$. Thus, suppose that $R(a,b) > 0$ and $R(c,d) > 0$.

If T is positive then $T(R(a,b), R(b,a)) = 0$ if and only if $\min(R(a,b), R(b,a)) = 0$ for all $a, b \in A$. Therefore, $R(b,a) = 0$ and $R(d,c) = 0$.

By negative S-transitivity of R we have

$$\begin{aligned}
T(R(a,b), R(c,d)) &\le T[S(R(a,d), R(d,b)), S(R(c,b), R(b,d))] \\
&\le T[S(R(a,d), S(R(d,c), R(c,b))), S(R(c,b), S(R(b,a), R(a,d)))] \\
&= T[S(R(a,d), R(c,b)), S(R(a,d), R(c,b))] \\
&\le S(R(a,d), R(c,b)),
\end{aligned}$$

since $R(b,a) = R(d,c) = 0$ and $T(x,x) \le x$.

Thus, we proved that R is T-S-Ferrers when T is a positive t-norm.

Suppose now that T is a continuous Archimedean t-norm with zero divisiors. Therefore, there is an automorphism φ of the unit interval such that

$$T(x,y) = \varphi^{-1}(\max\{\varphi(x) + \varphi(y) - 1, 0\}).$$

Thus, it sufficies to prove that

$$\varphi(R(a,b)) + \varphi(R(c,d)) - 1 \le \varphi(R(a,d)) + \varphi(R(c,b))$$

holds for all $a, b, c, d \in A$.

We have by negative S-transitivity that

$$\begin{aligned}
&\varphi(R(a,b)) + \varphi(R(c,d)) - 1 \\
&\leq [\varphi(R(a,d)) + \varphi(R(d,b))] + [\varphi(R(c,b)) + \varphi(R(b,d))] - 1 \\
&= \varphi(R(a,d)) + \varphi(R(c,b)) + [\varphi(R(d,b)) + \varphi(R(b,d)) - 1] \\
&\leq \varphi(R(a,d)) + \varphi(R(c,b)),
\end{aligned}$$

since R is T-asymmetric, i.e.,

$$\varphi(R(d,b)) + \varphi(R(b,d)) - 1 \leq 0.$$

To prove the statement for any continuous t-norm T and its N-dual t-conorm S, use their ordinal sum representations.

By the previously proved cases (i.e., min and any continuous Archimedean t-norm and their N-dual t-conorms), using similar argumentation, the statement follows.

■

2.5.12 Linearity

Definition 2.16 A valued binary relation R on A is called
(a) *left-linear* if for every $a, b, c, d \in A$ we have

$$R(d,a) < R(d,b) \Rightarrow R(c,a) \leq R(c,b);$$

(b) *right-linear* if for every $a, b, c, d \in A$ we have

$$R(a,d) > R(b,d) \Rightarrow R(a,c) \geq R(b,c);$$

(c) *linear* if it is both left- and right-linear.

Theorem 2.9 *A valued binary relation R on A is*

(a) *left-linear if and only if its left-trace is strongly complete;*

(b) *right-linear if and only if its right-trace is strongly complete;*

(c) *linear if and only if both of its traces are strongly complete relations on A.*

Proof. We prove only part (a).

Let R be left-linear and suppose $R^\ell(a,b) < 1$ for some $a, b \in A$. Then there exists $d \in A$ such that $R(d,a) > R(d,b)$ (otherwise $R^\ell(a,b) = 1$). R is left-linear, so for every $c \in A$ we have $R(c,a) \geq R(c,b)$. This means that

$$R^\ell(b,a) = \inf_{c\in A} I_T^{\rightarrow}(R(c,b), R(c,a)) = \inf_{c\in A} 1 = 1,$$

that is, R^ℓ is strongly complete on A.

Suppose now that R^ℓ is strongly complete and let $a, b \in A$ such that $R^\ell(a,b) = 1$. Then we have $R(c,a) \leq R(c,b)$ for every $c \in A$. Thus, there are no four elements $a, b, c, d \in A$ such that $R(c,a) < R(c,b)$ and $R(d,a) > R(d,b)$. That is, R is left-linear.

Parts (b) and (c) can be proved analogously.

■

Chapter 3

Valued preference modelling

Preference modelling is a fundamental part of several applied fields but at the same time it has its own interesting theoretical problems. Our aim in this chapter is to propose an axiomatic approach to the definition of valued strict preference, indifference and incomparability associated with a valued (weak) preference relation so that most of the relationships (corresponding to the classical connections) among these valued relations are preserved.

A special emphasis is given to the study of transitivity of the strict preference associated with a transitive weak preference relation.

3.1 Basic notions of preference modelling

In the classical theory of preference modelling (see e.g.Roubens and Vincke (1985)), a binary relation R with respect to each pair of alternatives (a, b) of a given set A is considered as a *weak preference relation*:

$$aRb \iff \text{“}a \text{ is not worse than } b\text{”}.$$

This definition implies that R is a *reflexive* relation, i.e., aRa holds for any $a \in A$.

Three binary relations corresponding to the given preference relation R: *strict preference* P, *indifference* I and *incomparability* J are also defined as follows:

$$\begin{aligned} aPb &\iff aRb \text{ and not } bRa, \\ aIb &\iff aRb \text{ and } bRa, \\ aJb &\iff \text{not } aRb \text{ and not } bRa. \end{aligned}$$

Using set-theoretic operations, the previous verbal expressions can be translated as

$$P = R \cap R^d \tag{3.1}$$

$$I = R \cap R^{-1} \tag{3.2}$$

$$J = R^c \cap R^d. \tag{3.3}$$

These relations form a *preference structure* (P, I, J) and are linked together:

$$P \cup I = R, \tag{3.4}$$

$$P \cap I = \emptyset, \tag{3.5}$$

$$P \cap J = \emptyset, \tag{3.6}$$

$$I \cap J = \emptyset, \tag{3.7}$$

$$P \cup I \cup P^{-1} = R \cup R^{-1}. \tag{3.8}$$

$R(a,b)$	$R(b,a)$	$P(a,b)$	$I(a,b)$	$J(a,b)$
0	0	0	0	1
0	1	0	0	0
1	0	1	0	0
1	1	0	1	0

Table 3.1: Monotony of P, I, J

It is worth mentioning that (3.4) is equivalent to

$$P \cup J = R^d. \tag{3.9}$$

Moreover, I and J are *symmetric*, P is *asymmetric*, i.e., $I = I^{-1}$, $J = J^{-1}$ and $P \cap P^{-1} = \emptyset$. Finally,

$$P \cup P^{-1} \cup I \cup J = A \times A. \tag{3.10}$$

If we introduce the usual *valuation*:

$$R(a,b) = \begin{cases} 1 & \text{if } aRb \\ 0 & \text{otherwise} \end{cases},$$

then J is nonincreasing, I and P are nondecreasing functions of $R(a,b)$, while I is nondecreasing, J and P are nonincreasing functions of $R(b,a)$, as one can see by Table 3.1.

3.2 Axiomatics for valued preference relations

From now on we consider problems in which preferences between alternatives are described by a *valued preference relation* R such that the value $R(a,b)$, which lies between 0 and 1 for convenience, is understood as the degree to which the proposition "a is not worse than b" is true. We investigate the problem of defining P, I and J in terms of R and introducing models for set-theoretic (fuzzy logical) operations that preserve as many of the classical properties as possible. This chapter is based mainly on the results obtained by Fodor and Roubens (1994a). For other results on the same topic, see Fodor (1991e,1992b), Ovchinnikov and Roubens (1991,1992) and the overview paper Fodor and Roubens (1991a).

3.2.1 Do we need axiomatics at all ?

One might think that we are in an easy situation : all we need is using t-norms in (3.1), (3.2), (3.3). Let us see what happens if we try. Let (T, S, n) be any continuous De Morgan triple with a strict negation n.

According to (3.1), (3.2) and (3.3), we have

$$\begin{aligned} P(a,b) &= T(R(a,b), n(R(b,a))), && (3.11)\\ I(a,b) &= T(R(a,b), T(b,a)), && (3.12)\\ J(a,b) &= T(n(R(a,b)), n(R(b,a))). && (3.13) \end{aligned}$$

Unfortunately, the following negative result can be proved easily (see also Alsina (1985)).

Proposition 3.1 *There is no De Morgan triple (T, S, n) such that (3.4) holds with P, I defined by (3.11), (3.12).*

Proof. Denoting $x = R(a,b)$, $y = R(b,a)$ for short, equality (3.4) has the following form now :

$$S(T(x, n(y)), T(x, y)) = x. \tag{3.14}$$

Taking $x = 1$, we must have $S(n(y), y) = 1$, or equivalently, $T(z, n(z)) = 0 \; \forall z \in [0,1]$.

Let $y = x$ in (3.14) such that $0 < x \leq n(x)$. Then we have

$$\begin{aligned} x &= S(T(x, n(x)), T(x,x)) = S(0, T(x,x)) \\ &= T(x,x) \leq T(x, n(x)) = 0, \end{aligned}$$

a contradiction.

■

Therefore, we have to find appropriate definitions for P, I and J in order to preserve classical relationship $R = P \cup I$ and even more. This will be carried out successfully by establishing the following axiomatic approach.

3.2.2 Axioms for defining (P, I, J)

To define valued binary relations P, I and J, we propose the following general axioms:

(IA) Independence of Irrelevant Alternatives:

- *For any two alternatives a, b the values of $P(a,b)$, $I(a,b)$ and $J(a,b)$ depend only on the values $R(a,b)$ and $R(b,a)$.*

According to (IA), there exist three functions p, i, j from $[0,1]^2$ to $[0,1]$ such that

$$\begin{aligned} P(a,b) &= p(R(a,b), R(b,a)), \\ I(a,b) &= i(R(a,b), R(b,a)), \\ J(a,b) &= j(R(a,b), R(b,a)). \end{aligned}$$

(PA) Positive Association Principle:

- *The functions $p(x, n(y))$, $i(x,y)$, $j(n(x), n(y))$ are nondecreasing with respect to both arguments.*

(S) Symmetry:

- *$i(x,y)$ and $j(x,y)$ are symmetric functions.*

(PA) can be justified by Table 3.1, (S) is a natural assumption, supported also by the classical case.

It might be surprising that no asymmetry has been assumed as an axiom. The reason is twofold : first, P is always T-asymmetric (see Theorem 3.2 later); second, asymmetry (with respect to minimum) is a very restrictive condition (see Theorem 3.4).

3.3 The system of functional equations

We start with properties (3.4) and (3.9) (which are equivalent in the crisp case). These can be translated now in the following way:

$$S(P(a,b), I(a,b)) = R(a,b), \tag{3.15}$$
$$S(P(a,b), J(a,b)) = R^d(a,b), \tag{3.16}$$

for all $a, b \in A$. Denoting $x = R(a,b), y = R(b,a)$ for short, we write conditions (3.15) and (3.16) as a system of functional equations:

$$S(p(x,y), i(x,y)) = x, \tag{3.17}$$
$$S(p(x,y), j(x,y)) = n(y) \tag{3.18}$$

for all $x, y \in [0,1]$.

3.3.1 Properties of the solutions

According to the previous section, our models are described by $< p, i, j, T, S, n >$, where p, i, j are functions based on (IA) and (T, S, n) is a De Morgan triple such that T is a continuous t-norm and n is a strict negation (whence S is a continuous t-conorm). In the first proposition we prove some boundary conditions for p, i, j.

Lemma 3.1 *If (3.17) and (3.18) are fulfilled then*

(a) $i(0,y) = p(0,y) = 0$,

(b) $j(x,1) = p(x,1) = 0$,

(c) $i(x,1) = p(x,0) = x$,

(d) $j(0,y) = p(1,y) = n(y)$

hold for every $x, y \in [0,1]$.

Proof. (a) Let $x = 0$ in (3.17). $S(u,v) = 0 \iff u = v = 0$.
(b) Let $y = 1$ in (3.18).
(c) Let $y = 1$ in (3.17). Then we have that

$$x = S(p(x,1), i(x,1)) = S(0, i(x,1)) = i(x,1),$$

by (b). If $y = 0$ in (3.17) then

$$x = S(p(x,0), i(x,0)) = S(p(x,0), 0) = p(x,0),$$

by (a) and (S).
(d) Let $x = 0$ in (3.18). Thus

$$n(y) = S(p(0,y), j(0,y)) = S(0, j(0,y)) = j(0,y),$$

by (a). If $x = 1$ in (3.18) then

$$n(y) = S(p(1,y), i(1,y)) = S(p(1,y),0) = p(1,y),$$

by (b) and (S).

■

In the next lemma we prove that T must be a φ-transform of the Łukasiewicz t-norm W.

Lemma 3.2 *If (3.17) and (3.18) are satisfied then there exists an automorphism φ of the unit interval such that*

$$T(x,y) = \varphi^{-1}(\max\{\varphi(x) + \varphi(y) - 1, 0\}). \tag{3.19}$$

Proof. First we prove that T fulfils the law of contradiction. Indeed, let $x = 1$ in (3.17). Using Lemma 3.1 we obtain that $1 = S(p(1,y), i(1,y)) = S(n(y), y)$, or equivalently,

$$T(n(z), z) = 0$$

for every $z \in [0,1]$ with $z = n(y)$. This means that T has zero divisors. On the other hand, T is continuous, thus it is Archimedean (see Lemma 1.1). Hence, there exists an automorphism φ of the unit interval such that (3.19) holds.

■

Now we prove that n is the strong negation defined by the same φ as in the previous lemma.

Lemma 3.3 *If (3.17) and (3.18) are satisfied then*

$$n(x) = \varphi^{-1}(1 - \varphi(x)) \tag{3.20}$$

with the same automorphism φ as in (3.19).

Proof. Using (3.19), equation (3.17) has the following equivalent form:

$$\max\{\varphi(n(p(x,y))) + \varphi(n(i(x,y))) - 1, 0\} = \varphi(n(x)).$$

If $x < 1$ then this equation is equivalent to

$$\varphi(n(p(x,y))) + \varphi(n(i(x,y))) - 1 = \varphi(n(x)),$$

that is,

$$i(x,y) = n^{-1}\{\varphi^{-1}\{\varphi(n(x)) - \varphi(n(p(x,y))) + 1\}\}.$$

This last equation implies, by symmetry of i, that

$$\varphi(n(x)) - \varphi(n(p(x,y))) = \varphi(n(y)) - \varphi(n(p(y,x))). \tag{3.21}$$

Let $y = 1$ in (3.21). Then we obtain the following equality:

$$\varphi(n(x)) - 1 = -\varphi(n(n(x))),$$

or introducing $z = n(x)$,

$$\varphi(z) + \varphi(n(z)) = 1.$$

Thus our lemma is proved.

■

Therefore, we have obtained that (T, S, n) should be a strong (Łukasiewicz-like) De Morgan triple depending only on the automorphism φ of the unit interval. In other words, as soon as we have chosen φ, the logical connectives (T, S, n) are uniquely determined by our functional equations:

$$\begin{aligned} T(x,y) &= W_\varphi(x,y) = \varphi^{-1}(\max\{\varphi(x) + \varphi(y) - 1, 0\}, \\ S(x,y) &= W'_\varphi(x,y) = \varphi^{-1}(\min\{\varphi(x) + \varphi(y), 1\}, \\ n(x) &= N_\varphi(x) = \varphi^{-1}(1 - \varphi(x)). \end{aligned}$$

Keeping this fact in mind, a solution $< p, i, j, T, S, n >$ of our system can be denoted simply by $< p, i, j >_\varphi$.

Although the logical connectives are uniquely determined (up to an automorphism φ), this is not the case for the generator functions p, i, j of strict preference, indifference and incomparability, respectively. In general, we can obtain the following inequalities for $< p, i, j >_\varphi$.

Theorem 3.1 *If (3.17) and (3.18) are fulfilled by* $< p, i, j >_\varphi$ *then the following inequalities are true:*

$$W_\varphi(x, N_\varphi(y)) \leq \ p(x,y) \ \leq \min\{x, N_\varphi(y)\}, \tag{3.22}$$
$$W_\varphi(x,y) \leq \ i(x,y) \ \leq \min\{x, y\}, \tag{3.23}$$
$$W_\varphi(N_\varphi(x), N_\varphi(y)) \leq \ j(x,y) \ \leq \min\{N_\varphi(x), N_\varphi(y)\}. \tag{3.24}$$

Proof. By the previous lemmas we know that (T, S, n) is given by $(W_\varphi, W'_\varphi, N_\varphi)$. Using (PA) and Lemma 3.1, it follows that

$$p(x,y) \leq p(1,y) = N_\varphi(y) \quad \text{and} \quad p(x,y) \leq p(x,0) = x,$$

whence the upper bound in (3.22) is obtained.

On the other hand,

$$\begin{aligned} x &= S(p(x,y), i(x,y)) \leq S(p(x,y), i(1,y)) \\ &= S(p(x,y), y) \\ &= N_\varphi[W_\varphi(N_\varphi(p(x,y)), N_\varphi(y))]. \end{aligned}$$

This is true if and only if $N_\varphi(x) \geq W_\varphi(N_\varphi(p(x,y)), N_\varphi(y))$, or equivalently, if and only if $p(x,y) \geq W_\varphi(x, N_\varphi(y))$, which implies the lower bound in (3.22).

The proof of (3.23) and (3.24) can be done in a similar way.

■

These general inequalities imply the following result.

Theorem 3.2 *For any solution* $< p, i, j >_\varphi$ *of (3.17) and (3.18) and for every* $x, y \in [0, 1]$ *the following equalities hold:*

$$\begin{array}{ll} (a) & W_\varphi(p(x,y), p(y,x)) = 0, \\ (b) & W_\varphi(p(x,y), i(x,y)) = 0, \\ (c) & W_\varphi(p(x,y), j(x,y)) = 0, \\ (d) & W_\varphi(i(x,y), j(x,y)) = 0, \\ (e) & W'_\varphi(W'_\varphi(p(x,y), p(y,x)), W'_\varphi(i(x,y), j(x,y))) = 1. \end{array}$$

Proof. (a) $W_\varphi(p(x,y), p(y,x)) \le W_\varphi(\min\{x, N_\varphi(y)\}, \min\{y, N_\varphi(x)\}) \le W_\varphi(x, N_\varphi(x)) = 0$.

(b) $W_\varphi(p(x,y), i(x,y)) \le W_\varphi(\min\{x, N_\varphi(y)\}, \min\{x, y\}) \le W_\varphi(N_\varphi(y), y) = 0$.

(c) and (d) can be proved similarly.

(e)

$$\begin{aligned} & W'_\varphi[W'_\varphi(p(x,y), p(y,x)), W'_\varphi(i(x,y), j(x,y))] = \\ &= W'_\varphi[W'_\varphi(p(x,y), i(x,y)), W'_\varphi(p(y,x), j(y,x))] \\ &= W'_\varphi(x, N_\varphi(x)) = 1. \end{aligned}$$

■

As a consequence of this theorem, we obtain the following corollary (compare with the classical properties (3.5)–(3.7) and (3.10)).

Corollary 3.1 *For any preference structure (P,I,J) defined via solutions of (3.17), (3.18) we have*

$$\begin{array}{ll} (a) & P \text{ is } W_\varphi\text{-asymmetric}, \\ (b) & P \cap_{W_\varphi} I = \emptyset, \\ (c) & P \cap_{W_\varphi} J = \emptyset, \\ (d) & I \cap_{W_\varphi} J = \emptyset, \\ (e) & P \cup_{W'_\varphi} P^{-1} \cup_{W'_\varphi} I \cup_{W'_\varphi} J = A \times A. \end{array}$$

Although the previous results are rather general ones, unique formulas can be obtained for (P, I, J) when R is strongly complete.

Proposition 3.2 *Suppose that R is strongly complete. Then we have for any $a, b \in A$*

$$\begin{aligned} P(a,b) &= R^d(a,b) \\ I(a,b) &= \min(R(a,b), R(b,a)), \\ J(a,b) &= 0. \end{aligned}$$

Proof. Let $a, b \in A$. Suppose first that $R(a,b) = 1$. Then, $P(a,b) = N_\varphi(R(b,a)) = R^d(a,b)$, $I(a,b) = R(b,a)$ and $J(a,b) = 0$, by (3.22), (3.23) and (3.24), respectively.

If $R(b,a) = 1$ then, also by (3.22), (3.23) and (3.24), we obtain that

$$P(a,b) = 0, \quad I(a,b) = R(a,b), \quad J(a,b) = 0.$$

■

3.3.2 Characterization of some particular solutions

In addition to the previous general results, in this section some important particular solutions of system (3.17), (3.18) are characterized. Since logical connectives are uniquely determined (up to an automorphism φ of the unit interval), we can obtain particular forms of the functions p, i and j only.

Due to negative results in Section 3.2, define first

$$\begin{aligned} p(x,y) &= T_1(x, N_\varphi(y)), \\ i(x,y) &= T_2(x,y) \end{aligned}$$

with continuous t-norms T_1, T_2.

Theorem 3.3 *Assume that $p(x,y) = T_1(x, N_\varphi(y))$ and $i(x,y) = T_2(x,y)$, where T_1 and T_2 are continuous t-norms. Then $< p, i, j >_\varphi$ satisfies (3.17) and (3.18) if and only if there exists a number $s \in [0, \infty]$ such that*

$$\begin{aligned} T_1(x,y) &= \varphi^{-1}(T^s(\varphi(x), \varphi(y))), \\ T_2(x,y) &= \varphi^{-1}(T^{1/s}(\varphi(x), \varphi(y))), \end{aligned}$$

where T^s and $T^{1/s}$ belong to the Frank family.

Proof. Suppose that (3.17) and (3.18) are satisfied. Then, because

$$S(x,y) = W'_\varphi(x,y) = \varphi^{-1}(\min\{\varphi(x) + \varphi(y), 1\})$$

and

$$N_\varphi(x) = \varphi^{-1}(1 - \varphi(x)),$$

equation (3.4) is equivalent to the following one:

$$\min\{\varphi(T_1(x, N_\varphi(y))) + \varphi(T_2(x,y)), 1\} = \varphi(x),$$

or introducing

$$T'_k(x,y) = \varphi(T_k(\varphi^{-1}(x), \varphi^{-1}(y))) \tag{3.25}$$

for $k = 1, 2$ and using $\varphi(x) = u$, $\varphi(y) = v$ then the last equation implies

$$T'_1(u, 1 - v) + T'_2(u, v) = u. \tag{3.26}$$

Applying this equation again for the values $1 - v$ and u, we obtain that

$$T'_1(1 - v, u) + T'_2(1 - v, 1 - u) = 1 - v,$$

that is,

$$T'_1(1 - v, u) = 1 - T'_2(1 - v, 1 - u) - v.$$

Using that T'_1 is commutative and substituting this form of $T'_1(1 - v, u)$ into (3.26), we get

$$T'_2(u,v) + 1 - T'_2(1 - u, 1 - v) = u + v, \tag{3.27}$$

which is exactly the same as that of (1.43). Thus, any solution T_2 of (3.4) fulfils (3.27). This means in the Archimedean case that

$$T_2(x,y) = \varphi^{-1}(T^s(\varphi(x), \varphi(y))$$

for some $s \in (0, \infty]$, as it follows from Theorem 1.11, and $s = 0$ if $T_2' = \min$.

It is easy to see that in this case

$$T_1(x, y) = \varphi^{-1}(T^{1/s}(\varphi(x), \varphi(y)).$$

One can verify immediately that these pairs (T_1, T_2) are indeed solutions of (3.4) for any $s \in [0, \infty]$.

■

Notice that this pair of solution (T_1, T_2) is supported also by fuzzy matrix logic, see Fodor (1993a). As a trivial consequence, the following result of Alsina (1985) is obtained.

Corollary 3.2 *Assume that* $p(x, N_\varphi(y)) = i(x, y) = j(N_\varphi(x), N_\varphi(y)) = T_1(x, y)$, *where* T_1 *is a continuous t-norm. Then* $< p, i, j >_\varphi$ *fulfils (3.17) and (3.18) if and only if*

$$T_1(x, y) = \varphi^{-1}\{\varphi(x)\varphi(y)\}.$$

We have proved in Theorem 3.2 that, for any solution p of our system, the strict preference P is W_φ-asymmetric, i.e., $W_\varphi(P(a, b), P(b, a)) = 0$ for any $a, b \in A$. However, it is also reasonable to expect that if $P(a, b) > 0$ then $P(b, a) = 0$, i.e., P is *asymmetric*, as in the crisp case. In the next theorem we prove that asymmetry for P is equivalent to a particular solution of (3.17) and (3.18). In this solution p reaches its lower bound in (3.22) while i and j are on their upper bounds obtained in (3.23) and (3.24), respectively.

Theorem 3.4 *A solution* $< p, i, j >_\varphi$ *of (3.17), (3.18) is such that* P *is asymmetric if and only if*

$$\begin{aligned} p(x, y) &= W_\varphi(x, N_\varphi(y)), \\ i(x, y) &= \min\{x, y\}, \\ j(x, y) &= \min\{N_\varphi(x), N_\varphi(y)\}. \end{aligned}$$

Proof. Using a result of Ovchinnikov and Roubens (1991),

$$P \text{ is asymmetric} \iff p(x, x) = 0 \quad \forall x \in [0, 1].$$

So let $y \geq x$ in (3.17). This implies that $i(x, y) = x$ for all $x \in [0, 1]$, $x \leq y$ i.e., $i(x, y) = \min\{x, y\}$. Similarly, $y \geq x$ in (3.18) implies that $j(x, y) = N_\varphi(y)$ for all $x \in [0, 1]$ and $y \geq x$, whence $j(x, y) = \min\{N_\varphi(x), N_\varphi(y)\}$. The rest of the proof immediately follows by the Theorem of Ovchinnikov and Roubens (1992).

■

It is easy to see that another particular solution is obtained if p is given by the upper bound in (3.22), i and j are given by lower bounds in (3.23) and (3.24), respectively. This corresponds to property (3.8), as we prove now.

Theorem 3.5 *A solution $< p, i, j >_\varphi$ of (3.17), (3.18) is such that $P \cup I \cup P^{-1} = R \cup R^{-1}$ is fulfilled by (P, I, J) defined via $< p, i, j >_\varphi$ if and only if*

$$\begin{aligned} p(x,y) &= \min\{x, N_\varphi(y)\}, \\ i(x,y) &= W_\varphi(x,y), \\ j(x,y) &= W_\varphi(N_\varphi(x), N_\varphi(y)). \end{aligned}$$

Proof. In fact, property (3.8) is equivalent to

$$S(p(x,y), i(x,y), p(y,x)) = S(x,y) \tag{3.28}$$

for all $x, y \in [0,1]$. Using (3.17), equation (3.28) is equivalent to

$$S(p(x,y), y) = S(x,y),$$

that is,

$$\varphi^{-1}(\min\{\varphi(p(x,y)) + \varphi(y), 1\}) = \varphi^{-1}(\min\{\varphi(x) + \varphi(y), 1\}). \tag{3.29}$$

Assume first that $\varphi(x) + \varphi(y) < 1$. In this case, by (3.29), $p(x,y) = x$.

If $\varphi(x) + \varphi(y) \geq 1$ then (3.29) implies that $p(x,y) \geq \varphi^{-1}(1 - \varphi(y))$. Taking into account inequality (3.22), we immediately obtain that now $p(x,y) = N_\varphi(y)$. In other words, $p(x,y) = \min\{x, N_\varphi(y)\}$.

Consider now equation (3.17) and let $x > N_\varphi(y)$, i.e., $\varphi(x) + \varphi(y) > 1$. Then (3.17) is read as follows:

$$1 - \varphi(y) + \varphi(i(x,y)) = \varphi(x),$$

i.e., $i(x,y) = W_\varphi(x,y)$ for $x > N_\varphi(y)$. On the other hand, $x \leq N_\varphi(y)$ and (3.17) together imply that $i(x,y) = 0$. Thus our theorem is proved.

■

One can think by the previous particular solutions of system (3.17), (3.18) that $i(x,y) = j(N_\varphi(x), N_\varphi(y))$ holds for any solution. The following example shows that this is not the case in general.

Assume that $T(x,y) = W(x,y) = \max\{x + y - 1, 0\}$ and $N(x) = 1 - x$. Define

$$\begin{aligned} p(x,y) &= \begin{cases} x^2(1-y) & \text{if } x \leq y \\ x - y + y^2(1-x) & \text{if } x > y \end{cases}, \\ i(x,y) &= x - p(x,y), \\ j(x,y) &= 1 - y - p(x,y). \end{aligned}$$

Obviously, equations (3.17) and (3.18) are satisfied by thus defined $< p, i, j, W, W', N >$.

Proposition 3.3 *Thus defined $< p, i, j, W, W', N >$ fulfils (PA) and (S). Moreover,*

$$i(x,y) \neq j(N(x), N(y)).$$

The proof, which can be carried out by elementary calculus, is left to the reader. In fact, we can prove the following result on having $i(x,y) = j(N_\varphi(x), N_\varphi(y))$.

Proposition 3.4 *Let $< p, i, j >_\varphi$ be any solution of the system (3.17), (3.18). Then $i(x,y) = j(N_\varphi(x), N_\varphi(y))$ holds if and only if $p(x,y) = p(N_\varphi(y), N_\varphi(x))$.*

Proof. By equations (3.17), (3.18), i and j are expressed as follows :

$$\begin{aligned} i(x,y) &= \varphi^{-1}(\varphi(x) - \varphi(p(x,y))), \\ j(x,y) &= \varphi^{-1}(1 - \varphi(y) - \varphi(p(x,y))). \end{aligned}$$

Taking into account that j is symmetric, we obtain that

$$j(x,y) = \varphi^{-1}(1 - \varphi(x) - \varphi(p(y,x))).$$

Therefore, $i(x,y) = j(N_\varphi(x), N_\varphi(y))$ holds if and only if

$$\varphi^{-1}(\varphi(x) - \varphi(p(x,y))) = \varphi^{-1}(1 - (1 - \varphi(x) - \varphi(p(N_\varphi(y), N_\varphi(x))),$$

or equivalently, if and only if $p(x,y) = p(N_\varphi(y), N_\varphi(x))$.

■

3.3.3 Strict preference and implications

In this section we investigate connections between strict preference and fuzzy implications (see also Fodor (1991e)). Suppose that $< p, i, j >_\varphi$ is a solution of the system (3.17), (3.18) for a given automorphism φ of the unit interval. Define a function $I^{\rightarrow}$ by

$$I^{\rightarrow}(x,y) = N_\varphi(p(x,y)). \tag{3.30}$$

It can be proved that thus defined $I^{\rightarrow}$ is a fuzzy implication (in the sense of Definition 1.15), having additional properties.

Proposition 3.5 *If $< p, i, j >_\varphi$ is a solution of the system (3.17), (3.18) then $I^{\rightarrow}(x,y) = N_\varphi(p(x,y))$ is a fuzzy implication such that*

$$\begin{aligned} I^{\rightarrow}(1,x) &= x \quad \forall\, x \in [0,1], \\ I^{\rightarrow}(x,0) &= N_\varphi(x) \quad \forall\, x \in [0,1] \end{aligned}$$

are also true.

Proof. $I^{\rightarrow}(x,y)$ is nonincreasing in the first place and is nondecreasing in the second place since the positive association principle holds for p. Moreover,

$$\begin{aligned} I^{\rightarrow}(0,y) &= N_\varphi(p(0,y)) = N_\varphi(0) = 1 && \text{by Lemma 3.1 (a).} \\ I^{\rightarrow}(x,1) &= N_\varphi(p(x,1)) = N_\varphi(0) = 1 && \text{by Lemma 3.1 (b).} \\ I^{\rightarrow}(1,x) &= N_\varphi(p(1,x)) = N_\varphi(N_\varphi(x)) = x && \text{by Lemma 3.1 (d).} \\ I^{\rightarrow}(x,0) &= N_\varphi(p(x,0)) = N_\varphi(x) && \text{by Lemma 3.1 (c).} \end{aligned}$$

■

Therefore, we can conclude that strict preference relation P can be defined in a consistent way by using a fuzzy implication $I^{\rightarrow}$ satisfying the following inequality

$$\max(N_\varphi(x), y) \leq I^{\rightarrow}(x,y) \leq W'_\varphi(N_\varphi(x), y)$$

so that

$$P(a,b) = N_\varphi[I^{\rightarrow}(R(a,b), R(b,a))].$$

We would like to recall an alternative approach from Fodor (1992b) which also supports the use of implications in the definition of strict preferences.

Assume that axioms (IA), (PA) and (S) hold. Moreover, assume that the strict preference P is asymmetric: $\min(P(a,b), P(b,a)) = 0$ for all $a, b \in A$. Asymmetry of P is equivalent to $p(x,x) = 0$ for all $x \in [0,1]$ (see Ovchinnikov and Roubens (1991)).

Starting from the classical properties (3.4), (3.6) and observing (by (3.2) and (3.3)) that the crisp case implies $j(x,y) = i(n(x), n(y))$, we can study a new system of functional equations. The following result was proved by Fodor (1992b).

Theorem 3.6 *Suppose that (T, S, n) is a De Morgan triple such that T is a continuous t-norm, S is a continuous t-conorm and n is a strict negation. Assume that axioms (IA), (PA), (S) hold and P is asymmetric. Then $< p, i, j, T, S, n >$ satisfies the system of functional equations*

$$\begin{aligned} S(p(x,y), i(x,y)) &= x, \\ T(p(x,y), j(x,y)) &= 0, \\ j(x,y) &= i(n(x), n(y)) \end{aligned}$$

if and only if

(i) T is a continuous Archimedean t-norm with zero divisors;
(ii) n is a strict negation such that $n(x) \le I_T^{\rightarrow}(x, 0)$;
(iii) $S(x,y) = n^{-1}(T(n(x), n(y)))$;
(iv) $p(x,y) = n^{-1}[I_T^{\rightarrow}(n(y), n(x))]$;
(v) $i(x,y) = \min(x,y)$;
(vi) $j(x,y) = \min(n(x), n(y))$.
Moreover, $n(x) = I_T^{\rightarrow}(x, 0)$ if and only if $p(x,y) = T(x, n(y))$.

¿From a semantical point of view, the connection between strict preferences and implications is supported by the interrelation

$$\textbf{NOT } aPb \quad \Longleftrightarrow \quad aRb \text{ implies } bRa.$$

3.4 Transitivity of strict preference relations

In the theory of classical preference modelling, a strict preference relation associated with a transitive preference relation is always transitive.

The corresponding result for valued preferences has been established by Ovchinnikov and Roubens (1991). We cite that result now.

First note that transitivity in the valued case is understood as transitivity with respect to the t-norm 'min'. In addition, it is assumed that the valued strict preference P is asymmetric. That is, $\min\{P(a,b), P(b,a)\} = 0$ holds for all $a, b \in A$. Using generator functions p of strict preferences P satisfying the positive association principle (PA), it was proved by Ovchinnikov and Roubens (1991) that $P(a,b) = p(R(a,b), R(b,a))$ is asymmetric if and only if $p(x,x) = 0$ for all $x \in [0,1]$.

We need the following lemma (for the proof see Ovchinnikov and Roubens (1991)).

Lemma 3.4 *Let R be a transitive valued binary relation on A such that*

$$R(c,a) < R(a,c) \quad \text{and} \quad R(b,c) < R(c,b)$$

hold for some $a, b, c \in A$. Then we have

$$R(b,a) = \min\{R(b,c), R(c,a)\}.$$

The main theorem of this section is formulated as follows.

Theorem 3.7 *An asymmetric valued strict preference relation P associated with a transitive valued preference relation R is a transitive valued preference relation.*

Proof. We have to prove that

$$\min\{P(a,c), P(c,b)\} \leq P(a,b),$$

for all $a, b, c \in A$. This inequality is trivially satisfied if $P(a,c) = 0$ or $P(c,b) = 0$.

Thus, suppose that $P(a,c) > 0$ and $P(c,b) > 0$. Then we have $R(c,a) < R(a,c)$ and $R(b,c) < R(c,b)$, by asymmetry of P. This implies, by the previous lemma, that $R(b,a) = \min\{R(b,c), R(c,a)\}$. Therefore, we have

$$\begin{aligned}
\min\{P(a,c), P(c,b)\} &= \min\{p(R(a,c), R(c,a)), p(R(c,b), R(b,c))\} \\
&\leq \min\{p(R(a,c), R(b,a)), p(R(c,b), R(b,a))\} \\
&= p(\min\{R(a,c), R(c,b)\}, R(b,a)) \\
&\leq p(R(a,b), R(b,a)) \\
&= P(a,b).
\end{aligned}$$

■

Remark that it was not supposed that function p is a part of a solution $< p, i, j >_\varphi$ of our functional equations. Moreover, asymmetry of P is a necessary condition to ensure transitivity of P by transitivity of R, as it is indicated by the following example.

Example. Let $A = \{a, b, c\}$ and R is defined by the following matrix

$$R = \begin{bmatrix} 1 & 0.5 & 0.5 \\ 0.9 & 1 & 0.7 \\ 0.8 & 0.5 & 1 \end{bmatrix}.$$

It is easy to check that R is a transitive and reflexive valued binary relation on A. Let $p(x,y) = \min(x, 1-y)$. Then $p(x,x) \neq 0$ for $0 < x < 1$. That is, the strict preference relation

$$P(a,b) = \min\{R(a,b), 1 - R(b,a)\}$$

is not asymmetric on A.

One can compute that

$$\begin{aligned}
P(a,c) &= \min\{R(a,c), 1 - R(c,a)\} = 0.2, \\
P(c,b) &= \min\{R(c,b), 1 - R(b,c)\} = 0.3, \\
P(a,b) &= \min\{R(a,b), 1 - R(b,a)\} = 0.1.
\end{aligned}$$

Therefore, we have

$$\min\{P(a,c),P(c,b)\} > P(a,b),$$

whence P is not transitive.

■

Therefore, the only transitivity preserving generator function p of strict preferences among the solutions of our functional equations is given by

$$p(x,y) = W_\varphi(x,N_\varphi(y)) = \varphi^{-1}(\max\{\varphi(x)-\varphi(y),0\}),$$

see Theorem 3.4.

The following example shows that W-transitivity is not preserved in general by using $P(a,b) = \max\{R(a,b) - R(b,a),0\}$.

Example. Let $A = \{a,b,c\}$ and R be giben by

$$R = \begin{bmatrix} 1 & 0.9 & 0.8 \\ 0.05 & 1 & 0.9 \\ 0.15 & 0.05 & 1 \end{bmatrix}.$$

Then P is obtained as

$$P = \begin{bmatrix} 0 & 0.85 & 0.65 \\ 0 & 0 & 0.85 \\ 0 & 0 & 0 \end{bmatrix}.$$

It is easy to check that R is W-transitive. However, we have

$$\max\{P(a,b)+P(b,c)-1,0\} = 0.7 > 0.65 = P(a,c).$$

Therefore, W-transitivity of R does not imply W-transitivity of P in general.

■

Chapter 4

Similarity relations and valued orders

In the classical theory of binary relations probably the two most important and most frequently applied classes consist of equivalence relations (with associated partitions) and different kinds of orders.

Equivalence relations (i.e., reflexive, symmetric and transitive binary relations) and partitions are mathematical models for the concept of 'sameness'.

Generally speaking, an order is simply a transitive relation. To distinguish between equivalences (which are also transitive relations) and orders, a kind of asymmetry and completeness is assumed in the second case.

In this chapter we study valued counterparts of equivalence relations and several important classes of orders in the light of the general representation of T-transitive valued binary relations.

4.1 Covers and proximity relations

Similarity relations are particular cases of *proximity* relations (sometimes called *tolerance* relations). In fact, a similarity relation is a transitive proximity relations. First we study proximity relations and valued covers. The way of presentation in this section follows the paper of Ovchinnikov (1991). The results can also be found in Yeh and Bang (1975) and in Ovchinnikov and Riera (1982).

Definition 4.1 Let A be a given set. A family of mappings $\Sigma = \{C\}$ with $C : A \to [0,1]$ is called a *valued cover* of A if

$$A = \bigcup_{C \in \Sigma} C.$$

Notice that condition $A = \bigcup_{C\in\Sigma} C$ means that

$$\sup_{C\in\Sigma} C(a) = 1 \text{ for all } a \in A.$$

For a given valued cover $\Sigma = \{C\}$ of A we define a valued binary relation R_Σ associated with Σ in the following way :

$$R_\Sigma(a,b) = \sup_{C\in\Sigma} \min\{C(a), C(b)\}. \tag{4.1}$$

Obviously, R_Σ is a reflexive and symmetric valued binary relation on A.

Definition 4.2 A valued binary relation R defined on A is said to be a *proximity relation* on A if R is reflexive and symmetric.

Using this terminology, a valued binary relation on A associated with a valued cover of A is a proximity relation on A.

Definition 4.3 Let R be a valued binary relation on A. A function $K : A \to [0,1]$ is called a *pre-class* of R if

$$\min\{K(a), K(b)\} \leq R(a,b)$$

holds for all $a, b \in A$. The set of all maximal pre-classes (with respect to $\subseteq$) of R is denoted by Σ_R.

In the following theorem it is shown that for any proximity relation R on A there exists a valued cover of A such that R is associated with that cover.

Theorem 4.1 *Let R be a proximity relation on A. Then there exists a valued cover Σ of A such that $R = R_\Sigma$.*

Proof. Define a family $\{K_{\{a,b\}}\}$ of functions from A to $[0,1]$ by

$$K_{\{a,b\}}(x) = \begin{cases} R(a,b) & \text{if } x \in \{a,b\}, \\ 0 & \text{otherwise} \end{cases},$$

for all $a, b \in A$. We prove that $K_{\{a,b\}}$ is a pre-class of R for all $a, b \in A$.
Indeed, we have

$$\begin{aligned} \min\{K_{\{a,b\}}(a), K_{\{a,b\}}(a)\} &= R(a,b) \leq 1 = R(a,a), \\ \min\{K_{\{a,b\}}(a), K_{\{a,b\}}(b)\} &= R(a,b), \\ \min\{K_{\{a,b\}}(b), K_{\{a,b\}}(a)\} &= R(a,b) = R(b,a), \\ \min\{K_{\{a,b\}}(b), K_{\{a,b\}}(b)\} &= R(a,b) \leq 1 = R(b,b), \\ \min\{K_{\{a,b\}}(x), K_{\{a,b\}}(y)\} &= 0 \text{ if } x \notin \{a,b\} \text{ and } y \notin \{a,b\}. \end{aligned}$$

Let Σ be a set of classes such that for any $\{a,b\}$ there is a class in Σ containing $K_{\{a,b\}}$. Then, for any $K \in \Sigma$, $\min\{K(a), K(b)\} \leq R(a,b)$ and there exists K such that $\min\{K(a), K(b)\} = R(a,b)$.
Indeed, let $K \in \Sigma$ be a class containing $K_{\{a,b\}}$. Then

$$\begin{aligned} R(a,b) &= \min\{K_{\{a,b\}}(a), K_{\{a,b\}}(b)\} \leq \min\{K(a), K(b)\} \\ &\leq R(a,b). \end{aligned}$$

Hence

$$\sup_{K \in \Sigma} \min\{K(a), K(b)\} = R(a,b) \text{ for all } a, b \in A.$$

Since R is a proximity relation on A, we have

$$1 = R(a,a) = \sup_{K \in \Sigma} K(a),$$

that is, Σ is a valued cover of A and $R_\Sigma = R$.

■

Corollary 4.1 *If $\Sigma = \Sigma_R$ is the set of maximal pre-classes of R then $R_{\Sigma_R} = R$.*

Let us consider an example from Ovchinnikov (1991). Let $A = \{a, b, c\}$ and let two valued covers of A be given by

$$\Sigma_1 = \begin{array}{c|ccc} & C_1 & C_2 & C_3 \\ \hline a & 1 & \alpha & \alpha \\ b & \alpha & 1 & \gamma \\ c & \beta & \gamma & 1 \end{array}, \quad \Sigma_2 = \begin{array}{c|ccc} & C_1 & C_2 & C_3 \\ \hline a & 1 & \alpha & \beta \\ b & \alpha & 1 & \alpha \\ c & \beta & \gamma & 1 \end{array}$$

with $\alpha < \beta < \gamma$. It is easy to verify that

$$R_{\Sigma_1} = R_{\Sigma_2} = \begin{bmatrix} 1 & \alpha & \beta \\ \alpha & 1 & \gamma \\ \beta & \gamma & 1 \end{bmatrix}.$$

$R = R_{\Sigma_1} = R_{\Sigma_2}$ is clearly a proximity relation with the following maximal pre-classes :

$$\Sigma_R = \begin{array}{c|cccc} & C_1 & C_2 & C_3 & C_4 \\ \hline a & 1 & \alpha & \alpha & \beta \\ b & \alpha & 1 & \gamma & \alpha \\ c & \beta & \gamma & 1 & 1 \end{array}.$$

Therefore, there exist at least three different valued covers Σ_1, Σ_2 and Σ_R such that $R = R_{\Sigma_1} = R_{\Sigma_2} = R_{\Sigma_R}$.

4.2 Similarity relations and valued partitions

Similarity relations have been introduced and investigated by Zadeh (1971) (see also Ovchinnikov (1981,1991)).

Let (T, S, N) be a De Morgan triple such that T is a left-continuous t-norm and N is a strong negation (whence S is a right-continuous t-conorm).

Definition 4.4 A valued binary relation R on A is called T-*similarity relation* on A if R is reflexive, symmetric and T-transitive.

In other words, T-similarity relations on A are T-transitive proximity relations on A. Recall that, since R is reflexive, T-transitivity means $R \circ_T R = R$.

There is a nice characterization and representation of T-similarity relations published by Valverde (1985). We cite this result first. Remark that Valverde used the term "*indistinguishability*" instead of similarity.

Theorem 4.2 *Let R be a valued binary relation on A. Then R is a T-similarity relation on A if and only if there exists a family $\{h_\gamma\}_{\gamma \in \Gamma}$ of functions from A to $[0,1]$ so that for all $a, b \in A$*

$$R(a,b) = \inf_{\gamma \in \Gamma} I_T^{\rightarrow}(\max\{h_\gamma(a), h_\gamma(b)\}, \min\{h_\gamma(a), h_\gamma(b)\}), \tag{4.2}$$

where $I_T^{\rightarrow}$ is the R-implication defined by T.

Proof. Using Theorem 2.2 on the representation of T-transitive valued binary relations, formula (2.14) implies that a T-transitive valued binary relation is reflexive if and only if $h_\gamma = k_\gamma$ for all $\gamma \in \Gamma$. That is, R is reflexive and T-transitive if and only if there exists a family $\{h_\gamma\}_{\gamma\in\Gamma}$ such that

$$R(a,b) = \inf_{\gamma\in\Gamma} I_T^{\rightarrow}(h_\gamma(b), h_\gamma(a))$$

holds for all $a, b \in A$. By symmetry of R, the result follows easily.

■

Using the traces of a valued binary relation, the following result can be proved.

Theorem 4.3 *The following statements are equivalent for any valued binary relation R on A:*
(a) R is a T-similarity relation on A;
(b)

$$R(a,b) = R^\ell(a,b) = R^\ell(b,a); \tag{4.3}$$

(c) $R(a,b) = R^r(a,b) = R^r(b,a)$
holds for all $a, b \in A$.

Proof. (a) $\iff$ (b): By Propositions 2.7 and 2.20, R is a reflexive and T-transitive valued binary relation on A if and only if $R = R^\ell$. Thus, from the definition of symmetry we obtain the statement.
(a) $\iff$ (c) can be proved in the same way.

■

This theorem has an important consequence on the maximal number of functions $\{h_\gamma\}_{\gamma\in\Gamma}$ defining a T-similarity relation.

Corollary 4.2 *Any T-similarity relation on A can be defined by (4.2) with at most as many functions h_γ as the number of elements in A.*

Proof. By Theorem 4.3, representation (4.3) holds. Therefore, $\Gamma = A$ and $h_c(a) = R(c,a)$ is an appropriate choice.

■

Now we introduce T-pre-classes and T-classes of a valued binary relation R.

Definition 4.5 Let R be a valued binary relation on A. A function $K : A \longrightarrow [0,1]$ is called a *T-pre-class* of R if

$$T(K(a), K(b)) \leq R(a,b) \tag{4.4}$$

holds for all $a, b \in A$. A T-pre-class K of R is called a *T-class* of R if there exists an $a_0 \in A$ such that $K(a_0) = 1$ and we have

$$T(K(a), R(a,b)) \leq K(b) \tag{4.5}$$

for all $a, b \in A$.

This definition has been proposed by Höhle (1988), see also Dubois and Prade (1991c). Remark that Höhle (1988) works in a more general framework (reflexivity of R is weakened, equality in A is redefined by a special similarity relation, and a more abstract algebraic structure is used instead of the unit interval).

Inequality (4.4) requires that R should contain the Cartesian product (with respect to T) of any T-pre-class by itself. The existence of an element $a_0 \in A$ means that a T-class K is nonempty. Condition (4.5) indicates that elements interrelating with b should be in the T-class of b.

In case of $T = \min$ there is a close connection between thus defined T-classes and maximal pre-classes (which have been called classes by Ovchinnikov (1991)) of a valued binary relation R, as we state in the following proposition.

Proposition 4.1 *Suppose $T = \min$ and R is a valued binary relation on A. A function $K : A \to [0,1]$ such that $K(a_0) = 1$ for some $a_0 \in A$ is a maximal pre-class of R if and only if*

$$\min(K(a), K(b)) \leq R(a, b)$$

and

$$\min(K(a), R(a, b)) \leq K(b)$$

hold for all $a, b \in A$.

Proof. Suppose K is a maximal pre-class of R and there exist $c, d \in A$ such that

$$\min(K(c), R(c, d)) > K(d).$$

Define a function K' by

$$K'(x) = \begin{cases} R(c,d) & \text{if } x = d \\ K(x) & \text{otherwise} \end{cases}.$$

Obviously, thus defined K' is also a pre-class of R such that $K'(d) > K(d)$. Hence, K cannot be a maximal pre-class of R.

Suppose now that $K : A \to [0,1]$ is a function with $K(a_0) = 1$ for some $a_0 \in A$ such that (4.4) and (4.5) hold. Let $a = a_0$. Then we have that $K(b) \leq R(a_0, b) \leq K(b)$, i.e., $R(a_0, b) = K(b)$ for all $b \in A$.

Let K' be a pre-class containing K. Then we should have $K'(a_0) = K(a_0) = 1$. Hence,

$$K(b) \leq K'(b) \leq R(a_0, b) \leq K(b),$$

i.e., $K'(b) = K(b)$ holds for all $b \in A$. This proves that K is a maximal pre-class of R. ∎

Following Zadeh (1971), similarity classes of a T-similarity relation consist of elements being close to each other and formally are defined as follows.

Definition 4.6 Let R be a T-similarity relation on A. For any given $a \in A$, a *similarity class* of a is a function $R[a] : A \to [0,1]$ defined by $R[a](c) = R(a, c)$ for all $c \in A$.

Following Dubois and Prade (1991c), we prove that the T-classes of a T-similarity relation R on A are exactly similarity classes of R. This is shown via a sequence of lemmas.

Lemma 4.1 *Any similarity class $R[a]$ of a T-similarity relation R is a T-class of R.*

Proof. Reflexivity of R implies that $R[a]$ is normalized. Using symmetry of R, both conditions (4.4) and (4.5) require the T-transitivity of R.

■

Lemma 4.2 *Let R be a relexive valued binary relation on A. Then we have*

$$\sup_{c\in A} T(K(c), R(c,a)) = K(a) \quad \forall a \in A \tag{4.6}$$

for any T-class K of R.

Proof. We have the following inequality for any R and any T-class K of R

$$\sup_{c\in A} T(K(c), R(c,a)) \leq K(a),$$

by (4.5).

On the other hand, reflexivity of R implies that

$$\sup_{c\in A} T(K(c), R(c,a)) \geq T(K(a), R(a,a)) = K(a).$$

■

Lemma 4.3 *If K and K' are two T-classes of the T-similarity relation R such that $K(a_0) = K'(a_0) = 1$ for some $a_0 \in A$ then $K = K'$.*

Proof. Let $a \in A$ be any element. Then

$$\begin{aligned} K(a) &= T(K(a), \sup_{c\in A} T(K'(c), K'(c))) \qquad \text{(since } K' \text{ is normalized)} \\ &= \sup_{c\in A} T(K(a), T(K'(c), K'(c))) \\ &= \sup_{c\in A} T(T(K(a), K'(c)), K'(c)) \\ &\leq \sup_{c\in A} T(R(c,a), K'(c)) \qquad \text{(by condition (4.5))} \\ &= K'(a) \qquad \text{(by Lemma 4.2).} \end{aligned}$$

By symmetry, we can show that $K'(a) \leq K(a)$ for all $a \in A$, so we obtain that $K = K'$.

■

This lemma implies that T-classes of a T-similarity relation cannot be ordered via fuzzy set inclusion, as it is pointed out by Dubois and Prade (1991c). Indeed, if K and K' are two T-classes then $K \subseteq K'$ implies $K = K'$. Moreover, since similarity classes $R[a]$ of a T-similarity relation R are T-classes of R, no fuzzy subsets of these similarity classes can belong to T-classes of R.

Lemma 4.4 *Suppose that C is a normalized fuzzy set on A for which there exists no $a \in A$ such that $C \subseteq R[a]$. Then C is not a T-class of R.*

Proof. By the conditions of this lemma, for all $a \in A$ there exists a $c \in A$ such that $R(a,c) < C(c)$. Let a be such that $C(a) = 1$. Then, for some c we have $R(a,c) < K(c) = T(K(a), K(c))$. This contradicts to condition (4.4).

■

So far we have proved that a T-class of a T-similarity relation cannot be strictly contained by some $R[a]$ and at the same time should be contained by some $R[a]$. This proves the following statement.

Theorem 4.4 *The set of T-classes of a T-similarity relation R coincides with the set of similarity classes of R.*

■

It may happen that $R[a] = R[b]$ for different elements $a, b \in A$. The following proposition characterizes this case.

Proposition 4.2 $R[a] = R[b]$ *holds if and only if* $R(a,b) = 1$.

Proof. Suppose $R[a] = R[b]$. Then

$$R(a,b) = R[a](b) = R[b](b) = R(b,b) = 1.$$

Conversely, suppose that $R(a,b) = 1$. Then, by T-transitivity of R,

$$R[a](c) = R(a,c) \geq T(R(a,b), R(b,c)) = R(b,c) = R[b](c).$$

Similarly, $R[b](c) \geq R[a](c)$ for all $c \in A$.
This proves the statement. ■

Corollary 4.3 *If* $R[a] \neq R[b]$ *then* $h(R[a] \cap_T R[b]) < 1$.

Proof. If the height of the intersection is 1, then there exists $c \in A$ such that

$$R[a](c) = R[b](c) = 1.$$

Then,

$$R(a,b) \geq T(R(a,c), R(c,b)) = 1,$$

and thus $R[a] = R[b]$.

■

Following Ovchinnikov (1991) further, from now on in this section, we suppose that $T = \min$ and A is a finite set.

Let Σ be a cover of A and R_Σ be a proximity relation associated with Σ. In the following theorem, necessary and sufficient condition is given to ensure that R_Σ is a similarity relation.

Theorem 4.5 *Let Σ be a cover of A. R_Σ is a similarity relation on A if and only if for any C', $C'' \in \Sigma$ there exists $C \in \Sigma$ such that*

$$\min\{h(C', C''), C'(a), C''(b)\} \leq \min\{C(a), C(b)\} \tag{4.7}$$

holds for all $a, b \in A$, where

$$h(C', C'') = \sup_{a \in A} \min\{C'(a), C''(a)\}$$

is the height of $C' \cap C''$.

Proof. Suppose Σ satisfies (4.7). It suffices to prove transitivity of R_Σ. We have

$$\begin{aligned} \min\{R_\Sigma(a,c), R_\Sigma(c,b)\} &= \min\left\{\sup_{C\in\Sigma}\min\{C(a),C(c)\}, \sup_{C\in\Sigma}\min\{C(c),C(b)\}\right\} \\ &= \sup_{C',C''\in\Sigma} \min\{C'(a), C'(c), C''(c), C''(b)\} \\ &\leq \sup_{C',C''\in\Sigma} \min\{h(C',C''), C'(a), C''(b)\} \\ &\leq \sup_{C',C''\in\Sigma} \min\{C(a), C(b)\} = R_\Sigma(a,b). \end{aligned}$$

To prove the converse, suppose that R_Σ is a similarity relation on A. Then,

$$\min\{R_\Sigma(a,c), R_\Sigma(c,b)\} \leq R_\Sigma(a,b)$$

holds for all $a, b, c \in A$, which implies

$$\sup_{C',C''\in\Sigma} \min\{C'(a), C'(c), C''(c), C''(b)\} \leq \sup_{C\in\Sigma} \min\{C(a), C(b)\}.$$

Hence, for any given $C', C'' \in \Sigma$ there eixsts $C \in \Sigma$ such that

$$\min\{C'(a), C'(c), C''(c), C''(b)\} \leq \min\{C(a), C(b)\}$$

for any $c \in A$. Therefore,

$$\begin{aligned} \min\{h(C',C''), C'(a), C''(b)\} &= \sup_{c\in A}\min\{C'(a), C'(c), C''(c), C''(b)\} \\ &\leq \min\{C(a), C(b)\}. \end{aligned}$$

This means that (4.7) is fulfilled. ∎

Notice that a classical equivalence relation is reflexive, symmetric and transitive. Thus, similarity relations are natural generalizations of equivalence relations in the valued case. Clearly, each λ-cut of a similarity relation is an equivalence relation, as one can check easily.

If an equivalence relation is given on A, then its equivalence classes form a partition of A, that is, A can be represented as the union of disjoint nonempty equivalence classes. In the valued case, although different similarity classes could have a nonempty intersection, their λ-cuts form a partition of A for a given λ. These observations lead us to the following definition.

Definition 4.7 A valued cover Σ of A is said to be a *valued partition* of A if there exists a similarity relation R on A such that Σ is the set of all distinct similarity classes of R.

Let Π_Σ be the set of all different 1.0-cuts of elements of Σ. If Π_Σ is a partition of A, then for any $a \in A$ there exists an element $[a] \in \Sigma$ such that $a \in [a]_{1.0}$, i.e., $[a](a) = 1$.

Theorem 4.6 *Let Σ be a finite cover of A. Σ is a valued partition of A if and only if*

(i) Π_Σ *is a partition of A, and*

(ii) $h([a] \cap [b]) = \min\{[a](b), [b](a)\}$ *for all $a, b \in A$.*

Proof. (a) Suppose that Σ is a valued partition of A. That is, Σ is the set of similarity classes of some similarity relation R. Then Π_Σ is the set of all distinct 1.0-cuts of similarity classes of R. Since $R[a](a) = 1$ for any $a \in A$, elements of Π_Σ are nonempty sets with union A. By Corollary 4.3, intersections of different sets in Π_Σ are empty. Therefore, Π_Σ is a partition of A.

For any element $a \in A$, we have $[a] = R[a]$. Indeed, $[a]$ is a similarity class of R. Therefore, $[a] = R[c]$ for some $c \in A$. We have $R(c, a) = R[c](a) = [a](a) = 1$, and, by Proposition 4.2, $R[c] = R[a]$. Further, using symmetry of R and $R \circ R = R$, we have

$$\begin{aligned} h([a] \cap [b]) &= \sup_{c \in A} \min\{[a](c), [b](c)\} \\ &= \sup_{c \in A} \min\{R[a](c), R[b](c)\} \\ &= \sup_{c \in A} \min\{R(a, c), R(b, c)\} \\ &= R(a, b) = \min\{R(a, b), R(b, a\} \\ &= \min\{[a](b), [b](a)\}. \end{aligned}$$

(b) Let Σ be a valued cover satisfying conditions (i) and (ii). Define a valued binary relation R on A by $R(a, b) = [a](b)$. We have to prove that R is a reflexive, symmetric and transitive valued binary relation on A.

Obviously, $R(a, a) = [a](a) = 1$ for all $a \in A$, thus R is reflexive.

We have the following inequality by (ii) :

$$\min\{[a](c), [b](c)\} \leq \min\{[a](b), [b](a)\}$$

for all $c \in A$.

Substituting $c = a$ and $c = b$, we obtain $[a](a) \leq [a](b)$ and $[a](b) \leq [b](a)$, respectively. Therefore, $[a](b) = [b](a)$ and hence R is symmetric.

Finally, by symmetry of R and condition (ii), we have

$$\min\{R(a, c), R(c, b)\} = \min\{[a](c), [b](c)\} \leq \min\{[a](b), [b](a)\} = R(a, b).$$

This means that R is transitive.

Thus our theorem is proved. ■

Let R be a similarity relation on A. Then for any given $\lambda \in [0, 1]$, R_λ is an equivalence relation on A and $aR_\lambda b$ holds if and only if $R(a, b) \geq \lambda$.

For a given $a \in A$, we have

$$\begin{aligned} R[a]_\lambda &= \{c \mid R[a](c) \geq \lambda\} = \{c \mid R(a,c) \geq \lambda\} \\ &= \{c \mid aR_\lambda c\}. \end{aligned}$$

Therefore, λ-cuts of similarity classes of R are exactly equivalence classes of the λ-cut of R.

Closing this section, consider the following example given by Zadeh (1971).
Let R be a valued binary relation on the set $A = \{a_1, a_2, \ldots, a_6\}$ defined by

$$R = \begin{bmatrix} 1 & 0.2 & 1 & 0.6 & 0.2 & 0.6 \\ 0.2 & 1 & 0.2 & 0.2 & 0.8 & 0.2 \\ 1 & 0.2 & 1 & 0.6 & 0.2 & 0.6 \\ 0.6 & 0.2 & 0.6 & 1 & 0.2 & 0.8 \\ 0.2 & 0.8 & 0.2 & 0.2 & 1 & 0.2 \\ 0.6 & 0.2 & 0.6 & 0.8 & 0.2 & 1 \end{bmatrix}.$$

One can check easily that R is a similarity relation of A. There are five distinct similarity classes of R :

$$R[a_1] = R[a_3], R[a_2], E[a_4], R[a_5], R[a_6].$$

Since we have only four different values of $R(a,b)$, there are only four different λ-cuts to consider : $R_{0.2}, R_{0.6}, R_{0.8}$ and $R_{1.0}$. Clearly, $R[a]_{\lambda'} \subseteq R[a]_\lambda$ for $\lambda' \geq \lambda$. Therefore, a partition generated by $R_{\lambda'}$ is a refinement of the partition generated by R_λ. A nested family of partitions generated by λ-cuts of R can be represented in the form of partition tree, as it is shown in Figure 4.1.

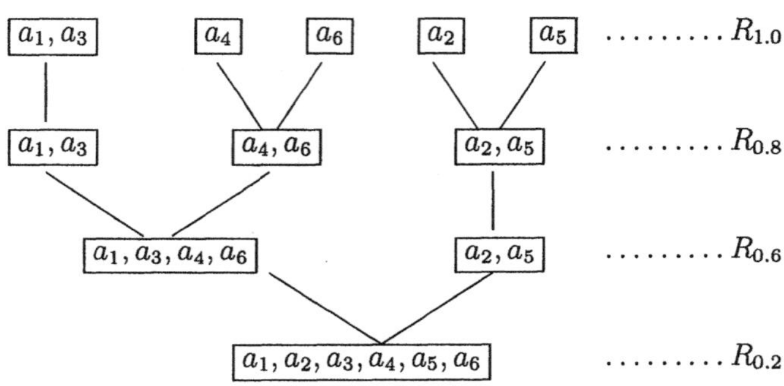

Figure 4.1. Partition tree

The notion of a partition tree may be considered as a generalization of the concept of a quotient set with respect to an equivalence relation.

4.3 Preorders

Several particular classes of crisp binary relations have been introduced at the end of Section 2.1. To define corresponding valued classes, suppose that (T, S, N) is a De Morgan triple with a left-continuous t-norm T, a strong negation N and A is the set of alternatives.

Definition 4.8 A valued binary relation R on A is called
– *partial T-preoder* (or a T-quasiorder) if R is reflexive and T-transitive;
– *total T-preorder* (or a linear T-quasiorder) if R is strongly complete and T-transitive.

Valued T-preorders can be characterized and represented by using the representation theorem of T-transitive valued binary relations. Remark that the original idea of representation comes from Ovchinnikov (1984). The general representation theorem was published first by Valverde (1985).

Theorem 4.7 *A valued binary relation R on A is a partial T-preorder if and only if there exists a family $\{h_\gamma\}_{\gamma\in\Gamma}$ of functions from A to $[0,1]$ such that*

$$R(a,b) = \inf_{\gamma\in\Gamma} I_T^{\rightarrow}(h_\gamma(b), h_\gamma(a)) \tag{4.8}$$

holds for all $a, b \in A$.

Proof. By Theorem 2.2, a valued binary relation R on A is T-transitive if and only if there exist two families $\{h_\gamma\}_{\gamma\in\Gamma}$, $\{k_\gamma\}_{\gamma\in\Gamma}$ of functions from A to [0,1] such that $k_\gamma \leq h_\gamma$ for all $\gamma \in \Gamma$ and (2.14) holds.

Choosing $b = a$, a relation R represented by (2.14) is reflexive if and only if

$$I_T^{\rightarrow}(h_\gamma(a), k_\gamma(a)) = 1$$

for all $a \in A$ and $\gamma \in \Gamma$. This condition is equivalent to $h_\gamma \leq k_\gamma$ since axiom I8 is satisfied by $I_T^{\rightarrow}$ (see Chapter 1). Thus, we can conclude that $k_\gamma = h_\gamma$ in (2.14). This proves the theorem. ∎

Another characterization of partial T-preorders can be given in terms of the traces.

Theorem 4.8 *A valued binary relation R on A is a partial T-preorder if and only if $R = R^\ell = R^r$.*

Proof. By Proposition 2.7, R is reflexive if and only if $R^\ell \subseteq R$, if and only if $R^r \subseteq R$. Similarly, by Proposition 2.20, R is T-transitive iff $R \subseteq R^\ell$ iff $R \subseteq R^r$. Hence we obtain the statement. ∎

Similarly to the crisp case, a close connection between linear partial T-preorders and total T-preorders can be established as follows.

Theorem 4.9 *A valued binary relation R on A is a total T-preorder if and only if R is a linear partial T-preorder.*

Proof. Assume that R is a total T-preorder on A. That is, by Theorem 4.7, representation (4.8) holds for R and

$$\max(R(a,b), R(b,a)) = 1 \qquad \forall a, b \in A. \tag{4.9}$$

Using Theorem 4.8, $R = R^\ell = R^r$. Therefore, (4.9) implies that $\max(R^\ell(a,b), R^\ell(b,a)) = 1$ and $\max(R^r(a,b), R^r(b,a)) = 1$ hold for all $a, b \in A$. Thus, by Theorem 2.9, R is linear.

To prove the converse statement, suppose that R is linear. Then both R^ℓ and R^r are strongly complete relations, by Theorem 2.9. Since R is a partial T-preorder, $R = R^\ell = R^r$. Thus, R is strongly complete and T-transitive, i.e., R is a total T-preorder. ■

In the rest of this subsection only *positive* valued binary relations are considered on a finite set A. Moreover, let T be a continuous Archimedean t-norm with additive generator f. Then $I_T^{\rightarrow}$ is defined by (1.55). It is easy to see that $x \geq y \geq z$ imply

$$T(I_T^{\rightarrow}(x,y), I_T^{\rightarrow}(y,z)) = I_T^{\rightarrow}(x,z). \tag{4.10}$$

Indeed, we have

$$\begin{aligned}
&T(I_T^{\rightarrow}(x,y), I_T^{\rightarrow}(y,z)) \\
&= f^{-1}(\min\{\max\{f(y) - f(x), 0\} + \max\{f(z) - f(y), 0\}, f(0)\}) \\
&= f^{-1}(\max\{f(y) - f(x), f(z) - f(y), f(z) - f(x), 0\}) \\
&= f^{-1}(\max\{f(z) - f(x), 0\}) \\
&= I_T^{\rightarrow}(x,z).
\end{aligned}$$

Definition 4.9 A valued binary relation R on A is a *strong T-preorder* if and only if $R(a,b) > 0$ for all $a, b \in A$, R is strongly complete on A and

$$\begin{array}{ccc}
R(a,b) = 1 & \text{and} & R(b,c) = 1 \\
& \text{imply} & \\
R(a,c) = 1 & \text{and} & R(c,a) = T(R(c,b), R(b,a))
\end{array}$$

for all $a, b, c \in A$.

Notice that the last condition can be regarded as a particular form of T-transitivity property. In fact, any strong T-preorder is a total T-preorder. The proof is left to the reader.

Since strong T-preorders are T-transitive and reflexive valued binary relations, the representation (4.8) is valid. We prove now that $|\Gamma| = 1$ (see Fodor and Ovchinnikov (1992)).

Theorem 4.10 *A valued binary relation R on A is a strong T-preorder if and only if there exists a function $h : A \to (0,1]$ such that*

$$R(a,b) = I_T^{\rightarrow}(h(b), h(a)) \tag{4.11}$$

holds for all $a, b \in A$.

Proof. a) Suppose R is given by (4.11). Then, by (1.44) and by property I6 of $I_T^{\rightarrow}$,

$$R(a,b) = I_T^{\rightarrow}(h(b), h(a)) \geq I_T^{\rightarrow}(1, h(a)) = h(a) > 0$$

and, by I8,

$$\max(R(a,b), R(b,a)) = 1.$$

Suppose $R(a,b) = 1$ and $R(b,c) = 1$ for some $a, b, c \in A$. Then, by I8, $h(a) \geq h(b) \geq h(c)$. Hence $R(a,c) = 1$. We have, by (4.10), that

$$R(c,a) = I_T^{\rightarrow}(h(a), h(c)) = T[I_T^{\rightarrow}(h(a), h(b)), I_T^{\rightarrow}(h(b), h(c))] = T(R(c,b), R(b,a)).$$

Therefore, R is a strong T-preorder.

b) Suppose R is a strong T-preorder. Then $R_{1.0}$ (1.0-level set of R) is a crisp total preorder. Let $a_{\max}$ be a maximal element of A with respect to $R_{1.0}$. Then, by definition of $R_{1.0}$, $R(a_{\max}, a) = 1$ for all $a \in A$. We define $h(a) = R(a, a_{\max})$. Obviously, $h(a) > 0$. To prove (4.10), consider two cases.

i) $R(a,b) = 1$. Since R is a strong T-preorder and $R(a_{\max}, a) = 1$, we have

$$R(b, a_{\max}) = T(R(b,a), R(a, a_{\max}))$$

or, equivalently,

$$h(b) = T(R(b,a), h(a)) \leq h(a).$$

By I8, $I_T^{\rightarrow}(h(b), h(a)) = 1 = R(a,b)$.

ii) $R(a,b) < 1$. Then $R(b,a) = 1$. Since R is a strong T-preorder and $R(a_{\max}, b) = 1$, we have

$$R(a, a_{\max}) = T(R(a,b), R(b, a_{\max}))$$

or, equivalently,

$$0 < h(a) = T(R(a,b), h(b)).$$

By Lemma 1.4, $R(a,b) = I_T^{\rightarrow}(h(b), h(a))$. ■

Now we characterize any positive partial T-preorder as a finite intersection of strong T-preorders.

Theorem 4.11 *A valued binary relation R on A is a positive partial T-preorder if and only if it is a finite intersection of strong T-preorders.*

Proof. a) Suppose R is a finite intersection of strong T-preorders. Then, by Theorem 4.10, there is a finite family $\{h_i\}$ of functions $h_i : A \to (0,1]$ such that

$$R(a,b) = \min_i \{I_T^{\rightarrow}(h_i(b), h_i(a))\}.$$

Obviously, $R(a,a) = 1$ for all $a \in A$. Since T is a non-decreasing function,

$$\begin{aligned} T(R(a,b), R(b,c)) &= T(\min_i\{I_T^{\rightarrow}(h_i(b), h_i(a))\}, \min_i\{I_T^{\rightarrow}(h_i(c), h_i(b))\}) \\ &\leq \min_i\{T(I_T^{\rightarrow}(h_i(b), h_i(a)), I_T^{\rightarrow}(h_i(c), h_i(c)))\} \\ &\leq \min_i\{I_T^{\rightarrow}(h_i(c), h_i(a))\} = R(a,c), \end{aligned}$$

by Proposition 1.10. Hence, R is a partial T-preorder.

b) Suppose R is a partial T-preorder. We define $h_c(a) = R(a,c)$ for each $a \in A$. By Theorem 4.10, all valued binary relations

$$R_c(a,b) = I_T^{\rightarrow}(h_c(b), h_c(b)) = I_T^{\rightarrow}(R(b,c), R(a,c))$$

are strong T-preorders. We prove that $R(a,b) = \min_c\{R_c(a,b)\}$. Indeed, by reflexivity of R

$$\min_c\{R_c(a,b)\} \leq R_b(a,b) = I_T^{\rightarrow}(R(b,b), R(a,b)) = R(a,b).$$

On the other hand, $T(R(a,b), R(b,c)) \leq R(a,c)$ (T-transitivity property), which implies

$$R(a,b) \leq I_T^{\rightarrow}(R(b,c), R(a,c)) = R_c(a,b).$$

Hence, $R(a,b) \leq \min_c\{R_c(a,b)\}$. We conclude that $R(a,b) = \min_c\{R_c(a,b)\}$, i.e., R is an intersection of strong T-preorders. ■

4.4 Orders

In this section we investigate different valued ordering relations. Suppose that (T, S, N) is a De Morgan triple such that T is a left-continuous t-norm and N is a strong negation (whence S is a right-continuous t-conorm).

All of the orderings in this section are T-transitive valued binary relations. However, reflexivity does not hold in general. Instead, antisymmetry is supposed to be satisfied. In addition, some other properties can be required as it is formulated in the following definition.

Definition 4.10 A valued binary relation R on A is called a
– *partial T-order* if R is antisymmetric and T-transitive;
– *strict partial T-order* if R is asymmetric and T-transitive;
– *total T-order* (or a linear T-order) if R is a complete partial T-order (that is, R is antisymmetric, complete and T-transitive);
– *strict total T-order* if R is a complete strict partial T-order (that is, R is asymmetric, complete and T-transitive).

In our study we start from the general representation theorem of T-transitive valued binary relations. That is, by Theorem 2.2, a valued binary relation R defined on a finite set A is T-transitive on A if and only if there exists two finite families $\{h_\gamma\}_{\gamma\in\Gamma}$ and $\{k_\gamma\}_{\gamma\in\Gamma}$ of functions from A to $[0,1]$ with $h_\gamma \geq k_\gamma$ such that

$$R(a,b) = \min_{\gamma\in\Gamma} I_T^{\rightarrow}(h_\gamma(b), k_\gamma(a)). \tag{4.12}$$

Antisymmetry of R implies further restrictions on the families $\{h_\gamma\}$ and $\{k_\gamma\}$. Recall that a valued binary relation R is antisymmetric on A if and only if

$$\min\{R(a,b), R(b,a)\} = 0 \quad \text{for all } a, b \in A,\ a \neq b.$$

Therefore, a T-transitive valued binary relation R represented by (4.12) is antisymmetric if and only if for $a \neq b$ we have

$$\min\{\min_{\gamma\in\Gamma} I_T^{\rightarrow}(h_\gamma(b), k_\gamma(a)), \min_{\gamma\in\Gamma} I_T^{\rightarrow}(h_\gamma(a), k_\gamma(b))\} = 0,$$

or equivalently, if and only if

$$\min_{\gamma\in\Gamma} I_T^{\rightarrow}(h_\gamma(b), k_\gamma(a)) > 0 \quad \text{implies} \quad \min_{\gamma\in\Gamma} I_T^{\rightarrow}(h_\gamma(a), k_\gamma(b)) = 0. \tag{4.13}$$

We have to distinguish two cases: first we consider t-norms such that the law of contradiction holds with the strong negation N, then the case of positive t-norms is investigated.

We will see that (strict) partial T-orders have representations with functions $\{h_\gamma\}$, $\{k_\gamma\}$ owning different properties depending on the class where the t-norm T belongs to. However, total T-orders and strict total T-orders are identical in both cases of t-norms: total T-orders differ from a crisp total order only in the values of $R(a,a)$, while strict total T-orders are always crisp strict total orders.

4.5 Orders under the law of contradiction

Suppose that (T,S,N) is a De Morgan triple (T is a left-continuous t-norm, N is a strong negation) such that the law of contradiction holds:

$$T(x, N(x)) = 0 \quad \text{for all } x \in [0,1].$$

This class includes the following particular cases (φ is an automorphism of the unit interval):

(i) Łukasiewicz-like De Morgan triples:

$$\begin{aligned} T(x,y) &= W_\varphi(x,y) = \varphi^{-1}(\max\{\varphi(x)+\varphi(y)-1, 0\}), \\ S(x,y) &= W'_\varphi(x,y) = \varphi^{-1}(\min\{\varphi(x)+\varphi(y), 1\}), \\ N(x) &= N_\varphi(x) = \varphi^{-1}(1-\varphi(x)). \end{aligned}$$

(ii) De Morgan triples based on the nilpotent minimum and maximum:

$$\begin{aligned} T(x,y) &= \min_{0,\varphi}(x,y) = \begin{cases} \min(x,y) & \text{if } \varphi(x)+\varphi(y) > 1 \\ 0 & \text{otherwise} \end{cases}, \\ S(x,y) &= \max_{1,\varphi}(x,y) = \begin{cases} 1 & \text{if } \varphi(x)+\varphi(y) \geq 1 \\ \max(x,y) & \text{otherwise} \end{cases}, \\ N(x) &= N_\varphi(x) = \varphi^{-1}(1-\varphi(x)). \end{aligned}$$

In any case, we can prove the following lemma.

Lemma 4.5 *Suppose that (T,S,N) is a De Morgan triple such that $T(x,N(x)) = 0$ holds for all $x \in [0,1]$ with a strong negation N. Then*

$$I_T^{\rightarrow}(x,y) = 0 \quad \Longleftrightarrow \quad x = 1, y = 0.$$

Proof. Obviously we have $I_T^{\rightarrow}(1,0) = 0$ for any implication.

To prove the converse in the present case, notice that $I_T^{\rightarrow}(x,y) = 0$ implies that $I_T^{\rightarrow}(x,y) \geq N(x)$ (since we have $T(x,N(x)) = 0$). $N(x) > 0$ follows from $x < 1$, thus we must have $x = 1$. Hence $I_T^{\rightarrow}(1,y) = y$ implies $y = 0$.

■

Corollary 4.4 *Suppose that $T(x,N(x)) = 0$ is satisfied. Then*

$$I_T^{\rightarrow}(x,y) > 0 \quad \Longleftrightarrow \quad \max\{1-x,y\} > 0.$$

Proof. By the previous lemma we have

$$I_T^{\rightarrow}(x,y) > 0 \quad \Longleftrightarrow \quad x < 1 \text{ or } y > 0,$$

which is equivalent to $\max\{1-x,y\} > 0$.

■

4.5.1 Partial T-orders

Now we are able to prove the following result concerning the representation of partial T-orders in the present case.

Theorem 4.12 *Suppose (T,S,N) is a De Morgan triple with a left-continuous t-norm T and a strong negation N such that $T(x,N(x)) = 0$ holds for all $x \in [0,1]$. A valued binary relation R defined on a finite A is a partial T-order if and only if there exist two finite families $\{h_\gamma\}_{\gamma\in\Gamma}$, $\{k_\gamma\}_{\gamma\in\Gamma}$ of functions from A to $[0,1]$ with the following properties:*
i) $h_\gamma \geq k_\gamma$ for all $\gamma \in \Gamma$,
ii) for all $a \neq b$ there exists $\delta \in \Gamma$:

$$\min\{\max\{1-h_\delta(b),k_\delta(a)\},\max\{1-h_\delta(a),k_\delta(b)\}\} = 0$$

such that

$$R(a,b) = \min_{\gamma\in\Gamma} I_T^{\rightarrow}(h_\gamma(b),k_\gamma(a)).$$

Proof. By Corollary 4.4, we have $I_T^{\rightarrow}(x,y) > 0$ if and only if $\max\{1-x,y\} > 0$. Therefore, for $a \neq b$

$$R(a,b) > 0 \quad \Rightarrow \quad R(b,a) = 0$$

if and only if

$$[\forall\gamma \ \max\{1-h_\gamma(b),k_\gamma(a)\} > 0] \quad \Rightarrow \quad \exists\delta \ : \ \max\{1-h_\delta(a),k_\delta(b)\} = 0,$$

or in an equivalent form, for $a \neq b$

$$\min\{R(a,b),R(b,a)\} = 0 \iff \min\{\max\{1-h_\delta(b),k_\delta(a)\},\max\{1-h_\delta(a),k_\delta(b)\}\} = 0$$

for some $\delta \in \Gamma$.

■

It may happen that R is a partial T-order being also reflexive. In other words, R is an antisymmetric T-preorder. Then we have the following statement.

Theorem 4.13 *R is a reflexive partial T-order on a finite A if and only if there exists a finite family* $\{h_\gamma\}_{\gamma\in\Gamma}$ *of functions from A to* $[0,1]$ *with property*

$$\forall\ a \neq b\ \exists\ \delta \in \Gamma\ :\ \max\{h_\delta(a), h_\delta(b)\} = 1,\ \min\{h_\delta(a), h_\delta(b)\} = 0$$

such that

$$R(a,b) = \min_{\gamma\in\Gamma} I_T^{\rightarrow}(h_\gamma(b), h_\gamma(a)).$$

Proof. It follows easily by Theorem 4.12 and by reflexivity of R.

■

4.5.2 Strict partial T-orders

By Definition 4.10, a strict partial T-order is an irreflexive partial T-order on A. Therefore, we can apply Theorem 4.12 and add the condition of irreflexivity of R:

$$R(a,a) = 0 \quad \text{for all } a \in A,$$

which is equivalent to

$$\min_{\gamma\in\Gamma} I_T^{\rightarrow}(h_\gamma(b), k_\gamma(a)) = 0 \quad \text{for all } a \in A.$$

Hence, by Lemma 4.5, for all $a \in A$ there exists $\gamma_a \in \Gamma$ such that

$$h_{\gamma_a}(a) = 1,\ k_{\gamma_a}(a) = 0.$$

Therefore, we have proved the following result.

Theorem 4.14 *Suppose* (T, S, N) *is a De Morgan triple with a left-continuous t-norm T and a strong negation N such that* $T(x, N(x)) = 0$ *holds for all* $x \in [0,1]$. *A valued binary relation R defined on a finite A is a strict partial T-order if and only if there exist two finite families* $\{h_\gamma\}_{\gamma\in\Gamma}$, $\{k_\gamma\}_{\gamma\in\Gamma}$ *of functions from A to* $[0,1]$ *with the following properties:*
i) $h_\gamma \geq k_\gamma$ *for all* $\gamma \in \Gamma$,
ii) for all $a, b \in A$ *there exists* $\delta \in \Gamma$:

$$\min\{\max\{1 - h_\delta(b), k_\delta(a)\}, \max\{1 - h_\delta(a), k_\delta(b)\}\} = 0$$

such that

$$R(a,b) = \min_{\gamma\in\Gamma} I_T^{\rightarrow}(h_\gamma(b), k_\gamma(a)).$$

4.5.3 Total T-orders

By Definition 4.10, a total T-order on A is a complete partial T-order on A. That is, we must have

$$\begin{aligned}\min\{R(a,b), R(b,a)\} &= 0 \quad \text{for } a \neq b,\\ \max\{R(a,b), R(b,a)\} &= 1 \quad \text{for } a \neq b.\end{aligned}$$

Theorem 4.15 *A valued binary relation R defined on a finite A is a total T-order if and only if there exists a crisp total order Q such that*

$$R(a,b) = \begin{cases} 1 & \text{if } aQb \\ 0 & \text{if not } aQb \end{cases}, \quad \text{for } a \neq b,$$

and $R(a,a) \in [0,1]$ is arbitrary for each $a \in A$.

Proof. Suppose R is a valued total T-order. Completeness and antisymmetry together imply that $R(a,b) \in \{0,1\}$ for $a \neq b$. T-transitivity is reduced to transitivity of crisp binary relations when different alternatives $a, b, c \in A$ are considered. Therefore, the crisp binary relation Q defined by aQa for all $a \in A$ and

$$aQb \iff R(a,b) = 1, \; a \neq b$$

is a crisp total order that corresponds to the statement.

Now suppose that Q is a crisp total order and define $R(a,b)$ according to the statement. The only thing that we have to check is T-transitivity when identical elements occur:

$$T(R(a,a), R(a,b)) \leq R(a,b)$$

follows immediately since $T(x,y) \leq y$. Finally,

$$T(R(a,b), R(b,a)) \leq R(a,a)$$

since $\min\{R(a,b), R(b,a)\} = 0$.

■

According to this theorem, total T-orders are almost identical to a crisp total order: the only difference can be in the values of $R(a,a)$.

4.5.4 Strict total T-orders

By Definition 4.10, a strict total T-order is a complete strict partial T-order, or in other words: an irreflexive total T-order. Therefore, the following result is immediate by Theorem 4.15.

Theorem 4.16 *A valued binary relation R on A is a strict total T-order on A if and only if it is a crisp strict total order on A.*

Proof. R is an irreflexive total T-order, thus we can apply Theorem 4.15 and use irreflexivity of R.

■

By this last theorem, it is impossible to define a really valued strict total T-order: it is always a crisp strict total order.

4.6 Orders when the underlying t-norm is positive

Suppose that (T, S, N) is a positive left-continuous t-norm and N is a strong negation. This class includes the following particular cases (φ is an automorphism of the unit interval):

(i) $T(x,y) = \min(x,y)$, $S(x,y) = \max(x,y)$, $N(x) = N_\varphi(x) = \varphi^{-1}(1-\varphi(x))$.

(ii) Product-like De Morgan triples:

$$\begin{aligned} T(x,y) &= \Pi_\varphi(x,y) = \varphi^{-1}(\varphi(x)\varphi(y)), \\ S(x,y) &= \Pi'_\varphi(x,y) = \varphi^{-1}(\varphi(x)+\varphi(y)-\varphi(x)\varphi(y)), \\ N(x) &= N_\varphi(x) = \varphi^{-1}(1-\varphi(x)). \end{aligned}$$

First we prove the following lemma.

Lemma 4.6 *Let T be a positive t-norm. Then*

$$I_T^{\to}(x,y) = 0 \quad \Longleftrightarrow \quad x > 0, y = 0.$$

Proof. If $y > 0$ then $I_T^{\to}(x,y) \geq y > 0$ is a contradiction.
If $x = 0$ then $I_T^{\to}(0,y) = 1$ is a contradiction, too.
On the other hand, $I_T^{\to}(x,0) = 0$ for $x > 0$, see Proposition 1.12. ■

Corollary 4.5 *Let T be a positive t-norm. The*

$$I_T^{\to}(x,y) > 0 \quad \Longleftrightarrow \quad x = 0 \text{ or } y > 0.$$

4.6.1 Partial T-orders

By using the previous lemma and its corollary, we can prove the following result.

Theorem 4.17 *Suppose T is a positive left-continuous t-norm. A valued binary relation R defined on a finite A is a partial T-order on A if and only if there exist two finite families $\{h_\gamma\}_{\gamma\in\Gamma}$ and $\{k_\gamma\}_{\gamma\in\Gamma}$ of functions from A to $[0,1]$ with the properties*
i) $h_\gamma \geq k_\gamma$ for all $\gamma \in \Gamma$,
ii) for $a \neq b$ there exists $\delta \in \Gamma$: $h_\delta(a) > k_\delta(b) = 0$ or $h_\delta(b) > k_\delta(a) = 0$
such that

$$R(a,b) = \min_{\gamma\in\Gamma} I_T^{\to}(h_\gamma(b), k_\gamma(a)).$$

Proof. By Corollary 4.5, $I_T^{\to}(x,y) > 0$ iff $x = 0$ or $y > 0$. Therefore, for $a \neq b$

$$R(a,b) > 0 \quad \Rightarrow \quad R(b,a) = 0$$

if and only if

$$\forall\, \gamma \in \Gamma,\ h_\gamma(b) = 0 \text{ or } k_\gamma(a) > 0 \quad \Rightarrow \quad \exists\, \delta \in \Gamma \ :\ h_\delta(a) > k_\delta(b) = 0,$$

or in another equivalent form,

$$\min\{R(a,b), R(b,a)\} = 0 \quad \text{for } a \neq b$$

if and only if there exists $\delta \in \Gamma$ such that $h_\delta(a) > k_\delta(b) = 0$ or $h_\delta(b) > k_\delta(a) = 0$.

■

If R is a reflexive partial T-order then we have the following statement.

Theorem 4.18 *R is a reflexive partial T-order on a finite A if and only if there exists a family $\{h_\gamma\}_{\gamma\in\Gamma}$ of functions from A to $[0,1]$ with property*

$$\forall\ a \neq b \quad \exists\ \delta \in \Gamma\ :\ \min\{h_\delta(a), h_\delta(b)\} = 0,\ \max\{h_\delta(a), h_\delta(b)\} > 0$$

such that

$$R(a,b) = \min_{\gamma\in\Gamma} I_T^{\rightarrow}(h_\gamma(b), h_\gamma(a)).$$

Proof. It follows easily from Theorem 4.17 and reflexivity of R.

■

4.6.2 Strict partial T-orders

By Definition 4.10, a strict partial T-order is an irreflexive partial T-order on A. Therefore, we can apply Theorem 4.17 and add the condition of irreflexivity:

$$R(a,a) = 0 \quad \forall\ a \in A.$$

This is equivalent to

$$\min_{\gamma\in\Gamma} I_T^{\rightarrow}(h_\gamma(b), k_\gamma(a)) = 0 \quad \forall\ a \in A,$$

that is, by Lemma 4.6, to

$$\forall\ a \in A \quad \exists\ \gamma_a \in \Gamma\ :\ h_{\gamma_a}(a) > k_{\gamma_a}(a) = 0.$$

Thus, the following result is true.

Theorem 4.19 *Suppose T is a positive left-continuous t-norm. A valued binary relation R defined on a finite A is a strict partial T-order on A if and only if there exist two finite families $\{h_\gamma\}_{\gamma\in\Gamma}$ and $\{k_\gamma\}_{\gamma\in\Gamma}$ of functions from A to $[0,1]$ with the properties*
i) $h_\gamma \geq k_\gamma$ for all $\gamma \in \Gamma$,
ii) for all $a, b \in A$ there exists $\delta \in \Gamma$: $h_\delta(a) > k_\delta(b) = 0$ or $h_\delta(b) > k_\delta(a) = 0$
such that

$$R(a,b) = \min_{\gamma\in\Gamma} I_T^{\rightarrow}(h_\gamma(b), k_\gamma(a)).$$

4.6.3 Total T-orders

By Definition 4.10, a total T-order on A is a complete partial T-order on A. That is, we must have

$$\begin{aligned}\min\{R(a,b),R(b,a)\} &= 0 \quad \text{for } a \neq b,\\ \max\{R(a,b),R(b,a)\} &= 1 \quad \text{for } a \neq b.\end{aligned}$$

Theorem 4.20 *A valued binary relation R defined on a finite A is a total T-order if and only if there exists a crisp total order Q such that*

$$R(a,b) = \begin{cases} 1 & \text{if } aQb \\ 0 & \text{if not } aQb \end{cases}, \qquad \text{for } a \neq b,$$

and $R(a,a) \in [0,1]$ is arbitrary for each $a \in A$.

Proof. See the proof of Theorem 4.15.

■

Here the same remark is valid as in case of total T-orders when the law of contradiction holds. That is, total T-orders are almost crisp total orders, except the values $R(a,a)$.

4.6.4 Strict total T-orders

By Definition 4.10, a strict total T-order is a complete strict partial T-order, or in other words: an irreflexive total T-order. Therefore, the following result is immediate by Theorem 4.20.

Theorem 4.21 *A valued binary relation R on A is a strict total T-order on A if and only if it is a crisp strict total order on A.*

Proof. R is an irreflexive total T-order, thus we can apply Theorem 4.20 and use irreflexivity of R.

■

Therefore, the class of valued strict total T-orders is identical to the class of crisp strict total orders.

Chapter 5

Aggregation operations

5.1 Basic problem

We consider m valued relations $R_1, \ldots, R_m$, $0 \leq R_i \leq 1$, and we are willing to substitute to the vector $(R_1, \ldots, R_m)$ one simple valued relation R using the *aggregation operator* M :

$$\bigcup_{m=1}^{\infty} [0,1]^m \to [0,1] \quad \text{such that} \quad 0 \leq R = M(R_1, \ldots, R_m) \leq 1.$$

In this Chapter, we suppose that the *condition of independence of irrelevant preferences* is satisfied. If a profile of valued relations $(R_1, \ldots, R_m)$ is modified in such a way that individual's paired comparisons among a pair of alternatives (a, b) are unchanged – $(R_1(x, y), \ldots, R_m(x, y))$ becomes $(R'_1(x, y), \ldots, R'_m(x, y))$ for all (x, y) belonging to $A \times A$ except for $(x, y) = (a, b)$ – the aggregation resulting from the original and modified profiles should be unchanged for the pair (a, b).

$R(a, b)$ depends only on $R_1(a, b), \ldots, R_m(a, b)$ and is a function of m arguments for every pair (a, b) of $A \times A$. We shortly write M as $M(x_1, \ldots, x_m)$.

Basically, we are searching for an operator which does not present any chaotic reaction to a small change of the arguments. In other words, $M(x_1, \ldots, x_m)$ should be *continuous* (C-operator).

The operator will be also *neutral* (N-operator), i.e. M is independent of the labels :

$$M(R_1, \ldots, R_m) = M(R_{i_1}, \ldots, R_{i_m})$$

if $(i_1, \ldots, i_m) = \sigma(1, \ldots, m)$ where σ represents a permutation operation.

In mathematical terms, $M(x_1, \ldots, x_m)$ should be symmetric.

Finally, M will present a non negative response to any increase of the arguments. This means that $M(x_1, \ldots, x_m)$ is *monotonic* (M-operator) and $x'_i > x_i$, implies

$$M(x_1, \ldots, x'_i, \ldots x_m) \geq M(x_1, \ldots, x_i, \ldots x_m), \quad i = 1, \ldots, m \tag{5.1}$$

Definition 5.1 The subclass of aggregation operators which are continuous, neutral and monotonic is called the class of CNM operators.

Definition 5.2 An operator is called *strict* (S-operator) if the second inequality of (5.1) is transformed into a strict inequality.

M is strict iff $x'_i > x_i$ implies

$$M(x_1, \ldots, x'_i, \ldots, x_m) > M(x_1, \ldots, x_i, \ldots, x_m), \quad i = 1, \ldots, m.$$

In the following sections we will characterize some particular CNM operators.

5.2 Idempotent CNM operators

It seems widely accepted to consider that an aggregation operator will satisfy the *unanimity* property, i.e. if all R_i are identical, $M(R_i, \ldots, R_i)$ restitutes the common valued relation.

Definition 5.3 An operator is called *idempotent* (I-operator) if it satisfies

$$M(x, \ldots, x) = x, \quad \text{for all } x \in [0,1] \tag{5.2}$$

An immediate property of an idempotent CNM operator corresponds to the compensative character of M.

Definition 5.4 M is *compensative* iff

$$\min_i(x_i) \leq M(x_1, \ldots, x_m) \leq \max_i(x_i). \tag{5.3}$$

The next proposition illustrates the correspondence between idempotence and compensation (or internness).

Proposition 5.1 *Idempotent monotonic operators and compensative monotonic operators are equivalent.*

Proof. M is an idempotent monotonic operator. Then

$$M(\min_i x_i, \ldots, \min_i x_i) \leq M(x_1, \ldots, x_m) \leq M(\max_i x_i, \ldots, \max_i x_i), \text{ (monotony)}$$

and

$$\min_i x_i \leq M(x_1, \ldots, x_m) \leq \max_i x_i, \text{ (idempotence)}$$

M is a compensative operator. Then

$$\min_i(x_i = x) \leq M(x, \ldots, x) \leq \max_i(x_i = x)$$

and

$$M(x, \ldots, x) = x.$$

■

5.3 Aggregation properties

There are at least two different ways to define $M^{(m)}(x_1, \ldots, x_m)$ in terms of $M^{(m-1)}(x_1, \ldots, x_{m-1})$ using associativity and decomposability.

Definition 5.5 *Associativity* property (A-operator) concerns aggregation of only two arguments which can be canonically extended to any finite number of arguments :

$$M^{(3)}(x_1, x_2, x_3) = M^{(2)}(x_1, M^{(2)}(x_2, x_3)) = M^{(2)}(M^{(2)}(x_1, x_2), x_3) \tag{5.4}$$

$$M^{(m)}(x_1, \ldots, x_m) = M^{(2)}(M^{(m-1)}(x_1, \ldots, x_{m-1}), x_m). \tag{5.5}$$

With the use of a graphical representation linked to clustering procedures, we obtain Figure 5.1

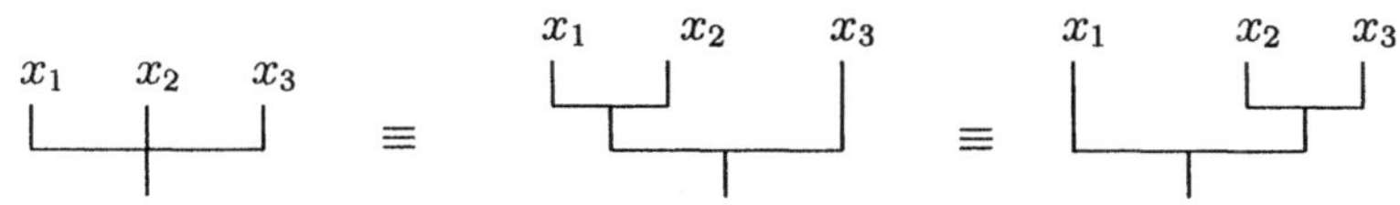

Fig. 5.1

Decomposability property was introduced by Kolmogorov (and independently by Nagumo) as early as 1930. He wrote a note in French which was published in an Italian mathematical journal with the title : "Sur la notion de la moyenne".

The famous russian author states that each element of a subgroup of elements to be aggregated can be substituted to its partial aggregation without change.

On a graphical basis, we obtain Figure 5.2

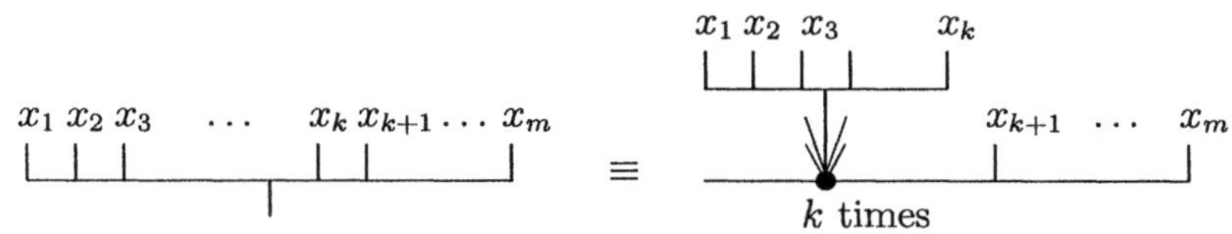

Fig. 5.2

Definition 5.6 An operator is called *decomposable (D-operator)* iff considering a sequence of functions :

$$M^{(1)}(x_1) = x_1, \quad M^{(2)}(x_1, x_2), \quad M^{(3)}(x_1, x_2, x_3), \ldots, M^{(m)}(x_1, \ldots, x_m) \ldots,$$

each function of the sequence has to satisfy the following condition

$$M^{(m)}(x_1, \ldots, x_k, x_{k+1}, \ldots, x_m) = M^{(m)}(M_k, \ldots, M_k, x_{k+1}, \ldots, x_m) \tag{5.6}$$

with $k = \{1, \ldots, m\}$, $M_k = M^{(k)}(x_1, \ldots, x_k)$.

If an idempotent and decomposable operator is considered, the concept of weighting clearly appears.

In fact, after Kolmogorov, we will prove that

Proposition 5.2

$$M(m.x, n.y) = M(m'.x, n'.y)$$

where M is an idempotent and decomposable CNM operator if

$$M(m.x, n.y) = M(\underbrace{x, \ldots, x}_{m\ times}, \underbrace{y, \ldots, y}_{n\ times}),$$

and

$$mn' = nm'.$$

Proof. We first prove that $M(x,y) = M(m.x, m.y)$

$$\begin{aligned} M(m.x, m.y) &= M(x,\ldots,x,y,\ldots,y) \quad \text{(definition)} \\ &= M(\underbrace{\underbrace{x,y}, \ldots, \underbrace{x,y}}_{m \text{ times}}) \quad \text{(commutativity)} \\ &= M(M(x,y),\ldots,M(x,y)) \quad \text{(decomposability)} \\ &= M(x,y) \quad \text{(idempotency).} \end{aligned}$$

We now prove that $M(m.x, m.y) = M(m'.x, n'.y)$ if $mn' = nm'$

$$\begin{aligned} M(m.x, n.y) &= M(mn'.x, nn'.y) \\ &= M(nm'.x, nn'.y) = M(m'.x, n'.y). \end{aligned}$$

■

This result already shows that decomposable and associative idempotent CNM operators are opposed in some sense.

In fact, $M(m.x, n.y) = M(x,y)$ when M is associative and idempotent for any m, n and $M(m.x, n.y) = M(x,y)$ if $m = n$ when M is decomposable and idempotent. However, min and max operators are idempotent, associative and decomposable CNM operators.

The results of the two following sections will show in an axiomatical way the differences between associativity on one side and decomposability, on the other side.

5.4 Associative idempotent CNM operators

The characterization of the subclass of associative and compensative (or equivalently, idempotent) CNM operators different from $\min_i x_i$ or $\max_i x_i$ was obtained by Fung and Fu (1975) and in a revisited way by Dubois and Prade (1984b).

Definition 5.7 An *averaging operator* is a member of the class of compensative CNM operators but different from $\min_i x_i$ or $\max_i x_i$.

Due to Proposition 5.1, an averaging operator is equivalently an idempotent CNM operator different from $\min_i x_i$ and $\max_i x_i$.

We already know from Proposition 1.4 that the extremes min and max are characterized in the following way : $\min_i x_i (\max_i x_i)$ corresponds to the class of associative, idempotent CNM such that $M(1,x) = x$ $(M(0,x) = x)$, $\forall x \in [0,1]$.

Following Dubois and Prade (1984b) and considering the last result, we now delete the properties $M(1,x) = x$ and $M(0,x) = x$, $\forall x \in [0,1]$, and we characterize the associative averaging operators (which are associative and idempotent CNM operators different from $\min_i x_i$ or $\max_i x_i$) :

Theorem 5.1 *An operator $M(x_1,\ldots,x_m)$ is an associative averaging operator iff*

$$M(x_1,\ldots,x_m) = median(\max_i x_i, \alpha, \min_i x_i),$$

where $\alpha \in (0,1)$.

Proof. (Necessity) We consider the case where $m = 2$. The proof is easily extended to any $m > 2$. We first prove that any associative, idempotent CNM operator satisfying $M(1,0) = 0$ ($M(0,1) = 1$) presents the property :

$$M(1,x) = x \; (M(0,x) = x), \quad \forall x \in [0,1] .$$

Associativity and idempotency imply :

$$M(1, M(1,x)) = M(M(1,1),x) = M(1,x).$$

When x increases from 0 to 1, $M(1,x)$ increases continuously form $M(1,0) = 0$ to $M(1,1) = 1$. This implies : $\forall y \in [0,1]$, $\exists x \in [0,1]$ such that $M(1,x) = y$. In other words,

$$\forall y \in [0,1] , \quad M(1,y) = M(1, M(1,x)) = M(1,x) = y.$$

We can easily prove, using the same arguments, the dual part of the proposal.

$M(x_1, \ldots, x_m)$ is an associative, idempotent CNM operator, different from $\min_i x_i$. This implies, with the use of Proposition 1.4, that : $\exists x : M(1,x) \neq x$.

From the first part of our proof, we obtain that $M(1,0)$ is different from 0. M is also different from max, which implies that $M(1,0) \neq 1$.

We consider now $M(0,1) = \alpha$, $\alpha \in (0,1)$.

Due to associativity, commutativity and idempotency, we have :

$$\begin{aligned} M(0,\alpha) &= M(0, M(0,1)) = M(M(0,0),1) = M(1,0) = \alpha \\ M(1,\alpha) &= M(1, M(0,1)) = M(M(1,1),0) = M(1,0) = \alpha \end{aligned}$$

Consequently, using monotonicity : $M(x,\alpha) = \alpha$, $\forall x \in [0,1]$.

We also have :

$$\begin{aligned} M(0,\alpha) \leq M(0,y) \leq M(x,y) \quad &, \quad \forall y \in [\alpha,1] , \; \forall x \in [0,1] \\ M(x,y) \leq M(x,1) \leq M(\alpha,1) \quad &, \quad \forall x \in [0,\alpha] , \; \forall y \in [0,1] . \end{aligned}$$

Finally,

$$M(x,y) = \alpha = \text{median } (x,y,\alpha) \; , \; \forall x \in [0,\alpha] , \; \forall y \in [\alpha,1] , \; \forall \alpha \in (0,1).$$

Two situations have still to be analyzed : $x, y \in [0,\alpha]$ and $x, y \in [\alpha,1]$. We first suppose, without restriction (due to commutativity), that

$$0 < \alpha \leq x \leq y \leq 1 \quad , \quad \alpha \neq 1.$$

Using associativity and idempotency, we obtain

$$M(M(x,y),y) = M(x, M(y,y)) = M(x,y).$$

If x increases from α to y, $M(x,y)$ increases continuously from $M(\alpha,y) = \alpha$ to $M(y,y) = y$. This implies that : $\forall z \in [\alpha,y]$, $\exists x \in [\alpha,y]$ such that $M(x,y) = z$.

In other words,

$$\forall \alpha \in (0,1), \quad \forall z \in [\alpha,y] : M(z,y) = M(M(x,y),y) = M(x,y) = z,$$

and

$$M(x,y) = x, \quad \forall x \in [\alpha,y] , \; \forall \alpha \in (0,1)$$

or

$$M(x, y) = x = \text{median } (x, y, \alpha), \quad \forall x, y \in [\alpha, 1], \ \forall \alpha \in (0, 1).$$

The case where : $0 \leq x \leq y \leq \alpha < 1$, $\alpha \neq 0$, gives, with the same arguments, that :

$$M(x, y) = y = \text{median } (x, y, \alpha), \quad \forall x, y \in [0, \alpha], \ \forall \alpha \in (0, 1).$$

Proof of sufficiency is immediate. ■

This aggregation operator $M(x_1, x_2)$ is an averaging operator which can be represented according to Figure 5.3

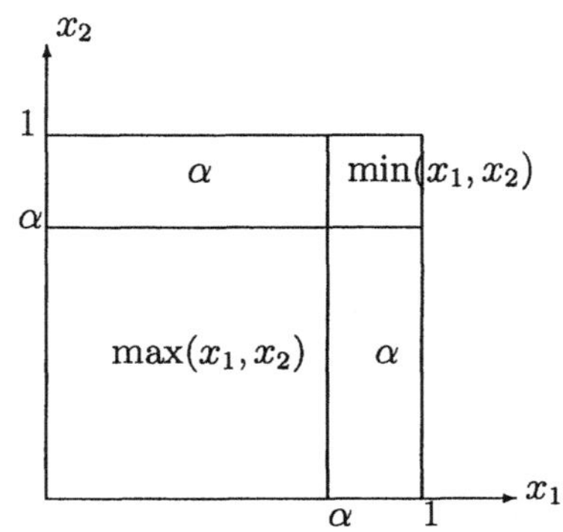

Fig. 5.3

This operator is non strict. This remark immediately implies that the subclass of strict, idempotent and associative CNM aggregation operators is empty.

5.5 Decomposable idempotent CNM operators

Kolmogorov (1930) and Nagumo (1930), in their pioneer work, consider the subclass of strict, idempotent and decomposable CNM operators on $[a, b]^m$ where $[a, b]$ is a closed interval of R. We restrict this interval to be $[0, 1]$ since we are dealing with binary valued relations.

The results of 1930 can be stated in the following way :

Theorem 5.2 *A CNM operator M is strict, idempotent and decomposable ((S&I&D)-CNM operator) iff*

$$M^{(m)}(x_1, \ldots, x_m) = f^{-1}\left[\frac{1}{m}\sum_i f(x_i)\right]$$

where f is any continuous strictly monotonic function on $[0, 1]$.

Proof. (Sufficiency). $M(x_1, \ldots, x_m) = f^{-1}\left[\frac{1}{m}\sum_i f(x_i)\right]$ is obviously continuous, strictly monotonic, commutative, idempotent and decomposable.

(Necessity). Suppose that z is a rational belonging to $[0, 1]$; $z = p/q$, where p and q are integers. We consider $\psi(z) = M(p.1, (q-p).0)$. ψ is univoquely defined; suppose

$z = p/q = p'/q'$, due to Proposition 5.2, $M(p.1, (q-p).0) = M(p'.1, (q'-p').0)$ because $p(q'-p') = (q-p)p'$.

ψ is a strictly increasing function; suppose $z' = p'/q > z = p/q$, due to strict monotonicity of M,

$$\psi(z') = M(p'.1, (q-p').0) > M(p.1, (q-p).0) = \psi(z).$$

Let us now consider $M(x_1, \ldots, x_m)$ where $x_i = \psi(z_i)$, z_i rationals belonging to $[0,1]$, $z_i = p_i/q$, $i = 1, \ldots, m$.

$$\begin{aligned} M(x_1, \ldots, x_m) &= M(q.x_1, \ldots, q.x_m), \text{ (Proposition 5.2)} \\ &= M(qM(p_1.1, (q-p_1).0), \ldots, qM(p_m.1, (q-p_m).0)) \\ &= M(p_1.1, (q-p_1).0, \ldots, p_m.1, (q-p_m).0) \end{aligned}$$

because $(p_i.1, (q-p_i).0)$ presents q terms and decomposability of M implies

$$M(\underbrace{p_1.1, (q-p_1).0}_{q \text{ elements}}, \ldots) = M(\underbrace{M(p_1.1, (q-p_1).0), \ldots, M(p_1.1, (q-p_1).0}_{q \text{ times}}, \ldots)$$

Finally,

$$\begin{aligned} M(x_1, \ldots, x_m) &= M((p_1 + \cdots + p_m).1,\ (nq - p_1, \ldots, -p_m).0), \text{ (commutativity)} \\ &= \psi\left(\frac{p_1 + \cdots + p_m}{nq}\right) = \psi\left(\frac{z_1 + \cdots + z_m}{m}\right). \end{aligned}$$

We will prove that ψ is continuous for each $y \in [0,1]$, y being rational or irrational. Suppose it is not true. Then

$$\psi(y-0) = u, \quad \psi(y+0) = v, \quad u < v.$$

However, for two rationals, z_1 and z_2,

$$M(\psi(z_1), \psi(z_2)) = \psi\left(\frac{z_1 + z_2}{2}\right).$$

If z_1 and z_2 converge at y, with $z_1 < y < z_2$,

$$\lim \psi\left(\frac{z_1 + z_2}{2}\right) = M(u, v) > M(u, u) = u.$$

However, we can do this, such that $\frac{z_1+z_2}{2}$ remains on the left side of y and $\lim \psi(\frac{z_1+z_2}{2}) = u$, a contradiction.

This implies that the formula

$$M(x_1, \ldots, x_m) = \psi\left[\frac{\varphi(x_1) + \cdots + \varphi(x_m)}{m}\right]$$

where φ is the inverse of ψ – that exists – can be extended to any $(x_1, \ldots, x_m) \in [0,1]^m$.

ψ is strictly increasing continuous function. If $\psi^{-1} = f$, the result can be transformed in

$$M(x_1, \ldots, x_m) = f^{-1}\left[\frac{f(x_1) + \cdots + f(x_m)}{m}\right].$$

If $\varphi = \psi^{-1} = -f$, f is a strictly decreasing continuous function and

$$\psi(0) = 0, \quad \psi(1) = 1, \quad \psi(-a) = f^{-1}(a)$$

$$M(x_1,\ldots,x_m) = f^{-1}\left[\frac{f(x_1)+\cdots+f(x_m)}{m}\right].$$

More generally, if $g = af + b$, $a \neq 0$, $b \in R$,

$$f^{-1}\left[\frac{f(x_1)+\cdots+f(x_m)}{m}\right] = g^{1}\left[\frac{g(x_1)+\cdots+g(x_m)}{m}\right],$$

(see Aczél (1948), p. 396). ■

The *generalized mean* $f^{-1}\left[\frac{1}{m}\sum_i f(x_i)\right]$ is an averaging operator and covers a wide spectrum of means including arithmetic, quadratic, geometric, harmonic and root-power means as it can be seen in Table 5.1.

$f(x)$	$M(x_1,\ldots,x_m)$	name
x	$\frac{1}{m}\sum_i x_i$	arithmetic
x^2	$\sqrt{\frac{1}{m}\sum_i x_i^2}$	quadratic
$\log x$	$\left(\prod_i x_i\right)^{\frac{1}{m}}$	geometric
x^{-1}	$\frac{1}{\frac{1}{m}\sum(\frac{1}{x_i})}$	harmonic
x^α, $\alpha \neq 0$ finite $\in R$	$\left(\frac{1}{m}\sum_i x_i^\alpha\right)^{\frac{1}{\alpha}}$	root-power

Table 5.1.

The *root-power mean* was extensively studied by Dujmovic (1974,1975) and corresponds to

$$M_m^{(\alpha)}(x_1,\ldots,x_m) = \left(\frac{1}{m}\sum_i x_i^\alpha\right)^{1/\alpha}, \quad 0 < |\alpha| < \infty.$$

The root-power mean includes

$$\begin{aligned} M_m^{(-1)}(x_1,\ldots,x_m) &= \left(\frac{1}{m}\sum_i\left(\frac{1}{x_i}\right)\right)^{-1}, \text{ the harmonic mean} \\ M_m^{(1)}(x_1,\ldots,x_m) &= \left(\frac{1}{m}\sum_i x_i\right), \text{ the arithmetic mean} \\ M_m^{(2)}(x_1,\ldots,x_m) &= \left(\frac{1}{m}\sum_i x_i^2\right)^{1/2}, \text{ the quadratic mean} \end{aligned}$$

and three limit cases :

$$\begin{aligned} M_m^{(0)}(x_1,\ldots,x_m) &= \left(\prod_i x_i\right)^{1/m}, \text{ the geometric mean} \\ M_m^{(\infty)}(x_1,\ldots,x_m) &= \max_i x_i \ , \text{ the disjunctive mean} \\ M_m^{(-\infty)}(x_1,\ldots,x_m) &= \min_i x_i \ , \text{ the conjunctive mean} \end{aligned}$$

with

$$\min \leq M_m^{(\alpha)} \leq \max .$$

Dujmovic also introduces the average value of a root-power mean as

$$\bar{M}_m^{(\alpha)} = \int_0^1 \ldots \int_0^1 M_m^{(\alpha)}(x_1,\ldots,x_m)dx_1 \ldots dx_m$$

$$\begin{aligned} \bar{M}_m^{(\infty)} = A_m = \frac{m}{m+1} \quad &, \quad \bar{M}_m^{(-\infty)} = a_m = \frac{1}{m+1}, \\ \bar{M}_m^{(0)} : \left(\frac{m}{m+1}\right)^m \quad &, \quad \bar{M}_m^{(1)} = \frac{1}{2}, \\ \bar{M}_2^{(2)} = .541 \quad &, \quad \bar{M}_3^{(2)} = .555 \\ \bar{M}_2^{(-1)} = .409 \quad &, \quad \bar{M}_3^{(-1)} = .363. \end{aligned}$$

A *degree of disjunction* d corresponds to

$$d = \frac{\bar{M}_m^{(\alpha)} - a_m}{A_m - a_m} = D_m(\alpha)$$

with $D_m(0) = .5$ for all m and is linguistically defined in table 5.2

Linguistic terms for basic multivariate logic

d value	name	symbol	operator
0	conjunctive (AND)	C	min
.125	strongly quasi-conjunctive	C^+	
.250	medium quasi-conjunctive	CA	
.375	weakly quasi-conjunctive	C^-	
.5	conjunctive-disjunctive	A	arith.mean
.625	weakly quasi-disjunctive	D^-	
.750	medium quasi-disjunctive	DA	
.875	strongly quasi-disjunctive	D^+	
1	disjunctive (OR)	D	max

Table 5.2

To these linguistic terms correspond the α-values, $D_m^{-1}(d)$ as in Table 5.3

α-values for basic many-valued logic
($\alpha = \infty$: max, $\alpha = 2$: square mean, $\alpha = 1$: arithmetic mean,
$\alpha = 0$: geometric mean, $\alpha = -1$: harmonic mean : $\alpha = -\infty$: min)

d	symbol	$m = 2$	3	4	5
0	C	$-\infty$	$-\infty$	$-\infty$	$-\infty$
.125	C^+	−3.51	−3.11	−2.82	−2.64
.250	CA	−0.72	−0.73	−0.71	−0.67
.375	C^-	0.26	0.20	0.17	0.18
.5	A	1	1	1	1
.625	D^-	2.02	2.19	2.30	2.36
.750	DA	3.92	4.45	4.83	5.05
.875	D^+	9.52	11.09	12.28	13.07
1	D	$+\infty$	$+\infty$	$+\infty$	$+\infty$

Table 5.3

Using different properties, Aczél and Alsina (1987) characterized the family of generalized means, called by them *quasi-arithmetic means*.

Definition 5.8 An operator is called *separable* (SE-operator) if

$$M(x_1, \ldots, x_m) = g(x_1) * \cdots * g(x_m) \tag{5.7}$$

where g is a continuous function which maps $[0,1]$ onto $[0,1]$ and $*$ is a continuous *associative* and *cancellative* operation mapping $[0,1]^2$ into $[0,1]$, i.e.

$$(u * v) * w = u * (v * w), \quad \text{for all } \; u, v, w \in [0,1] \tag{5.8}$$

$$\begin{cases} u * t = v * t, & \text{for all } t \in [0,1], \quad \text{implies } u = v \\ t * u = t * v, & \text{for all } t \in [0,1], \quad \text{implies } u = v. \end{cases} \tag{5.9}$$

Aczél and Alsina (1987) proved the following theorem.

Theorem 5.3 *An operator M is separable and idempotent (SE&I-operator) iff*

$$M(x_1, \ldots, x_m) = f^{-1}\left[\frac{1}{m}\sum_i f(x_i)\right] \tag{5.10}$$

where f corresponds to the function defined in Theorem 5.2.

Proof. From (5.8) and (5.9), following a result presented in Aczél's book (1966) it can be immediately deduced that

$$u * v = f^{-1}\left[f(u) + f(v)\right]. \tag{5.11}$$

Putting Equation (5.11) in (5.7), we obtain

$$M(x_1, \ldots, x_m) = f^{-1}\left[\sum_j f[g(x_j)]\right]. \tag{5.12}$$

If idempotency property holds, we obtain

$$x = M(x, \ldots, x) = f^{-1}\left[mf\left[g(x)\right]\right]$$

or

$$f\left[g(x)\right] = \frac{1}{m} f(x)$$

and Equation (5.12) is transformed in (5.10).

The operator given by Equation (5.10) presents obviously separability and idempotency properties.

■

5.6 Stable aggregation operators

We now characterize some aggregation operators defined from valued relations which are given according to some scale type as defined by Stevens (1946) – see also Coombs (1952) and Roberts (1979).

Let us suppose that admissible transformations related to the scale type are functions $\phi : [0,1] \to [0,1]$. Stability of the connective M is assumed if

$$M\left[\phi(x_1), \ldots, \phi(x_m)\right] = \phi M(x_1, \ldots, x_m)$$

where $(x_1, \ldots, x_m) \in [0,1]^m$.

Definition 5.9 An operator M is *ordinally stable* (SO-operator) if

$$M\left[\phi(x_1), \ldots, \phi(x_m)\right] = \phi M(x_1, \ldots, x_m)$$

for all continuous strictly increasing function $\phi : [0,1] \to [0,1]$. Denote that family by Φ.

Definition 5.10 An operator M is *stable for any admissible positive linear transformation* (SPL-operator) if

$$M(rx_1 + t, \ldots, rx_m + t) = rM(x_1, \ldots, x_m) + t,$$

$r > 0$, $rx_k + t \in [0,1]$ for all $k \in \{1, \ldots, m\}$ and $rM(x_1, \ldots, x_m) + t \in [0,1]$.

Definition 5.11 An operator M is *stable for any admissible similarity* (SSI-operator) if

$$M(rx_1, \ldots, rx_m) = rM(x_1, \ldots, x_m),$$

$r > 0$, $rx_k \in [0,1]$ for all $k \in \{1, \ldots, m\}$ and $rM(x_1, \ldots, x_m) \in [0,1]$.

Definition 5.12 An operator M is *stable for any admissible translation* (STR-operator) if

$$M\left[x_1 + t, \ldots, x_m + t\right] = M(x_1, \ldots, x_m) + t,$$

$x_k + t \in [0,1]$ for all $k \in \{1, \ldots, m\}$ and $M(x_1, \ldots, x_m) + t \in [0,1]$.

Definition 5.13 An operator M is *stable for the strong negation N* (SSN-operator) if

$$M\left[Nx_1, \ldots, Nx_m\right] = NM(x_1, \ldots, x_m).$$

Some representation theorems were obtained by Nagumo (1930) for STR- and SSI-stability and also by Silvert (1979) for SNN-stability. These results are resumed in the following propositions.

Proposition 5.3 (i) *An operator M is a STR-generalized mean iff*

$$M(x_1, \ldots, x_m) = \frac{1}{\alpha} \log \sum_i \frac{e^{\alpha x_i}}{m}, \quad \alpha \neq 0 \;\; or \;\; \frac{1}{m} \sum_i x_i$$

(ii) *An operator M is a SSI-generalized mean iff*

$$M(x_1, \ldots, x_m) = \left(\sum_i \frac{x_i^\alpha}{m} \right)^{1/\alpha}, \quad \alpha \neq 0 \;\; or \;\; \left(\prod_i x_i \right)^{1/m}$$

(iii) *A CNM operator such that $M(0, \ldots, 0) = 0$ and $M(1, \ldots, 1) = 1$ is stable for a given strong negation (generated by φ) iff*

$$M(x_1, \ldots, x_m) = \varphi^{-1} \left[\frac{g(x_1, \ldots, x_m)}{g(x_1, \ldots, x_m) + g(Nx_1, \ldots, Nx_m)} \right]$$

where g is a continuous, symmetrical positive mapping : $[0,1]^m \rightarrow [0,1]$ such that $g(0, \ldots, 0) = 0$ and φ is the generator corresponding to the given strong negation : $N(x) = \varphi^{-1}\left[1 - \varphi(x)\right]$.

Proof. For (i) and (ii), see Nagumo (1930) and Aczél and Alsina (1987) and for (iii), see Silvert (1979).

■

In order to subcharacterize the power mean, we present the following corollary:

Corollary 5.1 *The class of SSI-generalized means which presents the SSN-property for the particular negations $Nx = (1 - x^\alpha)^{1/\alpha}$, corresponds to the family of related power means.*

Proof. From Proposition 5.3 (ii), it is clear that the geometric mean fails to satisfy SSN-property for any negation $Nx = (1 - x^\alpha)^{1/\alpha}$, $\alpha \neq 0$. Moreover, the last property is true for the related power means.

■

We now characterize the families of ordinally stable and compensative neutral operators and associative or decomposable CNM operators which are stable for any admissible positive linear transformations (((D or A)&SPL)-CNM operators).

Theorem 5.4 *(i) Ordinally stable (SO) compensative and neutral operators are such that $M(x_1, \ldots, x_m) \in \{x_1, \ldots, x_m\}$.*

(ii) Ordinally stable (SO) compensative, neutral and continuous operators are characterized by the family of connectives $M(x_1, \ldots, x_m)$ equal to one order statistics for every $(x_1, \ldots, x_m) \in [0,1]^m$. (Marichal and Roubens (1993), Kamen and Ovchinnikov (1993)).

Proof. The sufficiency part of (ii) is evident. The necessary part is proved in two steps.

(i) Let us consider $(x_1, \ldots, x_m) \in [0,1]^m$. Due to neutrality,

$$M(x_1, \ldots, x_m) = M(x_{(1)}, \ldots, x_{(m)}),$$

with $x_{(1)} \leq \cdots \leq x_{(i)} \leq x_{(i+1)} \leq \cdots \leq x_{(m)}$.

Compensativeness means that

$$x_{(1)} \leq M(x_1, \ldots, x_m) = x_0 \leq x_{(m)}.$$

Suppose that $x_{(i)} < x_0 < x_{(i+1)}$ for one $i \in \{1, \ldots, m-1\}$. There exists $a, b \in [0,1]$ and a continuous and strictly increasing bijection ψ^* from $[x_{(i)}, x_{(i+1)}]$ onto itself such that

$$x_{(i)} < a < x_0 < b < x_{(i+1)}$$

$$\psi^*(x_{(i)}) = x_{(i)} \ , \ \psi^*(a) = b \ , \ \psi^*(x_{(i+1)}) = x_{(i+1)}.$$

Consider now ψ equal to ψ^* on $[x_{(i)}, x_{(i+1)}]$ and equal to the identity function on $[0,1] \setminus [x_{(i)}, x_{(i+1)}]$.

ψ is obviously a continuous, strictly increasing bijection of $[0,1]$ onto itself.

Now

$$b = \psi(a) < \psi(x_0) = \psi M(x_1, \ldots x_m) = M(\psi(x_1), \ldots, \psi(x_m)) = M(x_1, \ldots, x_m) = x_0$$

and $b < x_0$ which is a contradiction.

(ii) Consider now $z_1 < z_2 < \cdots < z_m \in (0,1)^m$. From (i), it is clear that there exists $r \in \{1, \ldots, m\}$ such that $M(z_1, \ldots, z_m) = z_r$.

Adding the continuity property, we will see that

$$M(x_1, \ldots, x_m) = x_r \ \text{ for every } \ (x_1, \ldots, x_m) \in [0,1]^m.$$

Due to neutrality, we suppose, without restriction, that $x_1 \leq x_2 \leq \cdots \leq x_m$.

Let us consider χ, a continuous non decreasing bijection from $[0,1]$ on itself such that $\chi(z_j) = x_j$, $\forall j \in \{1, \ldots, m\}$ (see Fig. 5.4).

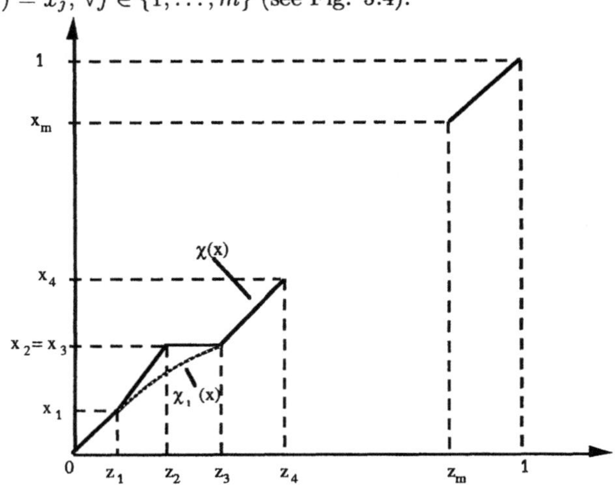

Fig. 5.4

It is always possible to build $\chi_i \in \Phi$ such that $\lim \chi_i(x) = \chi(x)$,for all $x \in [0,1]$.

M being compensative is idempotent (see proof in Proposition 5.1). We then have

$$z_r = M(z_r, \ldots, z_r) = M(z_1, \ldots, z_m).$$

Ordinal stability (SO) implies

$$\chi_i(z_r) = M(\chi_i(z_r), \ldots, \chi_i(z_r)) = M(\chi_i(z_1), \ldots, \chi_i(z_m))$$

and finally, with continuity,

$$\begin{aligned} x_r = \chi(z_r) &= \lim_{i\to\infty} \chi_i(z_r) = \lim_{i\to\infty} M(\chi_i(z_1), \ldots, \chi_i(z_m)) \\ &= M\left(\lim_{i\to\infty} \chi_i(z_1), \ldots, \lim_{i\to\infty} \chi_i(z_m)\right) \\ &= M(\chi(z_1), \ldots, \chi(z_m)) = M(x_1, \ldots, x_m). \end{aligned}$$

■

It is interesting to notice that (SO)-CNM operators are also characterized by one order statistics (see Marichal and Roubens, 1993) and that the results of Kamen and Ovchinnikov (1993) are obtained in the more general framework of ordered sets.

Corollary 5.2 *The class of ((A or D)&SO)-CNM operators equivalent to the class of ((A or D)&SO)-compensative and neutral operators corresponds to*

$$M(x_1, \ldots, x_m) = \min_i x_i \quad or \quad \max_i x_i.$$

Proof. Evident because $x_{(r)}$ is not associative nor decomposable for $r \neq 1, m$, if $x_{(1)} \leq x_{(2)} \leq \cdots \leq x_{(m)}$.

■

It is interesting to remember that operators min or max were previously characterized using the concept of idempotent extended t-norms and t-conorms (see Proposition 5.1 and Section 5.4).

The next theorem concerns the characterization of two classes of connectives : the $(A\&SPL)$-CNM and the $(D\&SPL)$-CNM operators (Marichal and Roubens (1993)). Let us first consider four lemmas.

Lemma 5.1 *The (SPL) operators are idempotent.*

Proof. We consider $r > 0$, $x = 0$ and $t = 0$:

$$M(0, \ldots, 0) = M(r0, \ldots, r0) = rM(0, \ldots, 0) \text{ and } M(0, \ldots, 0) = 0.$$

If $x = 0$, for any t, we obtain

$$M(t, \ldots, t) = t + M(0, \ldots, 0) = t.$$

■

Lemma 5.2 *For any (SPL)-operator and $m = 2$, we have $M(x_1, x_2) = \theta x_1 + (1 - \theta)x_2$ if $0 \leq x_1 \leq x_2 \leq 1$ and $\theta \in [0, 1]$.*

Proof. Consider $x_1 \leq x_2$,

$$M(x_1, x_2) - x_1 = M(0, x_2 - x_1) = (x_2 - x_1)M(0, 1) \quad \text{(stability)}.$$

Finally, $M(x_1, x_2) = \theta x_1 + (1 - \theta)x_2$, with $\theta = 1 - M(0, 1)$.

■

This result is very close to Theorem 1, p. 235 in Aczél (1966).

Lemma 5.3 *If A corresponds to the matrix*

$$A = \begin{pmatrix} \theta & \theta & 0 \\ 1-\theta & 0 & \theta \\ 0 & 1-\theta & 1-\theta \end{pmatrix}, \quad \theta \in [0, 1],$$

then

$$\lim_{i \to \infty} A^i = \frac{1}{D} \begin{pmatrix} \theta^2 & \theta^2 & \theta^2 \\ (1-\theta)\theta & (1-\theta)\theta & (1-\theta)\theta \\ (1-\theta)^2 & (1-\theta)^2 & (1-\theta)^2 \end{pmatrix} \textit{ with } D = \theta^2 + \theta(1-\theta) + (1-\theta)^2.$$

Proof. Suppose that $\theta \in (0, 1)$. The eigen values of A correspond to the solutions of $det(A - zI) = 0$ or $(z - 1)\left[\theta(1 - \theta) - z^2\right] = 0$.

Three distinct eigen values are obtained : $z_1 = 1$, $z_2 = \sqrt{\theta(1-\theta)}$, $z_3 = -\sqrt{\theta(1-\theta)}$, and A can be diagonalized :

$$S^{-1}AS = \begin{pmatrix} 1 & 0 & 0 \\ 0 & \sqrt{\theta(1-\theta)} & 0 \\ 0 & 0 & -\sqrt{\theta(1-\theta)} \end{pmatrix} = \Delta$$

We also have the following eigenvectors :

$$S_1 = \begin{pmatrix} s_{11} \\ s_{21} \\ s_{31} \end{pmatrix} = \begin{pmatrix} \theta^2 \\ \theta(1-\theta) \\ (1-\theta)^2 \end{pmatrix}, \quad S_2 = \begin{pmatrix} s_{12} \\ s_{22} \\ s_{32} \end{pmatrix} = \begin{pmatrix} \theta\left(1 - \theta - \sqrt{\theta(1-\theta)}\right) \\ -2\theta(1-\theta) + \sqrt{\theta(1-\theta)} \\ (1-\theta)\left(\theta - \sqrt{\theta(1-\theta)}\right) \end{pmatrix}$$

$$S_3 = \begin{pmatrix} s_{13} \\ s_{23} \\ s_{33} \end{pmatrix} = \begin{pmatrix} \theta\left(1 - \theta + \sqrt{\theta(1-\theta)}\right) \\ -2\theta(1-\theta) - \sqrt{\theta(1-\theta)} \\ (1-\theta)\left(\theta + \sqrt{\theta(1-\theta)}\right) \end{pmatrix}$$

A can be expressed under the form : $A = S\Delta S^{-1}$ and

$$A^i = S\Delta^i S^{-1}, \quad i \in N_0.$$

Finally,

$$\lim_{i \to \infty} A^i = S(\lim_{i \to \infty} \Delta^i)S^{-1} = (s_{11}^{-1}S_1, s_{12}^{-1}S_1, s_{13}^{-1}S_1).$$

We have to determine $s_{11}^{-1}, s_{12}^{-1}, s_{13}^{-1}$ such that

$$(s_{11}^{-1} s_{12}^{-1} s_{13}^{-1})S = (1\ 0\ 0)$$

and

$$s_{11}^{-1} = s_{12}^{-1} = s_{13}^{-1} = \frac{1}{D}.$$

When $\theta = 1$, A corresponds to the matrix

$$\begin{pmatrix} 1 & 1 & 0 \\ 0 & 0 & 1 \\ 0 & 0 & 0 \end{pmatrix}$$

and

$$A^i = A^2 = \left(\begin{array}{ccc} 1 & 1 & 1 \\ \hline 0 & 0 & 0 \\ 0 & 0 & 0 \end{array}\right) \quad \text{for } i > 2.$$

When $\theta = 0$, A corresponds to the matrix

$$\begin{pmatrix} 0 & 0 & 0 \\ 1 & 0 & 0 \\ 0 & 1 & 1 \end{pmatrix}$$

and

$$A^i = A^2 = \left(\begin{array}{ccc} 0 & 0 & 0 \\ 0 & 0 & 0 \\ \hline 1 & 1 & 1 \end{array}\right) \quad \text{for } i > 2.$$

■

Lemma 5.4 *If*

$$A_{\min,\, m\times m} = \left(\begin{array}{ccccc} 1 & \ldots & \ldots & 1 & 0 \\ \hline & & & & 1 \\ & & 0 & & \end{array}\right) \qquad A_{\max,\, m\times m} = \left(\begin{array}{ccccc} & & 0 & & \\ 1 & & & & \\ \hline 0 & 1 & \ldots & \ldots & 1 \end{array}\right)$$

$$A_{mean,\, m\times m} = \frac{1}{m-1}\begin{pmatrix} 1 & & 0 \\ & 0 & \\ 0 & & 1 \end{pmatrix}$$

then

$$\lim_{i\to\infty} A^i_{\min,\, m\times m} = \left(\begin{array}{ccc} 1 & \ldots & 1 \\ \hline & 0 & \end{array}\right) \qquad \lim_{i\to\infty} A^i_{\max,\, m\times m} = \left(\begin{array}{ccc} & 0 & \\ \hline 1 & \ldots & 1 \end{array}\right)$$

$$\lim_{i\to\infty} A^i_{mean,\, m\times m} = \frac{1}{m}\begin{pmatrix} 1 & \ldots & 1 \\ \vdots & & \vdots \\ 1 & \ldots & 1 \end{pmatrix}$$

Proof. The cases $A_{\min,\, m\times m}$ and $A_{\max,\, m\times m}$ are trivial.
Consider the case of $A_{mean,\, m\times m}$.

$$A_{mean,\, m\times m} = \frac{1}{m-1}(A_1 - A_2)$$

where

$$A_1 = \begin{pmatrix} 1 & \dots & 1 \\ \vdots & & \vdots \\ 1 & \dots & 1 \end{pmatrix} \text{ and } A_2 = \begin{pmatrix} 0 & & 1 \\ & 1 & \\ 1 & & 0 \end{pmatrix}$$

$A_1^i = m^{i-1}A_1$, $A_2^{2i} = I_{m\times m}$, $A_2^{2i+1} = A_2$, for $i \in N_0$ and $A_1A_2 = A_2A_1 = A_1$.
For $i \in N_0$,

$$\begin{aligned}
A^i_{mean,\, m\times m} &= \frac{1}{(m-1)^i}(A_2 - A_1)^i \\
&= \frac{1}{(m-1)^i}\sum_{k=0}^{i}\binom{i}{k}A_1^{i-k}A_2^k(-1)^k \\
&= \frac{1}{(m-1)^i}A_2^i(-1)^i + \frac{1}{(m-1)^i}\sum_{k=0}^{i-1}\binom{i}{k}m^{i-k-1}A_1A_2^k(-1)^k \\
&= \frac{1}{(m-1)^i}A_2^i(-1)^i + \frac{1}{m}A_1\left[\frac{1}{(m-1)^i}\sum_{k=0}^{i-1}\binom{i}{k}m^{i-k}(-1)^k\right] \\
&= \frac{1}{(m-1)^i}A_2^i(-1)^i + \frac{1}{m}A_1\left[1 - \frac{(-1)^i}{(m-1)^i}\right]
\end{aligned}$$

$$\lim A^i_{mean,\, m\times m} = \frac{1}{m}A_1.$$

■

The following theorem can now be proved.

Theorem 5.5 (i) *(A&SPL) operators correspond to the class of*

$$M(x_1, \dots, x_m) = \left[\min_i x_i\right] \text{ or } \left[\max_i x_i\right].$$

(ii) *(D&SPL) – CNM operators correspond to the class of*

$$M(x_1, \dots, x_m) = \left[\min_i x_i\right] \text{ or } \left[\max_i x_i\right] \text{ or } \left[\frac{1}{m}\sum_i x_i\right].$$

Proof. Sufficient part of the theorem is evident. Let us turn to the necessary part.

Let us consider first (i).
Associativity implies :

$$M(z_1, M(z_2, z_3)) = M(M(z_1, z_2), z_3).$$

If $z_1 \le z_2 \le z_3$, Lemma 5.2 gives

$$\theta z_1 + \theta(1-\theta)z_2 + (1-\theta)^2 z_3 = \theta^2 z_1 + \theta(1-\theta)z_2 + (1-\theta)z_3$$

or

$$\theta(1-\theta)(z_3 - z_1) = 0\,, \quad \text{for all } z_3 \ge z_1.$$

As a consequence, $\theta = 0$ or 1 and $M(x_1, x_2) = \min(x_1, x_2)$ or $\max(x_1, x_2)$.

The same values for θ are still obtained in a recurrent way for $m > 2$. ∎

We turn now to (ii)

(ii-1) : Let us first prove that for $m = 3$,

$$M(x_1,x_2,x_3) = \frac{\theta^2 x_1 + \theta(1-\theta)x_2 + (1-\theta)^2 x_3}{\theta^2 + \theta(1-\theta) + (1-\theta)^2}, \quad (x_1 \le x_2 \le x_3) \in [0,1]^3\,,\ \theta \in [0,1]$$

$$\begin{aligned} M(x_1,x_2,x_3) &= M(2.x_1, 2.x_2, 2.x_3) \quad \text{(Proposition 5.2)} \\ &= M(x_1, x_2, x_1, x_3, x_2, x_3) \quad \text{(commutativity)} \\ &= M(2.M(x_1,x_2), 2.M(x_1,x_3), 2.M(x_2,x_3)) \quad \text{(decomposability)} \\ &= M(\theta x_1 + (1-\theta)x_2, \theta x_1 + (1-\theta)x_3, \theta x_2 + (1-\theta)x_3) \quad \text{(Lemma 5.2)} \end{aligned}$$

$$\begin{aligned} M(x) &= M(x_1,x_2,x_3) = M(xA) \\ &= M\left[(x_1,x_2,x_3)\begin{pmatrix} \theta & \theta & 0 \\ 1-\theta & 0 & \theta \\ 0 & 1-\theta & 1-\theta \end{pmatrix}\right] = M(x^{(1)}). \end{aligned}$$

$x_1^{(1)} \le x_2^{(1)} \le x_3^{(1)}$ since $x_1 \le x_2 \le x_3$ and Lemma 5.2.

By iteration, $M(x) = M(xA^i) = M(x^{(i)})$ with $x_1^{(i)} \le x_2^{(i)} \le x_3^{(i)}$, $\forall i \in N_0$.

The diagonalization of A gives (see Lemma 5.3)

$$\lim_{i\to\infty} A^i = \frac{1}{D}\begin{pmatrix} \theta^2 & \theta^2 & \theta^2 \\ \theta(1-\theta) & \theta(1-\theta) & \theta(1-\theta) \\ (1-\theta)^2 & (1-\theta)^2 & (1-\theta)^2 \end{pmatrix}, \quad D = \theta^2 + \theta(1-\theta) + (1-\theta)^2.$$

Finally,

$$\begin{aligned} M(x) &= \lim_{i\to\infty} M(x^{(i)}) = M\left(x \lim_{i\to\infty} A^i\right) = M\left(3.\frac{\theta^2 x_1 + \theta(1-\theta)x_2 + (1-\theta)^2 x_3}{D}\right) \\ &= \frac{\theta^2 x_1 + \theta(1-\theta)x_2 + (1-\theta)^2 x_3}{D} \quad \text{(idempotency, see Lemma 5.1).} \end{aligned}$$ ∎

(ii-2) : We now prove that $\theta \in \{1, 0, 1/2\}$, i.e.

$$M(x_1,x_2) = \min(x_1,x_2) \ \text{ or } \ \max(x_1,x_2) \ \text{ or } \ \left(\frac{x_1+x_2}{2}\right).$$

Let us consider $0 \le z_1 \le z_2 \le z_3 \le 1$.

Decomposability implies

$$M(z_1,z_2,z_3) = M\left[M(z_1,z_3), M(z_1,z_3), z_2\right].$$

If $M(z_1, z_3) \leq z_2$,

$$\theta^2 z_1 + \theta(1-\theta) z_2 + (1-\theta)^2 z_3 = \theta^2 M(z_1, z_3) + \theta(1-\theta) M(z_1, z_3) + (1-\theta)^2 z_2$$

or $(1-\theta)(1-2\theta)(z_3 - z_2) = 0$.

As a consequence, $\theta = 1$ or $1/2$.

If $M(z_1, z_3) \geq z_2$,

$$\theta^2 z_1 + \theta(1-\theta) z_2 + (1-\theta)^2 z_3 = \theta^2 z_2 + \theta(1-\theta) M(z_1, z_3) + (1-\theta)^2 M(z_1, z_3),$$

or $\theta(1-2\theta)(z_2 - z_1) = 0$.

We can conclude that $\theta \in \{0, 1, 1/2\}$.

■

(ii-3) : We finally prove in a recurrent way that

$$M(x_1, \ldots, x_m) = \left[\min_i x_i\right] \text{ or } \left[\max_i x_i\right] \text{ or } \left[\frac{1}{m}\sum_i x_i\right].$$

Suppose $0 \leq x_1 \leq x_2 \leq \cdots \leq x_m \leq 1$, $m \geq 3$.

$$\begin{aligned} M(x_1, \ldots, x_m) &= M\left[(m-1).x_1, \ldots, (m-1).x_m\right] \\ &= M\left[x_1, x_2, \ldots, x_{m-1}, x_1, x_2, \ldots, x_{m-2}, x_m, \ldots, x_2, \ldots, x_m\right] \\ &= M\left[(m-1).M(x_1, \ldots, x_{m-1}), \ldots, (m-1).M(x_2, \ldots, x_m)\right] \\ &= M\left[M(x_1, \ldots, x_{m-1}), \ldots, M(x_2, \ldots, x_m)\right]. \end{aligned}$$

Using recurrency,

$$M(x_1, \ldots, x_m) = M\left[(x_1, \ldots, x_m) A_{m \times m}\right] = M(x_1^{(1)}, \ldots, x_m^{(1)})$$

where

$$A_{m \times m} = A_{\min,\, m \times m} \text{ or } A_{\max,\, m \times m} \text{ or } A_{mean,\, m \times m},$$

as defined in Lemma 5.4.

Due to monotonicity,

$$x_1^{(1)} = M(x_1, \ldots, x_{m-2}, x_{m-1}) \leq x_2^{(1)} = M(x_1, \ldots, x_{m-2}, x_m)$$

and $x_1^{(1)} \leq \cdots \leq x_m^{(1)}$.

By iteration,

$$\begin{aligned} M(x_1, \ldots, x_m) &= M\left[(x_1, \ldots, x_m) A_{m \times m}\right] = M\left[(x_1, \ldots, x_m) A^i_{m \times m}\right] \\ &= M(x_1^{(i)}, \ldots, x_m^{(i)}), \text{ for all } i \in N_0 \end{aligned}$$

with $x_1^{(i)} \leq \cdots \leq x_m^{(i)}$.

Consequently, due to the results of Lemma 5.4,

$$M(x_1, \ldots, x_m) = \lim_{i \to \infty} M\left[(x_1, \ldots, x_m) A^i_{m \times m}\right] = M\left[(x_1, \ldots, x_m) A^{\infty}_{m \times m}\right]$$

where

$$A^{\infty}_{m \times m} = \left[\lim_{i \to \infty} A^i_{\min,\, m \times m}\right] \text{ or } \left[\lim_{i \to \infty} A^i_{\max,\, m \times m}\right] \text{ or } \left[\lim_{i \to \infty} A^i_{mean,\, m \times m}\right]$$

■

Corollary 5.3 *The class of (D&S&SPL)-CNM operators corresponds to*

$$M(x_1, \ldots, x_m) = \frac{1}{m}\sum_i x_i.$$

Proof. Trivial consequence of Theorem 5.5. ∎

Stability can also be studied when x_i correspond to independent scales of measurement. Such extension immediately gives the following definitions which correspond to Definitions 5.9 to 5.11.

SOI-operator related to ordinal scales :

$$M(\phi_1(x_1), \ldots, \phi_m(x_m)) = \phi M(x_1, \ldots, x_m) \tag{5.13}$$

where ϕ_k and ϕ are continuous increasing functions which map $[0,1]$ onto $[0,1]$ and ϕ is a function of $\phi_1, \ldots, \phi_m$.

$SPLI$-operator related to interval scales :

$$M(r_1x_1 + t_1, \ldots, r_mx_m + t_m) = rM(x_1, \ldots, x_m) + t \tag{5.14}$$

$r > 0$, $r_k > 0$, $r_kx_k + t_k \in [0,1]$, for all $k \in \{1, \ldots, m\}$, $rM + t \in [0,1]$ and r and t are functions of $r_1, \ldots, r_m, t_1, \ldots, t_m$.

$SSII$-operator related to ratio scales :

$$M(r_1x_1, \ldots, r_mx_m) = rM(x_1, \ldots, x_m) \tag{5.15}$$

$r_k > 0$, $r_kx_k \in [0,1]$ for all $k \in \{1, \ldots, m\}$, $rM \in [0,1]$ and r is a function of $r_1, \ldots, r_m$.

Admissible transformations defined by Equation (5.15) were first studied by Luce (1964) and were clarified and extended to other transformations defined on B (an appropriate subset of the reals R) by Osborne (1970), Aczél (1984), Aczél, Roberts and Rosenbaum (1986), S. Kim (1988) and Aczél and Roberts (1989).

S. Kim has proved in a Technical Report (1988) the following result.

Theorem 5.6 *Continuous SOI-operators are characterized by* $M(x_1, \ldots, x_m) = g(x_j)$, *for some* $j \in \{1, \ldots, m\}$ *or a constant belonging to* $[0,1]$ *where* g *is a continuous strictly increasing function.*

Proof. See S. Kim (1988) and report by Aczél and Roberts (1989) in the general case where M is a function from R^m to R.

Corollary 5.4 *Continuous (SOI & I)-operators are characterized by*

$$M(x_1, \ldots, x_m) = x_j, \text{ for some } j \in \{1, \ldots, m\}.$$

Proof. Immediate. ∎

If commutativity is added to conditions of the previous Corollary, no solution can be found except when $m = 1$. This last negative result does not contradict Theorem 5.4 which is restricted to SO-operators.

5.7 Non compensative operators

The problem of choosing connectives for the logical combination of valued relations is a difficult one. In the previous sections, we considered the axiomatical point of view.

The proper choice of the functional representation to modelize expressions like ($R_1(a,b)$ *and/or* $R_2(a,b)$ *and/or* ...) is also linked to a semantic background.

In the fuzzy sets theory, there is a huge amount of effort to reconcile many-valued logic, which seems to ignore the existence of many connections between the AND and the OR, and the utility theory which does not handle disjunction and conjunction, as noticed by Dubois and Prade (1985) in a very nice review of fuzzy set aggregation connectives.

The fundamental attitudes in front of several goals – or several binary relations – related to different criteria are contemplated by these two authors. They essentially compare :

the conjunctive attitude : $M(x_1,\ldots,x_m) \le \min_i x_i$

the disjunctive attitude : $M(x_1,\ldots,x_m) \ge \max_i x_i$

the compromise attitude (we could call compensative) expressing a trade off between goals

$$\min_i x_i \le M(x_1,\ldots,x_m) \le \max_i x_i.$$

In the last case, M includes the class of averaging operators and the generalized mean seems to be the most appropriate. However, Zimmermann and Zysno (1983) report experiment in applied context (45 credit clerks of 5 different banks were asked to rate 50 fictious credit applications as to their credit worthiness) and concluded that both product operator and min operator had to be rejected to model the "and" used in human decisions. They defined a "compensatory AND" which had to be between the "logical AND" (which they modelized using the product) and the "inclusive OR" as

$$M(x_1,x_2,x_3,\ldots,x_m) = \left(\prod_i x_i\right)^{1-\gamma}\left(1-\prod_i(1-x_i)\right)^{\gamma}$$

where $\gamma \in [0,1]$.

This connective is another CNM operator which is not always compensative, non idempotent and covers a range from the product ($\gamma = 0$, conjunctive attitude) and the probabilistic sum ($\gamma = 1$, disjunctive attitude).

5.8 Weighted aggregation operators

We consider first the situation where each value x_i, $i = 1,\ldots,m$, is *weighted* by a non negative rational number (also called degree of significance of i) ω_i.

If these weights $\{\omega_i\}$ are given according to a ratio scale, ω_i are not univoqually estimated but are such that any other system of acceptable weights $\{\omega_i'\}$ corresponds to

$$\omega_i' = C\omega_i, \quad C \text{ a positive rational number.}$$

$\{\omega_i\}$ can be modified using a similarity transformation into

$$\omega_i' = \frac{\omega_i}{\sum_i \omega_i} \quad (\sum \omega_i' = 1)$$

or

$$p_i = q\omega_i',$$

where p_i and q ($\sum_i p_i = q$) are positive integers.

We now have to define weighted aggregation operators like :

$$\begin{array}{lll} M\left[(x_1,\omega_1),\ldots,(x_m,\omega_m)\right] & \text{with} & \max_i \omega_i = 1, \\ M\left[(x_1,\omega_1'),\ldots,(x_m,\omega_m')\right] & \text{with} & \sum_i \omega_i' = 1, \\ M\left[(x_1,p_1),\ldots,(x_m,p_m)\right] & \text{with} & p_i : \text{positive integers.} \end{array}$$

If the generalized mean is considered when (x_i, p_i) translates "x_i is obtained p_i times", then :

$$\begin{aligned} M((x_1,p_1),\ldots,(x_m,p_m)) &= M(p_1 \cdot x_1,\ldots,p_m \cdot x_m) \\ &= f^{-1}\left(\frac{1}{q}\sum_i p_i f(x_i)\right) \\ &= f^{-1}\left(\sum_i \omega_i' f(x_i)\right). \end{aligned}$$

Table 5.4 provides some particular cases of this type of mean.

$f(x)$	$M(x_1,\omega_1'),\ldots,(x_m,\omega_m')$	Name of weighted means
x	$\sum_i \omega_i' x_i$	arithmetic
$\log x$	$\left(\prod x_i^{\omega_i'}\right)$	geometric
x^α	$\left(\sum_i \omega_i' x_i^\alpha\right)^{\frac{1}{\alpha}}$	root-power

Table 5.4

The weighted quasi-arithmetic mean, also called *quasi-linear mean* was characterized by Aczél in 1948 using the property of bisymmetry replacing the Kolmogorov and Nagumo's decomposability.

Definition 5.14 An operator M is *bisymmetrical* iff

$$\begin{aligned} & M^{(m)}\{M^{(m)}(x_{11},\ldots,x_{1m}),\ldots,M^{(m)}(x_{m1},\ldots,x_{mm})\} \\ = \; & M^{(m)}\{M^{(m)}(x_{11},\ldots,x_{m1}),\ldots,M^{(m)}(x_{1m},\ldots,x_{mm})\}. \end{aligned}$$

In other words, we do not alter the aggregation operator if we replace x_{ij} by x_{ji} and vice versa.

Aczél (1948) obtained the following results :

Theorem 5.7 *(i) An operator $M^{(m)}$ is continuous, strictly monotonic, idempotent and bisymmetrical iff $M^{(m)}$ represents a quasi-linear mean, i.e.*

$$M^{(m)}(x_1,\ldots,x_m) = f^{-1}\left(\sum_i \omega_i' f(x_i)\right) \quad , \quad \omega_i' > 0 \, , \, \sum \omega_i' = 1.$$

(ii) An operator $M^{(m)}$ is a continuous, strictly monotonic and bisymmetrical iff $M^{(m)}$ represents a quasi-linear function, i.e.

$$M^{(m)}(x_1,\ldots,x_m) = f^{-1}\left(\sum_i a_i f(x_i) + b\right) \quad , \quad a_i > 0.$$

In both cases, f represents an increasing continuous function mapping $[0,1]$ into $[0,1]$.

Proof. See Aczél (1948) and Aczél and Dhombres (1989). ∎

Considering a *generalized separable operator* (GSE-operator)

$$M(x_1,\ldots,x_m) = g_1(x_1) * \cdots * g_m(x_m) \tag{5.16}$$

where g_j, $j = 1,\ldots,m$, are continuous functions which map $[0,1]$ onto $[0,1]$ and $*$ is a continuous, associative and cancellative operator, Aczél (1984) could characterize the weighted geometric and power means with the following theorem.

Theorem 5.8 *An operator M is separable (in the generalized sense), idempotent and stable for any admissible similarity (GSE & I & SSI-operator) iff*

$$M(x_1,\ldots,x_m) = \prod_k x_k^{\omega_k'} \text{ or } \left(\sum_k \omega_k' x_k^{\alpha}\right)^{\frac{1}{\alpha}}, \quad \alpha > 0, \ \sum_k \omega_k' = 1.$$

Proof. See Theorem 2 in Aczél (1984). ∎

If we use the characterization of a continuous, associative and cancellative operator as given by Aczél (1966), we immediately deduce that (GSE & I)-operators correspond to

$$M(x_1,\ldots,x_m) = f^{-1}\left[\sum_j f\left[g_j(x_j)\right]\right] = f^{-1}\left[\sum_j f_j(x_j)\right]$$

such that

$$f(x) = \sum_j f_j(x).$$

From the pioneer work of Luce (1964) (see also Aczél and Roberts (1989), case 4) we know that continuous $SSII$-operators from $\mathbb{R}_+^m$ to $\mathbb{R}_+$ are characterized by

$$M(x_1,\ldots,x_m) = a \prod_j x_j^{C_j} \quad , \quad a > 0, \ C_j \in \mathbb{R}.$$

When the aggregation operator M is a mapping from $(0,1]^m$ to $(0,1]$, we can prove the following result.

Theorem 5.9 *Non-constant SSII-operators $M : (0,1]^m \to (0,1]$ are characterized by*

$$M(x_1, \dots, x_m) = a \prod_{i=1}^{m} x_i^{c_i},$$

where $1 \geq a > 0$, $c_i \geq 0$ are constants for $i \in \{1, \dots, m\}$.

Proof. In the present case, $SSII$-property is expressed by

$$M(r_1 x_1, \dots, r_m x_m) = R(r_1, \dots, r_m) \cdot M(x_1, \dots, x_m) \ , \quad r_i, x_i \in (0,1]. \tag{5.17}$$

Using some arguments from Aczél et al. (1986), we obtain by (5.17) that

$$\begin{aligned} M(\underline{r} \cdot \underline{s} \cdot \underline{x}) &= R(\underline{r} \cdot \underline{s}) \cdot M(\underline{x}) \\ &= R(\underline{r}) \cdot R(\underline{s}) \cdot M(\underline{x}), \end{aligned} \tag{5.18}$$

where $\underline{r} = (r_1, \dots, r_m)$, $\underline{s} = (s_1, \dots, s_m)$, $\underline{x} = (x_1, \dots, x_m) \in (0,1]^m$.
We distinguish two cases.

Case a) : $R(\underline{r}) \not\equiv 1$.

That is, there exists $\underline{r}_0 > \underline{0}$ such that $R(\underline{r}_0) \neq 1$. Define $a = \frac{M(\underline{r}_0)}{R(\underline{r}_0)}$. Then, since we have

$$R(\underline{r}) \cdot M(\underline{x}) = R(\underline{x}) \cdot M(\underline{r}),$$

we obtain

$$M(\underline{x}) = a \cdot R(\underline{x}),$$

by substituting $\underline{r} = \underline{r}_0$.
On the other hand, we also have by (5.18) that

$$\begin{aligned} R(\underline{r}) &= R(r_1, \dots, r_m) = R(r_1, 1, \dots, 1) \cdot R(1, r_2, \dots, r_m) \\ & \quad R(r_1, 1, \dots, 1) \cdot R(1, r_2, 1, \dots, 1) \cdot \dots \cdot R(1, \dots, 1, r_m) \\ &= \prod_{i=1}^{m} R_i(r_i) \end{aligned}$$

with $R_i(rs) = R_i(r) R_i(s)$ for all $r, s \in (0,1]$, $i \in \{1, \dots, m\}$.
Considering now the functional equation

$$g(xy) = g(x) g(y) \quad , \quad x, y \in (0,1], \tag{5.19}$$

$g(x) \in (0,1]$ for all $x \in (0,1]$, we can introduce $x = e^{-u}$, $y = e^{-v}$, $u, v > 0$ and $h(u) = -\log g(e^{-u})$, $u > 0$. Then the functional equation (5.19) is equivalent to

$$h(u+v) = h(u) + h(v) \quad , \quad u, v > 0.$$

Now we have $h(u) > 0$ for $u > 0$, thus Theorem 1 (p.34) of Aczél (1966) can be applied. Hence we can conclude that

$$h(u) = c \cdot u \quad , \quad c > 0.$$

This implies that $g(x) = x^c$, $c > 0$.

Therefore, finally we have in case a) that

$$M(x_1,\ldots,x_m)=a\prod_{i=1}^{m}x_i^{c_i}\ ,\quad a\in(0,1],\ c_i>0.$$

(The constant $a>0$ shoud not be greater than 1 because $M(1,\ldots,1)=a\in(0,1]$.)

Case b) : $R(\underline{r})\equiv 1$.

This obviously implies that

$$M(\underline{r}\cdot\underline{x})=M(\underline{x})\quad\text{for all }\ \underline{r}\in(0,1]^m,$$

whence M is constant on $(0,1]^m$.

■

Corollary 5.5 *Continuous (SSII & I)-operators are characterized by the weighted geometric mean.*

Proof. Immediate. ■

This last result should be compared to Theorem 5.8 (the weighted power mean satisfies SSI – but not $SSII$-property) and Proposition 5.3 (ii).

The case $SPLI$ was studied by Luce in 1964 and by Aczél, Roberts and Rosenbaum (1986) when the same unit ($r_1=\cdots=r_m$) is considered for each interval scale of measurement ($SPLU$-operator).

Theorem 5.10 (i) *$SPLU$-operators are characterized by*

$$M(x_1,\ldots,x_m)=\sum_i a_i x_i+b$$

where a_k and b are arbitrary constants such that $M\in[0,1]$.

(ii) *$SPLI$-operators are characterized by*

$$M(x_1,\ldots,x_m)=ax_j+b,\quad \textit{for some } j\in\{1,\ldots,m\}$$

where a and b are arbitrary constants such that $M\in[0,1]$.

Proof. See Luce (1964) for (ii) when continuity is considered and Aczél, Roberts and Rosenbaum (1986) for general proof of (i) and (ii). ■

Corollary 5.6 *(SPLU & I)-operators are characterized by the weighted arithmetic mean.*

Proof. Immediate. ■

However the weighted sum $\sum_i a_i x_i$ is not identical to the weighted arithmetic mean, the constants a_i being arbitrary except that $\sum_i a_i=1$.

Corollary 5.7 *Neutral (SPLU & I)-operators are characterized by the arithmetic mean.*

Proof. Immediate. ■

A wide class of idempotent operators can be generated by implications and coimplications :

$$\begin{aligned} M_{I^{\rightarrow},\omega}(x_1,\ldots,x_m) &= \bigwedge_i I^{\rightarrow}(\omega_i, x_i) \\ M_{I_c^{\rightarrow},\omega}(x_1,\ldots,x_m) &= \bigvee_i I_c^{\rightarrow}(1-\omega_i, x_i) \end{aligned}$$

where $I^{\rightarrow}$ and $I_c^{\rightarrow}$ represent respectively implications and coimplications as defined in Chapter 1, $\max_i \omega_i = 1$.

These two families are obviously linked by the following relation :

$$M_{I_c^{\rightarrow},\omega}(x_1,\ldots,x_m) = 1 - M_{I^{\rightarrow},\omega}(1-x_1,\ldots,1-x_m),$$

if $I_c^{\rightarrow}(x,y) = 1 - I^{\rightarrow}(1-x,\ 1-y)$.

Proposition 5.4 *Under conditions (I-6) and (I-10) for fuzzy implications, $M_{I^{\rightarrow},\omega}$ and $M_{I_c^{\rightarrow},\omega}$ are idempotent and monotonic aggregation operators.*

Proof. If (I-6) and (I-10) are satisfied : $I^{\rightarrow}(1,x_i) = x_i$ and $I^{\rightarrow}(\omega_i, x) \geq x$. Knowing that $max_j \omega_j = 1$, we immediately obtain that

$$M_{I^{\rightarrow},\omega}(x,\ldots,x) = \min(I^{\rightarrow}(\omega_1,x),\ldots,x,\ldots) = x.$$

Moreover,

$$M_{I_c^{\rightarrow},\omega}(x,\ldots,x) = 1 - M_{I^{\rightarrow},\omega}(1-x,\ldots,1-x) = x.$$

Monotonicity derives from related property for implications and coimplications.

■

Corollary 5.8 *Under conditions (I-6) and (I-10),*

$M_{I^{\rightarrow},\omega}$ and $M_{I_c^{\rightarrow},\omega}$ are compensative.

Proof. Immediate because $M_{I^{\rightarrow},\omega}$ and $M_{I_c^{\rightarrow},\omega}$ are monotonic (see Proposition 5.1).

■

Two particular cases are of special interest.

If all $\omega_j = 1$,

$$\begin{aligned} M_{I^{\rightarrow},\omega}(x_1,\ldots,x_m) &= \min_j x_j \text{ (conjunction)}, \\ M_{I_c^{\rightarrow},\omega}(x_1,\ldots,x_m) &= \max_j x_j \text{ (disjunction)}. \end{aligned}$$

If $I^{\rightarrow}$ corresponds to the Kleene-Dienes implication (which satisfies (I-6) and (I-10)),

$$\begin{aligned} M_{I^{\rightarrow},\omega}(x_1,\ldots,x_m) &= \bigwedge_i [(1-\omega_i) \vee x_i] \\ M_{I_c^{\rightarrow},\omega}(x_1,\ldots,x_m) &= \bigvee_i [\omega_i \wedge x_i]. \end{aligned}$$

The two last aggregates correspond respectively to the *weighted minimum* and *maximum* studied by Dubois and Prade (1986) and they can be rewritten in terms of "weighted medians" (for proof, see Proposition 5.13 and Corollary 5.9 in the next Section).

Instead of using the compensatory "weighted sum" aggregator expressing a trade off, other compensatory aggregation operators will be proposed in Section 5.9 : the "weighted medians", which represent the qualitative counterparts of means.

If $\{\omega'_i\}$ are linked to the importances of the values $\{x_i\}$ where $(x_{(1)}, \omega'_1)$ means that ω'_1 is linked to the lowest value $x_{(1)}, \ldots, (x_{(m)}, \omega'_{(m)})$ means that ω'_m is linked to the greatest value $x_{(m)}$ and $x_{(1)} \leq \cdots \leq x_{(m)}$ the *OWA (ordered weighted averaging)* operator defined by Yager (1988) corresponds to

$$M\left[(x_{(1)}, \omega'_1), \ldots, (x_{(m)}, \omega'_m)\right] = \sum_i \omega'_i x_{(i)}.$$

This class of operators includes
$\min(x_1, \ldots, x_m)$, when $\omega'_1 = 1$,
$\max(x_1, \ldots, x_m)$, when $\omega'_m = 1$,
any order statistics $x_{(k)}$, when $\omega'_k = 1$,
the mean $(\sum_i x_i)/m$, when $\omega'_1 = \cdots = \omega'_m = 1/m$,
the median $(x_{(m/2)} + x_{(m/2)+1})/2$ when m is even and $\omega'_{(m/2)} = \omega'_{(m/2)+1} = 1/2$,
the median $x_{(\frac{m+1}{2})}$ when m is odd and $\omega'_{(\frac{m+1}{2})} = 1$,
the mean excluding the extremes as used by some jury of international olympic competitions, when $\omega'_1 = \omega'_m = 0$ and $\omega'_i = 1/(m-2)$, $i \neq 1, m$.

When $m = 2$, we can state that the OWA operators characterize (SPL) operators. Due to Lemma 5.2, we know that (SPL)-operators $(m = 2)$ correspond to

$$M(x_1, x_2) = \theta x_{(1)} + (1 - \theta) x_{(2)}. \tag{5.20}$$

Moreover, it is clear that $M(x_1, x_2)$ given by (5.20) is a (SPL) operator.

Let us extend this particular result.

Proposition 5.5 *The class of ordered weighted averaging operators corresponds to the aggregators which satisfy the properties of neutrality, monotonicity, idempotency and stability for positive linear transformations with the same unit, independent zeroes and ordered values (SPLU for ordered values).*

Proof. The conditions introduced in the proposition are trivially fulfilled by an OWA operator. To prove the converse, we start from the functional equation :

$$M(rx_{(1)} + t_{(1)}, \ldots, rx_{(m)} + t_{(m)}) = rM(x_{(1)}, \ldots, x_{(m)}) + T(t_{(1)}, \ldots, t_{(m)}) \tag{5.21}$$

where $x_{(1)} \leq \cdots \leq x_{(m)}$ and $t_{(1)} \leq \cdots \leq t_{(m)}$.

Due to idempotency, it is obvious that :

$$M(0, \ldots, 0) = 0$$

and

$$M(t_{(1)}, \ldots, t_{(m)}) = T(t_{(1)}, \ldots, t_{(m)}).$$

For solving Equation (5.21) with $M = T$, consider

$$f_i(x) = M(\underbrace{0,\ldots,0}_{(m-i)\ times}, \underbrace{x,\ldots,x}_{i\ times}).$$

Equation (5.21) with $r = x$, $x_{(1)} = \cdots = x_{(i-1)} = 0$, $x_{(i)} = \cdots = x_{(m)} = 1$, $t_{(1)} = \cdots = t_{(m)} = 0$, gives

$$\begin{aligned} f_i(x) &= xM(0,\ldots,0,1,\ldots,1) + M(0,\ldots,0) \\ &= c_i x,\ c_i \in [0,1],\ i = 1,\ldots,m. \end{aligned}$$

Idempotency implies $c_m = 1$ and monotonicity induces

$$c_1 \le c_2 \le \cdots \le c_m = 1.$$

The value of $M(x_{(1)},\ldots,x_{(m)}) = M(x_1,\ldots,x_m)$ (neutrality) is obtained now in a recursive way using Equation (5.21).

$$\begin{aligned} M(0,\ldots,0,x_{(m-1)},x_{(m)}) &= M(0,\ldots,0,x_{(m-1)},x_{(m-1)}) + M(0,\ldots,0,x_{(m)} - x_{(m-1)}) \\ &= c_2 x_{(m-1)} + c_1(x_{(m)} - x_{(m-1)}) \\ &= (c_2 - c_1)x_{(m-1)} + c_1 x_{(m)} \end{aligned}$$

$$M(x_{(1)},\ldots,x_{(m)}) = (c_m - c_{m-1})x_{(1)} + (c_{m-1} - c_{m-2})x_{(2)} + \cdots + c_1 x_{(m)}.$$

Denoting $\omega_i^{(m)} = c_{m-i+1} - c_{m-i}$, $i = 1,\ldots,m-1$

$$\omega_m^{(m)} = c_1,$$

we obtain that $M(x_{(1)},\ldots,x_{(m)}) = \sum_i \omega_i^{(m)} x_{(i)}$, $\omega_i^{(m)} \ge 0$, $\sum_i \omega_i^{(m)} = 1$, for $0 \le x_{(1)} \le \cdots \le x_{(m)}$.

■

Quasi-OWA operators correspond to

$$M(x_1,\ldots,x_m) = f^{-1}\left[\sum_i \omega_i' f(x_{(i)})\right]$$

where f represents any continuous strictly monotonic function on [0,1].

Proposition 5.6 *Any decomposable quasi-OWA operator corresponds to the* min *or* max *or quasi-arithmetic mean operators.*

Proof. Min, max and quasi-arithmetic mean operators are obviously decomposable quasi-OWA operators.

Let us now consider $M(x_1,\ldots,x_m) = f^{-1}\left[\sum_i \omega_i' f(x_{(i)})\right]$ where f is a continuous strictly monotonic function : $[0,1] \to [0,1]$. We suppose that M is decomposable. We will prove that

$$F(x_1,\ldots,x_m) = f\left[M(f^{-1}(x_1),\ldots,f^{-1}(x_m))\right] = \sum_i \omega_i' x_{(i)}$$

is also decomposable.

Indeed,

$$\begin{aligned}
&F(F(x_1,\ldots,x_k),\ldots,F(x_1,\ldots,x_k),x_{k+1},\ldots,x_m)\\
&= f\left[M\{M(f^{-1}(x_1),\ldots,f^{-1}(x_k)),\ldots,M(f^{-1}(x_1),\ldots,f^{-1}(x_k)),f^{-1}(x_{k+1}),\ldots,f^{-1}(x_m)\}\right]\\
&= f\left[M(f^{-1}(x_1),\ \ldots,f^{-1}(x_m))\right]\\
&= F(x_1,\quad ,x_m)\ \text{(due to decomposability of } M).
\end{aligned}$$

F being decomposable, is also continuous, commutative, monotonic and stable for positive linear transformations (obvious properties of OWA).

From Theorem 5.5 (ii), $F(x_1,\ldots,x_m)$ corresponds to min, max or arithmetic mean operators and $(\omega'_1,\ldots,\omega'_m) = (1,0,\ldots,0)$ or $(0,\ldots,0,1)$ or $(\frac{1}{m},\ldots,\frac{1}{m})$. M corresponds to min, max or quasi-arithmetic mean operators.

■

5.9 Weighted means and medians in terms of fuzzy integrals

Let us consider a fuzzy measure over a discrete set C (considered as the set of criteria : $C_1,\ldots,C_m$). This fuzzy measure satisfies three criteria :

(i) $\mu(A) \geq 0$, for every subset of C.

(ii) $\mu(C_1,\ldots,C_m) = \mu(C) = 1$, $\mu(\phi) = 0$.

(iii) $\mu(A) \leq \mu(B)$ if A and B are subsets of C such that $A \subseteq B$.

Fuzzy measure includes as particular cases :

- the *probability measure* which corresponds to

$$\mu(C_k) = \omega'_k \ \text{ and } \ \mu(C_\imath, C_k) = \omega'_\imath + \omega'_k, \text{ with } \sum_k \omega'_k = 1$$

- the *possibility measure* defined with

  $\mu(C_k) = \pi(C_k) = \omega_k$ and $\mu(C_\imath, C_k) = \max(\omega_\imath, \omega_k)$ such that $\max_k \omega_k = 1$.

- the *necessity measure* related to the possibility measure is the dual measure

$$\mathcal{N}(A) = 1 - \pi(\bar{A}) \ , \quad \bar{A} = \Omega \setminus A.$$

We now proceed to define fuzzy integrals using two main families (for theoretical developments, see Choquet (1953), Sugeno (1975,1977), Sugeno and Murofushi (1987), Grabisch, Murofushi and Sugeno (1992)).

Definition 5.15 The *Sugeno integral operator* corresponds to :

$$M_{S,\mu}(x_1,\ldots,x_m) = (S)\int x \circ \mu = \bigvee_{\imath=1}^{m} [x_\imath \wedge \mu(A_\imath)]$$

Definition 5.16 The *Choquet integral operator* corresponds to :

$$\begin{aligned} M_{C,\mu}(x_1,\ldots,x_m) &= (C)\int x \circ \mu = \sum_{i=1}^{m}(x_i - x_{i-1}).\mu(A_i) \\ &= \sum_{i=1}^{m} x_i \left[\mu(A_i) - \mu(A_{i+1})\right]. \end{aligned}$$

In both cases the elements of vector $\underline{x} : (x_1, \ldots, x_m)$ have been reordered such that

$$0 = x_0 \leq x_1 \leq x_2 \leq \cdots \leq x_m \leq x_{m+1} = 1$$

and $\mu(A_i) = \mu(C_i, \ldots, C_m)$. As an immediate consequence,

$$1 = \mu(C) = \mu(A_1) \geq \mu(A_2) \geq \cdots \geq \mu(A_m) = \mu(C_m) \geq \mu(A_{m+1}) = 0.$$

Proposition 5.7 *The Sugeno and Choquet integral aggregators belong to the class of monotonic and idempotent aggregation operators.*

Proof. Suppose that $x_i = x$. Then

$$\begin{aligned} M_{C,\mu}(x,\ldots,x) &= x\sum_i \left[\mu(A_i) - \mu(A_{i+1})\right] = x \\ M_{S,\mu}(x,\ldots,x) &= \bigvee_i \left[x \wedge \mu(A_i)\right] = x \wedge \left(\bigvee_i \mu(A_i)\right) = x. \end{aligned}$$

Monotonicity is obvious. ■

Proposition 5.8 *The Choquet integral aggregator reduces to the OWA for every additive measure (probability measure).*

Proof. If $\mu(A_i) = \sum_{k \geq i} \omega'_k$, obviously,

$$\sum_i x_i \left[\mu(A_i) - \mu(A_{i-1})\right] = \sum_i x_i \omega'_i.$$

■

Proposition 5.9 *Any OWA aggregator $M^{(m)}$ can be expressed in an equivalent way as a Choquet integral.*

Proof. Let us consider an OWA aggregator $M^{(m)}(x_1, \ldots, x_m)$. We first introduce

$$e_i^{(m)} = M^{(m)}(\underbrace{0,\ldots,0}_{(i)\text{ zeroes}}, \underbrace{1,\ldots,1}_{(m-i)\text{ ones}}), \quad i = 0,\ldots,m \tag{5.22}$$

We obviously have that ($M^{(m)}$ being monotonic and idempotent)

$$0 = e_m^{(m)} \leq e_i^{(m)} \leq e_0^{(m)} = 1 \quad , \quad i = 1,\ldots,m \tag{5.23}$$

$$e_{i-1}^{(m)} - e_i^{(m)} = \omega_i^{(m)} \quad , \quad i = 1,\ldots,m \tag{5.24}$$

Equation (5.24) indicates the equivalence between the knowledge of $(\omega_i^{(m)}, \sum \omega_i^{(m)} = 1, i = 1, \ldots, m)$ and $(e_i^{(m)}, i = 1, \ldots, m-1)$.

Finally,

$$M^{(m)}(x_1, \ldots, x_m) = \sum_{i=1}^{m} \omega_i^{(m)} x_i = \sum_{i=1}^{m} (e_{i-1}^{(m)} - e_i^{(m)}) x_i \quad (5.25)$$

$$= \sum_{i=1}^{m} e_{i-1}^{(m)} (x_i - x_{i-1}) \quad (5.26)$$

if $x_0 = 0$, by definition.

Let us now consider the subsets $A_j = \{x_j, \ldots, x_m\}$, $j = 1, \ldots, m$, $A_{m+1} = \phi$, and a fuzzy measure μ defined on these subsets :

$$\mu(A_j) = e_{j-1}^{(m)}.$$

We have

$$\mu(A_1) = \mu(x_1, \ldots, x_m) = e_0^{(m)} = 1$$

$$A_j \supset A_{j+1} \text{ and } \mu(A_j) \geq \mu(A_{j+1}) \ , \quad j = 1, \ldots, m$$

$$\mu(A_{m+1}) = \mu(\phi) = e_m^{(m)} = 0$$

and (5.25) can be rewritten as

$$M^{(m)}(x_1, \ldots, x_m) = \sum_{i=1}^{m} \mu(A_i)(x_i - x_{i-1})$$

which corresponds to the definition of a Choquet integral.

■

Proposition 5.10 *The Sugeno integral aggregator corresponds to a weighted median (Kandel and Byatt (1978)) :*

$$M_{S,\mu}(x_1, \ldots, x_m) = median(x_1, \ldots, x_m, \mu(A_2), \ldots, \mu(A_m)).$$

Proof. It can be easily proved that (see Proposition 1, Dubois and Prade (1986)) if $a_1 \leq \cdots \leq a_m$ and $1 = b_1 \geq b_2 \geq \cdots \geq b_m$, then

$$\bigvee_i [a_i \wedge b_i] = median(a_1, \ldots, a_m, b_2, \ldots, b_m).$$

$(x_1, \ldots, x_m)$ being ordered in such a way that $x_1 \leq x_2 \leq \cdots \leq x_m$, μ being monotonic :

$$1 = \mu(A_1) \geq \mu(A_2) = \mu(C_2, \ldots, C_m) \geq \cdots \geq \mu(A_m) = \mu(C_m),$$

and the result becomes obvious. ■

Proposition 5.11 *The Sugeno integral aggregator defined by a fuzzy measure μ corresponds to the dual of the Sugeno integral aggregator defined by the corresponding dual measure μ^* (Grabisch, Murofushi and Sugeno (1992)) :*

$$M_{S,\mu}(x_1, \ldots, x_m) = 1 - M_{S,\mu^*}(1 - x_1, \ldots, 1 - x_m)$$

when $\mu^*(A) = 1 - \mu(\bar{A})$.

Proof.

$$\begin{aligned} M_{S,\mu^*}(1-x_1,\ldots,1-x_m) &= \bigvee_j [(1-x_j) \wedge \mu^*(C_j,\ldots,C_1)] \\ &= \bigvee_j [(1-x_j) \wedge (1-\mu(A_{j+1}))] \\ &= 1 - \bigwedge_j [x_j \vee \mu(A_{j+1})] \end{aligned}$$

From a result presented in Dubois and Prade (1986) we know that if

$$c_1 \leq \cdots \leq c_m \quad \text{and} \quad d_1 \geq \cdots \geq d_{m-1} \geq d_m = 0$$

$$\bigwedge_i (c_i \vee d_i) = median(c_1,\ldots,c_m,d_1,\ldots,d_{m-1}).$$

Using that result and Proposition 5.10, it is then obvious that

$$\begin{aligned} 1 - \bigwedge_j [x_j \vee \mu(A_{j+1})] &= 1 - median(x_1,\ldots,x_m,\mu(A_2),\ldots,\mu(A_m)) \\ &= 1 - M_{S,\mu}(x_1,\ldots,x_m). \end{aligned}$$

■

Proposition 5.12 *The Choquet integral aggregator defined by a fuzzy measure μ corresponds to the dual of the Choquet integral aggregator defined by the corresponding dual measure μ^* (Grabisch, Murofushi and Sugeno (1992)) :*

$$M_{C,\mu}(x_1,\ldots,x_m) = 1 - M_{C,\mu^*}(1-x_1,\ldots,1-x_m)$$

where $\mu^*(A) = 1 - \mu(\bar{A})$.

Proof.

$$\begin{aligned} M_{C,\mu^*}(1-x_1,\ldots,1-x_m) &= \sum_{j=1}^{m}(1-x_j)\,[\mu^*(C_j,\ldots,C_1) - \mu^*(C_{j+1},\ldots,C_1)] \\ &= \sum_{j=1}^{m}(1-x_j)\,[\mu(C_j,\ldots,C_m) - \mu(C_{j+1},\ldots,C_m)] \\ &= \sum_{j=1}^{m}(1-x_j)\,[\mu(A_j) - \mu(A_{j+1})] \\ &= 1 - M_{C,\mu}(x_1,\ldots,x_m). \end{aligned}$$

■

Proposition 5.13 *The Sugeno integral aggregator reduces to the weighted maximum for a possibility measure (Dubois and Prade (1986)).*

Proof. Suppose that

$$\begin{aligned} \mu(C_i,\ldots,C_m) &= \pi(C_i,\ldots,C_m) = \bigvee_{k\geq i} \mu(C_k) \\ &= \bigvee_{k\geq i} \omega_k. \end{aligned}$$

We also have :

$$\bigvee_{i\leq k} x_i = x_k.$$

$$\begin{aligned}
M_{S,\pi}(x_1,\ldots,x_m) &= \bigvee_i [x_i \wedge \mu(C_i,\ldots,C_m)] \\
&= \bigvee_i \left[x_i \wedge \left(\bigvee_{k\geq i} \omega_k\right)\right] \\
&= \bigvee_i \left[\bigvee_{k\geq i} \{x_i \wedge \omega_k\}\right] \\
&= \bigvee_k \left[\left(\bigvee_{i\leq k} x_i\right) \wedge \omega_k\right] \\
&= \bigvee_k [x_k \wedge \omega_k]
\end{aligned}$$

■

Corollary 5.9 *The Sugeno integral aggregator reduces to the weighted minimum for an appropriate necessity measure (Dubois and Prade (1986)).*

Proof. If μ is a possibility measure π, we have that

$$M_{S,\mu}(x_1,\ldots,x_m) = M_{S,\pi}(x_1,\ldots,x_m) = \bigvee_i [x_i \wedge \omega_i].$$

From Proposition 5.11, we immediately obtain that

$$\begin{aligned}
M_{S,\mu^*}(x_1,\ldots,x_m) &= M_{S,\mathcal{N}}(x_1,\ldots,x_m) = \bigvee_i [x_i \wedge \mathcal{N}(A_i)] \\
&= 1 - M_{S,\pi}(1-x_1,\ldots,1-x_m) \\
&= 1 - \bigvee_i [(1-x_i) \wedge \omega_i] = \bigwedge_i [x_i \vee (1-\omega_i)].
\end{aligned}$$

■

From the previous results, the weighted maximum and minimum can be expressed in terms of weighted medians :

$$\begin{aligned}
M_{S,\pi}(x_1,\ldots,x_m) &= \bigvee_i [x_i \wedge \omega_i] \\
&= \textit{median}\,[x_1,\ldots,x_m,\pi(A_2),\ldots,\pi(A_m)]
\end{aligned}$$

$$M_{S,\pi}(x_1,\ldots,x_m) = \textit{median}\left[x_1,\ldots,x_i,\ldots,x_m,\bigvee_{k\geq 2}\omega_k,\ldots,\bigvee_{k\geq i}\omega_k,\ldots,\omega_m\right],$$

(recall that $\underline{x}$ must be ordered).

$$\begin{aligned}
&M_{S,\mathcal{N}}(x_1, \ldots, x_m) \\
&= \bigwedge_i [x_i \vee (1-\omega_i)] \\
&= median\,[x_1, \ldots, x_m, \mathcal{N}(A_2), \ldots, \mathcal{N}(A_m)] \\
&= median\,[x_1, \ldots, x_i, \ldots, x_m, 1-\pi(C_1), \ldots, 1-\pi(C_1, \ldots C_{i-1}), \ldots, 1-\pi(C_1, \ldots C_{m-1})], \\
&= median\left[x_1, \ldots, x_i, \ldots, x_m, 1-\omega_1, \ldots, 1-\bigvee_{k\leq i-1} \omega_k, \ldots, 1-\bigvee_{k\leq m-1} \omega_k\right]
\end{aligned}$$

$$\begin{aligned}
&M_{S,\mathcal{N}}(x_1, \ldots, x_m) \\
&= median\left[x_1, \ldots, x_i, \ldots, x_m, 1-\omega_1, \ldots, \bigwedge_{k\leq i-1} (1-\omega_k), \ldots, \bigwedge_{k\leq m-1} (1-\omega_k)\right]
\end{aligned}$$

(recall that $\underline{x}$ must be ordered).

5.10 Aggregation of transitive valued preference relations

Suppose that T is a continuous Archimedean t-norm (strict or with zero divisors). We know from a result of Ling (1965) that T can be expressed under the form :

$$T(x, y) = g^{-1}\left[\min(g(x) + g(y), g(0))\right]$$

where g is a continuous and strictly decreasing mapping : $[0,1] \to [0,\infty]$ with $g(1) = 0$. g is called an additive generator of T (see Theorem 1.3).

Assume that $R_1, \ldots, R_m$ are valued preference relations on a given set A which are T-transitive :

$$T\left[R_i(a,b), R_i(b,c)\right] \leq R_i(a,c), \quad \forall i = 1, \ldots, m, \quad \forall a, b, c \in A \tag{5.27}$$

Consider now the generalized mean generated by g :

$$R(a,b) = M(R_1(a,b), \ldots, R_m(a,b)) = g^{-1}\left[\frac{1}{m}\sum_i g(R_i(a,b))\right]. \tag{5.28}$$

We prove after Ovchinnikov (1992) that :

Theorem 5.11 *If the valued relations R_i, $i = 1, \ldots, m$ are T-transitive, where T is a continuous Archimedean t-norm with generator g, then $R = M(R_1, \ldots, R_m)$, an aggregation operator of the type "generalized mean" with the same generator g, is T-transitive.*

Proof. We first prove that transitivity condition (5.27) is identical to

$$gR_i(a,b) + gR_i(b,c) \geq gR_i(a,c). \tag{5.29}$$

In fact, (see also Lemma 2.3)

$$\begin{aligned} g^{-1}\left[\min(g(x)+g(y), g(0))\right] \leq z \quad &\Leftrightarrow \quad \min(g(x)+g(y), g(0)) \geq g(z) \text{ ; } g \text{ is decreasing} \\ &\Leftrightarrow \quad g(x)+g(y) \geq g(z) \text{ ; } g(0) \geq g(z), \ \forall x, z \in [0,1]. \end{aligned}$$

We now have to prove that

$$gR(a,b) + gR(b,c) \geq gR(a,c). \tag{5.30}$$

For each i, $gR_i(a,b) + gR_i(b,c) \geq gR_i(a,c)$, which implies

$$\frac{1}{m}\sum_i gR_i(a,b) + \frac{1}{m}\sum_i gR_i(b,c) \geq \frac{1}{m}\sum_i gR_i(a,c)$$

and (5.30) is a consequence of the definition (5.28). ∎

This proof can be extended to the Ferrers' property under the restriction that the t-norm, being continuous and Archimedean should have zero divisors.

We know, from Theorem 1.5 that a continuous and Archimedean t-norm with zero divisors can be expressed as a φ-transform of the Lukasiewicz t-norm

$$T(x,y) = \varphi^{-1}\left[\max(\varphi x + \varphi y - 1, 0)\right] \tag{5.31}$$

where φ is an automorphism from $[0,1]$ to $[0,1]$.

Assume that $R_1, \ldots, R_m$ are valued preference relations which are T-Ferrers with a strong negation generated by φ :

$$T\left[R_i(a,b), R_i(c,d)\right] \leq NT\left[NR_i(a,d), NR_i(c,b)\right] \tag{5.32}$$

with

$$Nx = \varphi^{-1}\left[1 - \varphi x\right].$$

We prove :

Theorem 5.12 *If the valued relations R_i, $i = 1, \ldots, m$, are T-Ferrers where T is a φ-transform of the Lukasiewicz t-norm, then the relation $R = M(R_1, \ldots, R_m)$, an aggregation operator of the type "generalized mean" with the same generator φ, is T-Ferrers.*

Proof. We first prove that Ferrers' condition (5.32) is identical to

$$\varphi R_i(a,b) + \varphi R_i(c,d) \leq \varphi R_i(a,d) + \varphi R_i(c,b) + 1. \tag{5.33}$$

In fact,

$$\begin{aligned} & T\left[x_i, y_i\right] \leq NT\left[Nu_i, Nv_i\right] \\ \Leftrightarrow \quad & \varphi^{-1}\left[\max(\varphi x_i + \varphi y_i - 1, 0\right] && \leq \varphi^{-1}\left[1 - \max(\varphi N u_i + \varphi N v_i - 1, 0)\right] \\ & && \leq \varphi^{-1}\left[1 - \max(1 - \varphi u_i - \varphi v_i, 0)\right] \\ & && \leq \varphi^{-1}\left[\min(\varphi u_i + \varphi v_i, 1)\right] \\ \Leftrightarrow \quad & \max(\varphi x_i + \varphi y_i - 1, 0) && \leq \min(\varphi u_i + \varphi v_i, 1) \\ \Leftrightarrow \quad & \varphi x_i + \varphi y_i - 1 && \leq \varphi u_i + \varphi v_i. \end{aligned}$$

We now have to prove that

$$\varphi R(a,b) + \varphi R(c,d) - 1 \leq \varphi R(a,d) + \varphi R(c,b). \tag{5.34}$$

For each i, $\varphi R_i(a,b) + \varphi R_i(c,d) - 1 \leq \varphi R_i(a,d) + \varphi R_i(c,b)$, which implies

$$\frac{1}{m}\sum_i \varphi R_i(a,b) + \frac{1}{m}\sum_i \varphi R_i(c,d) - 1 \leq \frac{1}{m}\sum_i \varphi R_i(a,d) + \frac{1}{m}\sum_i \varphi R_i(c,b)$$

and (5.34) is a consequence of the definition of a generalized mean with generator φ. ■

We can easily see that min-transitivity does not allow to state that the generalized mean preserves that type of transitivity.

Let us consider

$$R_1 = \begin{matrix} a \\ b \end{matrix} \underset{\displaystyle a \quad\; b}{\begin{bmatrix} 0 & 1 \\ 0 & 0 \end{bmatrix}} \quad \text{and } R_2 = \begin{bmatrix} 0 & 0 \\ 1 & 0 \end{bmatrix}. \text{ Obviously } R = \begin{bmatrix} 0 & .5 \\ .5 & 0 \end{bmatrix}$$

R_1 and R_2 are min-Ferrers because

$$\min\left[R_k(x,y), R_k(z,t)\right] \leq \max\left[R_k(x,t), R_k(z,y)\right]$$

for all x, y, z, t belonging to $\{a, b\}$ and for $k = 1, 2$.

If $R(a,a) = 0$, T-Ferrers implies T-transitivity. R_1 and R_2 are min-transitive but that property is preserved for R : $\min\left[R(a,b), R(b,a)\right] = .5 > (R(a,a) = 0)$.

Moreover R is not a min-Ferrers relation because R is not min-transitive and min-Ferrers' property is also not preserved for R.

Let us now consider the Π-Ferrers' property. The geometric mean preserves Π-transitivity because the product t-norm is continuous and Archimedean (see Theorem 5.11). We will show that Π-Ferrers is generally not preserved using the following example :

$$R_1 = \begin{matrix} a \\ b \end{matrix} \underset{\displaystyle a \qquad b}{\begin{bmatrix} .5 & .930 \\ .805 & .5 \end{bmatrix}} \quad R_2 = \begin{bmatrix} .5 & .805 \\ .930 & .5 \end{bmatrix} \quad R = \begin{bmatrix} .5 & .8652 \\ .8652 & .5 \end{bmatrix}$$

R_1 is Π-Ferrers, and particularly

$$R_1(a,b) \cdot R_1(b,a) \leq R(a,a) + R_1(b,b) - R_1(a,a) \cdot R_1(b,b)$$

$((.930)(.805) < .75)$. The same is true for R_2.

R is not Π-Ferrers because

$$R(a,b) \cdot R(b,a) = (.8652)^2 > R(a,a) + R(b,b) - R(a,a) \cdot R(b,b) = .75.$$

5.11 Nearest valued relation to the profile $(R_1, \ldots, R_m)$

An approach to determining an aggregation operator which is quite different from the previous ones consists in choosing the valued relation R which is at minimum distance from the profile $(R_1, \ldots, R_m)$.

If we consider the valued relations $R_i(a,b)$, $i = 1, \ldots, m$, $(a,b) \in A \times A$, one reasonable way to aggregate these relations is to calculate $R(a,b)$ such that the distance between R and any components of the vector $(R_1, \ldots, R_m)$ should be minimized :

$$\underset{R(a,b)}{[\text{MIN}]} \sum_i \sum_{a,b} \omega_i d(R_i(a,b), R(a,b)), \quad \sum_i \omega_i = 1, \tag{5.35}$$

where d represents a distance between two valued relations.

(5.35) is equivalent to

$$\underset{R(a,b)}{[\text{MIN}]} \sum_i \omega_i d(R_i(a,b), R(a,b)), \quad \forall a, b \in A \tag{5.36}$$

The optimal solution for (5.36) corresponds to :

- the *median valued relation* $\tilde{R}(a,b) = \text{median}\ (R_i(a,b); \omega_i)$ if d represents the Hamming distance : $d(R_i(a,b), R(a,b) = |R_i(a,b) - R(a,b)|$.
- the *mean valued relation* $\bar{R}(a,b) = \sum_i \omega_i R_i(a,b)$ if d stands for the quadratic distance : $d(R_i(a,b), R(a,b)) = [R_i(a,b) - R(a,b)]^2$.

If $R(a,b)$ represents the crisp result of a voting procedure, $\omega_i = \frac{1}{n}$, for all i, and

$$\begin{aligned} R_i(a,b) &= 1 \text{ if } a \text{ is preferred or indifferent to } b \\ &= 0 \text{ otherwise} \end{aligned}$$

and if (5.35) corresponds to the Kemeny (1959)'s group consensus function (see also Kemeny and Snell, 1962) :

$$\underset{R}{[MIN]} \sum_i \sum_{a,b} |R_i(a,b) - R(a,b)| \tag{5.37}$$

or

$$\underset{R}{[MIN]} \sum_i \sum_{a,b} [R_i(a,b)(1 - R(a,b)) + (1 - R_i(a,b))R(a,b)] \tag{5.38}$$

subject to $R(a,b) \in \{0,1\}$.

If we define

$c(a,b) = \sum_i R_i(a,b)$, the total number of voters in favour of the choice "a is not worse than b",

$\tilde{c}(a,b) = m - c(a,b)$, the total number of voters opposed to the choice "a is not worse than b", i.e. in favour of "b is preferred to a",

(5.38) is transformed in the following optimization function

$$\underset{R}{[\text{MAX}]} \sum_{a,b} (c(a,b) - \tilde{c}(a,b))R(a,b) \tag{5.39}$$

subject to $R(a,b) \in \{0,1\}$.

Obviously, the solution of this boolean program corresponds to the "majority rule" :

$$\text{If } \begin{cases} c(a,b) \geq \tilde{c}(a,b) & , \quad R(a,b) = 1 \ , \forall a \neq b \in A \\ c(a,b) < \tilde{c}(a,b) & , \quad R(a,b) = 0 \end{cases}$$

If R_i are associated to total crisp orders ($R_i(a,b) = 1 \Rightarrow R_i(b,a) = 0$), $c(a,b) + c(b,a) = m$, $\tilde{c}(a,b) = c(b,a)$, and (5.39) becomes

$$\left[\underset{R}{\text{MAX}}\right] \sum_{a,b} (c(a,b) - c(b,a)) R(a,b)$$

subject to $R(a,b) \in \{0,1\}$. In that case, the "majority rule" corresponds to the Condorcet tournaments if m is odd :

$$\text{If } c(a,b) > c(b,a) : R(a,b) = 1, \ R(b,a) = 0, \ \ \forall a \neq b \in A.$$

If m is even, any tie-breaking rule can be introduced.

Transitivity may be considered in the optimization procedure (median preorder of Kemeny) or in a post-optimization procedure (method of Slater (1961)).

The median preorder is a solution of the following boolean linear program (5.40) :

$$\left[\underset{R}{\text{MAX}}\right] \sum_{a,b} (c(a,b) - \tilde{c}(a,b)) R(a,b) \tag{5.40}$$

subject to :

$$\begin{cases} R(a,b) \in \{0,1\}, \ \ \forall a,b \in A \ \text{ (crisp constraint)} \\ R(a,b) + R(b,a) \geq 1, \ \ \forall a,b \in A \ \text{ (completeness)} \\ R(a,b) \geq R(a,c) + R(b,c) - 1, \ \ \forall a,b,c \in A \ \text{ (transitivity)} \end{cases}$$

This procedure is studied on the axiomatical level by Levenglick and Young (1978) and with a unifying view (algebraic, metric, geometrical, statistical and computational aspects) by Barthelemy and Monjardet (1981).

Slater's proposal is to take a linear order (total crisp order) minimizing the total number of inversions (or inconsistent responses) to the considered Condorcet tournament (first step).

If the Condorcet tournament R is transitive, R is the solution of the problem. If not, in a second step, consider the boolean program (5.41)

$$\left[\underset{R_0}{\text{MIN}}\right] \sum_{a,b} |R_0(a,b) - R(a,b)| \tag{5.41}$$

subject to :

$$\begin{cases} R_0(a,b) \in \{0,1\}, \\ R_0(a,b) + R_0(b,a) = 1, \ \ \forall a \neq b \in A \\ R_0(a,b) \geq R_0(a,c) + R_0(c,b) - 1, \ \ \forall a,b,c \in A \end{cases}$$

R_0 is the final solution.

Similar pathes have been investigated in the valued case in order to introduce max-min transitivity. The aggregation procedure is therefore linked to a ranking procedure (see Chapter 6).

Starting with a given aggregation rule M, we obtain $R = M(R_1, \ldots, R_m)$ and we consider the following quadratic programming problem (5.42) :

$$\left[\underset{R_q}{\text{MIN}}\right] \sum_{a,b} [R_q(a,b) - R(a,b)]^2 \tag{5.42}$$

subject to

$$\begin{cases} 0 \leq R_q(a,b) \leq 1, \quad \forall a, b \in A \\ R_q(a,a) = 1 \\ R_q(a,b) \geq \min(R_q(a,c), R_q(c,b)), \quad \forall a,b,c \in A \end{cases}$$

Montero and Tejada (1986) have suggested to consider an approximate solution to the problem (5.42) to reduce its complexity.

The idea is to preserve the ordinal properties of R : condition : $R(a,b) \geq R(c,d) \Rightarrow R_q(a,b) \geq R_q(c,d)$, $\forall \{a,b\} \cap \{c,d\} \neq \emptyset$, is substituted to the transitivity constraints ($2^{|A|(|A|-1)(|A|-2)}$ in number !).

Another possibility which is less time consuming consists in computing $\hat{R}$, the transitive closure of R :

$$\hat{R} = R^{|A|-1}$$

where $R^k = R^{k-1} \circ R$, $R^2(a,b) = \max_c \min [R(a,c), R(c,b)]$.

Properties of $\hat{R}$ have been obtained by Thomason (1977), Hashimoto (1983b) and Chakraborty and Das (1983). They are resumed in Section 2.5.6.

One reproach might be addressed to the transitive closure. It might be, in some cases, far (in the sense of the Hamming distance) from the aggregation operator R. Proposition 2.23 restrains this inconvenience. $\hat{R}$ is the nearest transitive relation to any relation R, that contains it.

One can also approach the aggregation relation R from the bottom. A maximal transitive relation $\check{R}$ contained in R can be obtained using the algorithm presented in Section 2.5.7.

5.12 Nearest crisp relation to the profile $(R_1, \ldots, R_m)$

When the idea of obtaining a ranking of the alternatives is clearly in mind before starting the aggregation procedure, two ways are offered if the Hamming distance is considered as an optimization tool,

(W1) : aggregate first (with the use of (5.36) and rank after (using methods proposed in Chapter 6).

(W2) : aggregate and rank at once.

If the second way is used, we are looking at a crisp relation close to the profile $(R_1, \ldots, R_m)$ such that (5.36) is optimized :

$$\left[\underset{R(a,b)}{\text{MIN}}\right] \sum_{a,b} \sum_{i} \omega_i \, |R_i(a,b) - R(a,b)| \tag{5.43}$$

subject to $R(a,b) \in \{0,1\}, \ \forall a,b \in A$.

(5.43) can be rewritten as

$$\left[\underset{R(a,b)\in\{0,1\}}{\text{MIN}}\right] \sum_{a,b} \left\{ \sum_i \omega_i R_i(a,b)(1-R(a,b)) + \sum_i \omega_i (1-R_i(a,b))R(a,b) \right\}$$

or in an equivalent way

$$\left[\underset{R(a,b)\in\{0,1\}}{\text{MIN}}\right] \sum_{a,b} \left\{1 - 2\bar{R}(a,b)\right\} R(a,b). \tag{5.44}$$

If no additional constraint is added, no guarantee of transitivity is offered but the program (5.44) is easily solved and corresponds to the "majority rule" :

$$\begin{array}{rl} \text{If } \bar{R}(a,b) \geq .5 & , \quad R(a,b) = 1, \\ \bar{R}(a,b) < .5 & , \quad R(a,b) = 0. \end{array}$$

If transitivity condition is added, the program to be solved corresponds to the boolean linear program (5.45)

$$\left[\underset{R(a,b)}{\text{MIN}}\right] \sum_{a,b} \left\{1 - 2\bar{R}(a,b)\right\} R(a,b) \tag{5.45}$$

under constraints :

$$\begin{array}{rl} R(a,b) \in \{0,1\} & , \quad \forall a,b \in A \\ R(a,b) \geq R(a,c) + R(c,b) - 1 & , \quad \forall a,b,c \in A \end{array}$$

which solution corresponds to a crisp "optimal" partial preorder.

If one wants to add completeness, the set of constraints should contain :

$$R(a,b) + R(b,a) \geq 1, \quad \forall a,b \in A. \tag{5.46}$$

If antisymmetry is required, (5.46) is replaced by

$$R(a,b) + R(b,a) \leq 1, \quad \forall a,b \in A.$$

The complexity of these programs (associativity is in n^3) is such that computational effort is rapidely insurmountable with n growing ($|A| = n$). Marcotorchino and Michaud (1979) reported on computational experience with programs of this type, up to $n = 80$.

Although the zero-one linear programs and the associated relaxed programs (crisp constraints are transformed in : $0 \leq R(a,b) \leq 1, \ \forall a,b \in A$) are not generally equivalent, the two authors obtain that the relaxed programs give a zero-one optimal solution, thus an optimal solution of the boolean linear programs.

The diagram shown in Figure 5.6 gives a short summary of some solutions proposed in Sections 5.11 and 5.12.

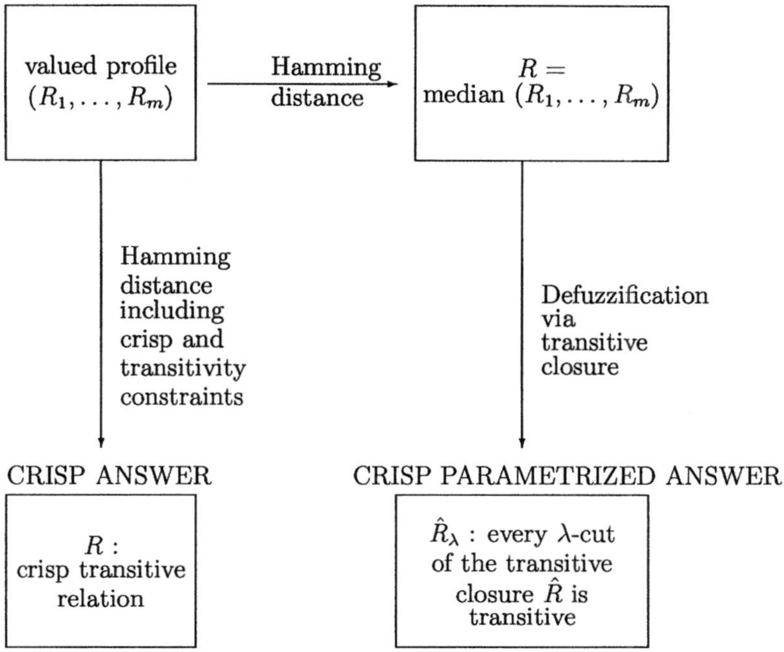
valued profile
$(R_1, \ldots, R_m)$
Hamming
distance
$R =$
median $(R_1, \ldots, R_m)$
Hamming
distance
including
crisp and
transitivity
constraints
Defuzzification
via
transitive
closure
CRISP ANSWER
R :
crisp transitive
relation
CRISP PARAMETRIZED ANSWER
$\hat{R}_\lambda$: every λ-cut
of the transitive
closure $\hat{R}$ is
transitive

Fig. 5.6

Chapter 6

Ranking procedures

6.1 Ranking by scoring

6.1.1 The problem of ranking

The problem of ranking deals with the definition of a preorder on a finite set of alternatives A based on a given valued preference relation $R : A^2 \rightarrow [0, 1]$.

A related problem concerns the ranking of the alternatives in a voting procedure. For each individual, the preference relation related to any pair of alternatives is supposed to be known and we want to aggregate the opinions to build a social ranking.

We mean by *complete ranking*, a complete and transitive crisp binary relation (called total preorder or linear quasiorder or weak order).

A complete ranking is not always required. If two complete rankings are competing, one might think on defining a *partial ranking* with the use of a reflexive and transitive crisp binary relation (called partial preorder or quasiorder) as the intersection of these weak orders.

Example 6.1 For a first example, let A = {Audi, BMW, Mercedes, Opel, Renault, Volvo} and R_1, R_2, R_2, R_4 be the rankings for four different criteria with respective weight equal to 4/12, 3/12, 3/12, 2/12

R_1	
Volvo	
Renault	
Opel	
BMW Mercedes	ex aequo
Audi	

R_2	
Audi BMW Mercedes Renault	ex aequo
Opel	
Volvo	

R_3	
Volvo Opel BMW	ex aequo
Audi	
Renault	
Mercedes	

R_4	
Audi Mercedes Opel Volvo	ex aequo
BMW Renault	ex aequo

The preference of one car a on another b is obtained as the sum of the weights related to criteria such that $(a > b)$ or $(a \sim b)$ with $R(a, a) = 1$.

The preferences are expressed as in Table 6.1.

Preference among cars (Example 6.1)

Entry a, b is the degree of preference of "car a is not worse than car b" multiplied by 12

	A	B	M	O	R	V
Audi (A)	12	5	8	5	8	5
BMW (B)	10	12	10	6	8	6
Mercedes (M)	9	9	12	5	5	5
Opel (O)	9	9	9	12	5	8
Renault (R)	7	9	10	7	12	3
Volvo (V)	9	9	9	9	9	12

Table 6.1

We want to rank the six cars.

Example 6.2 For a second example, we compare the taste of five Medoc wines. Table 6.2 shows as entry $R(a, b)$, the proportion of subjects who expressed a preference for the taste of a over the taste of b. This is a typical forced choice pair comparison tableau where $R(a, b) + R(b, a) = 1$, for all a, b in A. R is called in that case a *probabilistic* relation.

Probabilistic relation for Médoc wines (Example 6.2)

Entry a, b in the proportion of tasters who preferred wine a to wine b

	a	b	c	d	e
a	.50	.57	.57	.29	.67
b	.43	.50	.70	.52	.28
c	.43	.30	.50	.72	.48
d	.71	.48	.28	.50	.48
e	.33	.72	.52	.52	.50

Table 6.2

We want to rank the five wines.

6.1.2 The score functions

The most common and perhaps most natural procedure of ranking is in terms of a function, called a score, associated to every element of the set A of objects to be compared.

The idea of scoring was first proposed by Jean-Charles de Borda in 1781 when he wrote a paper proposing what will be called "the Borda count" for use in elections to the French Académie Royale des Sciences. The Borda count or the preference score is used in a voting procedure where each voter gives a complete preorder on the set of candidates.

de Borda assigns a merit $M_i(a) = k_1 + (m - r_j(a))k_\Delta$ to a candidate a every time it is ranked $r_i(a)$ by some voter i $(i = 1, \ldots, m)$. k_1 is the degree of merit of the last candidate and k_Δ represents the difference between the merits of two consecutive (non ex aequo) candidates for any voter.

The Borda rule gives a ranking $(\geq (B))$ such that

$$a \geq (B)b \text{ iff } \sum_{i=1}^{m} M_i(a) \geq \sum_{i=1}^{m} M_i(b)$$

which is equivalent to

$$a \geq (B)b \text{ iff } \sum_{i=1}^{m} r_j(a) \leq \sum_{i=1}^{m} r_i(b).$$

The principle of preference score can be extended in the valued case.

In general terms, a score $S(a, R)$ related to candidate a using the valued relation $R(x, y)$, $x, y \in A$, corresponds to

$$S(a, R) = F\left[R(a, b_1), \ldots, R(a, b_{n-1}), NR(b_1, a), \ldots, NR(b_{n-1}, a)\right], \quad b_i \neq a$$

where F is a non decreasing function of the 2 $(n-1)$ arguments if n represents the cardinality of A.

Let us consider that a candidate is associated to a vertex and let $R(a, b)$ the degree of preference of candidate a over candidate b be the value assigned to the arc which links the two candidates a and b.

In the valued directed graph $G(A, R)$, we can define for each vertex the scores which correspond to the entering flow, the leaving flow and the net flow :

$$S_E(a, R) = -\sum_{c \in A \setminus \{a\}} R(c, a) \tag{6.1}$$

$$S_L(a, R) = \sum_{c \in A \setminus \{a\}} R(a, c) \tag{6.2}$$

$$S_{L/E}(a, R) = \sum_{c \in A \setminus \{a\}} [R(a, c) - R(c, a)] = S_L(a, R) + S_E(a, R) \tag{6.3}$$

It seems natural to rank the candidates according to the decreasing order of the scores

$$\begin{array}{llll} a \geq_E b & \text{iff } S_E(a, R) & \geq & S_E(b, R) \\ a \geq_L b & \text{iff } S_L(a, R) & \geq & S_L(b, R) \\ a \geq_{L/E} b & \text{iff } S_{L/E}(a, R) & \geq & S_{L/E}(b, R) \end{array}$$

If we reconsider the first example of Section 6.1.1, the different scores assigned to each car are the following (see Table 6.3)

Scores related to each car (multiplied by 12)

Car	S_L	S_E	$S_{L/E}$
A	31	-44	-13
B	40	-41	-1
M	33	-46	-13
O	40	-32	8
R	36	-35	1
V	45	-27	18

Table 6.3.

To each score corresponds a complete ranking

$$\begin{array}{ll} S_L & \Rightarrow \; V > O \sim B > R > M > A \\ S_E & \Rightarrow \; V > O > R > B > A > M \\ S_{L/E} & \Rightarrow \; V > O > R > B > A \sim M \end{array}$$

These rankings can be represented with the use of directed graphs

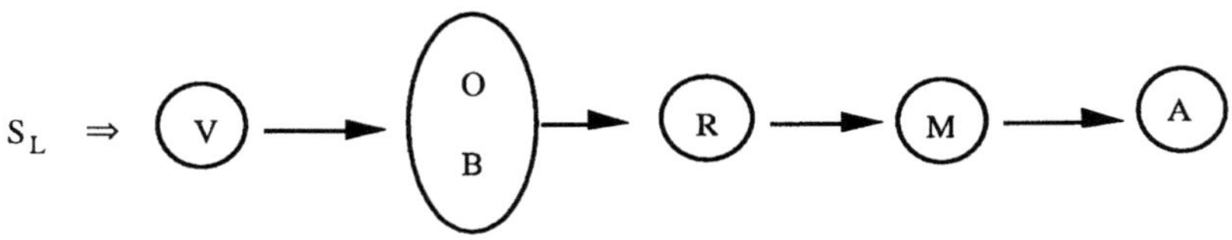

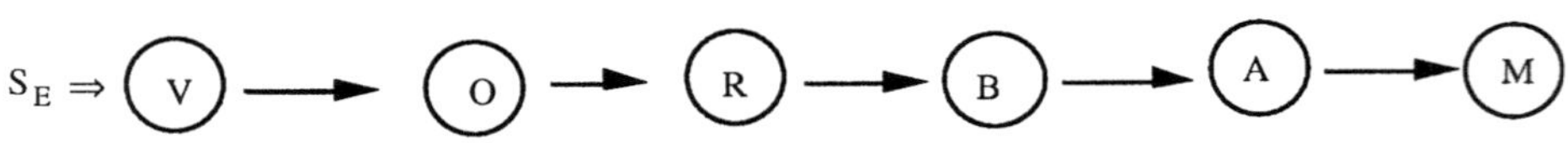

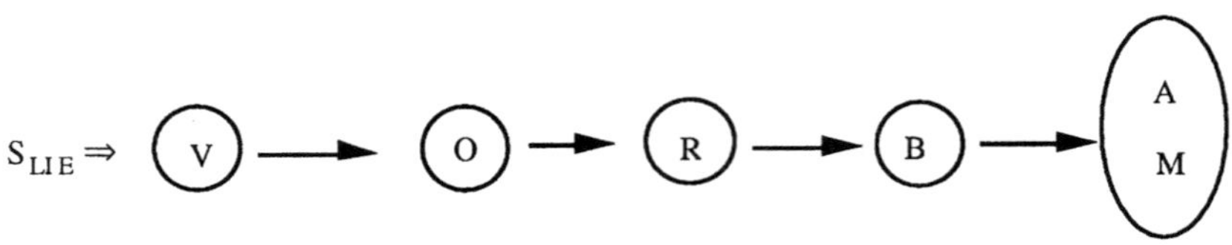

The intersection of S_L and S_E ($S_L \wedge S_E$) gives a partial ranking i.e. a quasiorder

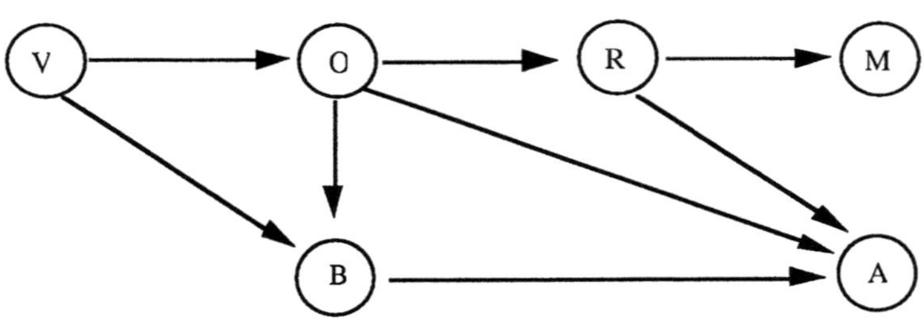

For the second example of Section 6.1.1, the scores are revealed in Table 6.4

Scores related to each wine

wine	S_L	S_E	$S_{L/E}$
a	2.10	-1.90	.20
b	1.93	-2.07	-.14
c	1.93	-2.07	-.14
d	1.95	-2.05	-.10
e	2.09	-1.91	.18

Table 6.4

The three scores $(S_E, S_L, S_{L/E})$ give the same complete ranking :

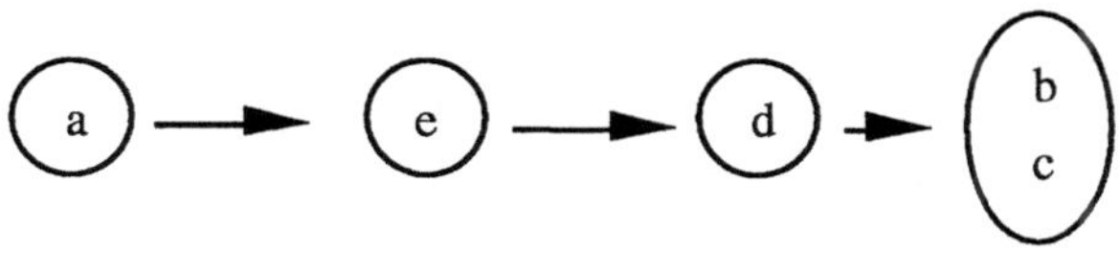

This is not a coincidence. In any probabilistic relation $(R(a,b) + R(b,a) = 1)$, we have, if n represents the number of candidates :

$$S_L(a,R) - S_E(a,R) = n - 1$$

$$\begin{aligned} S_{L/E}(a,R) &= (n-1) + 2S_E(a,R) \\ &= 2S_L(a,R) - (n-1) \end{aligned}$$

For general valued relations, we might think on the minimum leaving flow

$$S_{\text{MIN}}(a,R) = \underset{c \in A \setminus \{a\}}{\text{MIN}} R(a,c) \tag{6.4}$$

but also on the maximum entering flow

$$S_{\text{MAX}}(a,R) = -\underset{c \in A \setminus \{a\}}{\text{MAX}} R(c,a) \tag{6.5}$$

For example 6.1, we obtain

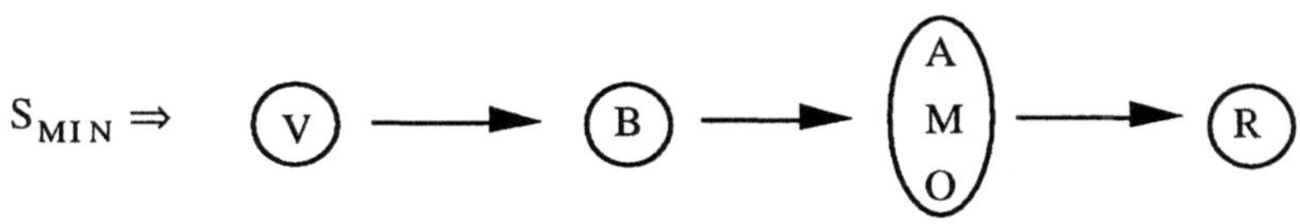

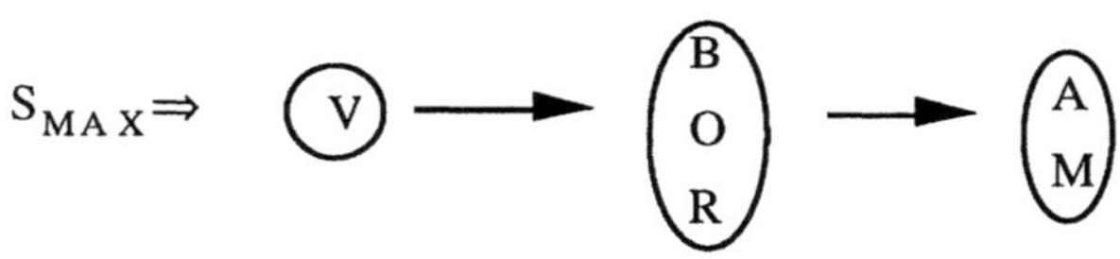

Coming back to example 6.2, we obtain

S_{MIN} and S_{MAX} $\Rightarrow$ 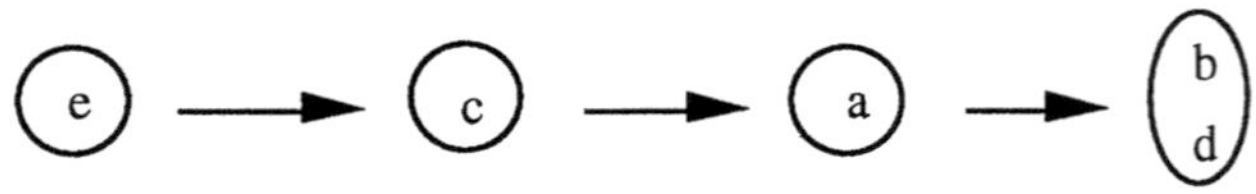

The identity of both maximum flow and minimum flow scores is another property for the probabilistic relation because

$$S_{\text{MIN}}(a, R) = 1 + S_{\text{MAX}}(a, R)$$

The concept of score was used for *selecting a winner in elections* in which voters can rank the candidates.

Suppose that $R(a, b)$ represents the proportion of voters that have a preferred to b (as in Example 6.2) and that R is a probabilistic relation.

The *social choice function* $C(A, R)$ is a function C from the set of candidates A into non empty subsets of A which represents the winners.

Fishburn (1977) has studied some social choice functions called Condorcet social choice functions (CSCF). Among them :

$$\begin{aligned}
\text{the Black's function } C_1(A, R) &= \{a \in A : S_L(a, R) \geq S_L(b, R),\ \forall b \neq a \in A\} \\
&= \{a \in A : a \geq (S_L)b,\ \forall b\} \\
&= \{a \in A : a \geq (S_E)b,\ \forall b\} \\
&= \{a \in A : a \geq (S_{L/E})b,\ \forall b\} \\
\text{the minimax function } C_6(A, R) &= \{a \in A : S_{\text{MIN}}(a, R) \geq S_{\text{MIN}}(b, R),\ \forall b \neq a \in A\} \\
&= \{a \in A : a \geq (S_{\text{MIN}})b,\ \forall b\} \\
&= \{a \in A : a \geq (S_{MAX})b,\ \forall b\}.
\end{aligned}$$

It is interesting to notice that both C_1 and C_6 satisfy the properties of *homogeneity* (if each voter suddenly splits into N voters, each of whom has the same preferrences as the original, the choice set will not change), *monotonicity* or nonnegative responsiveness (if a voter moves upward in his ranking and leaves the relative standing of the others unchanged, then candidate x will stand at least as well relative to each other candidate as before), *Pareto optimality* ($y \notin C(A, R)$, there is an $x \in A \setminus \{y\}$ such that $R(y, x) = 0$) but both procedures fail to satisfy *Condorcet transitivity* (if $y \in C(A, R)$ and $R(x, y) > R(y, x)$, then $x \in C(A, R)$).

A complete and valuable contribution to the literature of choice procedures can be found in a recent monograph written by Kitainik (1993).

A general extension of some of the previous scores gives

$$S_T(a, R) = \underset{c \in A\setminus\{a\}}{T} [R(a, c)] \tag{6.6}$$

$$S_S(a, R) = \underset{c \in A\setminus\{a\}}{S} N[R(c, a)] = N \underset{c \in A\setminus\{a\}}{T} [R(c, a)] \tag{6.7}$$

where $T(x_1, \ldots, x_n)$ represents the extended t-norm related to the classical t-norm $T(x, y)$ and $S(x_1, \ldots, x_n)$ is the related extended t-conorm

$$S(x_1, \ldots, x_n) = NT(Nx_1, \ldots, Nx_n).$$

The extensions (6.6) and (6.7) suggest two other scores : the product leaving and entering flow scores,

$$
\begin{aligned}
S_{\Pi^+}(a,R) &= \prod_{c\in A\setminus\{a\}} R(a,c) \\
S_{\Pi^-}(a,R) &= -\prod_{c\in A\setminus\{a\}} R(c,a).
\end{aligned}
$$

For example 6.1, we obtain

$S_{\Pi^+} \Rightarrow$ V → O → B → R → M → A

$S_{\Pi^-} \Rightarrow$ V → O → R → B → A → M

For example 6.2, the product flow scores gives the same ordering

$S_{\Pi^+}, S_{\Pi^-} \Rightarrow$ e → a → d → c → b

If we resume the results obtained for examples 6.1 and 6.2 in Tables 6.5 and 6.6, we can see that different rankings have been proposed

Ranking of six cars
(The hyphen between O and B represents the fact that they are tied in the ranking)

S_L	S_E	$S_{L/E}$	S_{MIN}	S_{MAX}	S_{Π^+}	S_{Π^-}
V	V	V	V	V	V	V
O-B	O	O	B	B-O-R	O	O
R	R	R	A-M-O	A-M	B	R
M	B	B	R		R	B
A	A	A-M			M	A
	M				A	M

Table 6.5

Ranking of five Médoc
(The hyphen between *b* and *c* represents the fact that they are tied in the ranking)

$S_L, S_E, S_{L/E}$	$S_{\text{MIN}}, S_{\text{MAX}}$	S_{Π^+}, S_{Π^-}
a	e	e
e	c	a
d	a	d
b-c	b-d	c
		b

Table 6.6

Under these circumstances, a possible approach is to list conditions which a "reasonable" preference score should satisfy and see if we can derive from them a particular score function. This is the axiomatical approach, which will be our purpose in the next section.

Before closing this section, we have to mention the meaningful character of the ranking procedures in relation with the score types which were used to define the valued relations R. An extensive study of the theory of meaningfulness can be found in the book of Roberts (1979) on Measurement Theory.

If the valuations R are given according to an *ordinal scale,* they should only preserve the order. The admissible transformations of R are monotonic increasing functions, i.e. functions $\phi(x)$ satisfying the condition that

$$R(x,y) \geq R(z,t) \Leftrightarrow \phi R(x,y) \geq \phi R(z,t),$$

with $\phi(0) = 0$ and $\phi(1) = 1$.

A score function $S(a, R)$ related to an ordinal scale R corresponds to a *meaningful procedure* $\geq (S)$ if

$$\geq (S) = \geq (S_\phi)$$

where

$$a \geq (S)b \text{ iff } S(a,R) \geq S(b,R)$$

and

$$a \geq (S_\phi)b \text{ iff } S(a,\phi R) \geq S(b,\phi R)$$

In the case of ordinal scaling, $S_E, S_L, S_{L/E}$ are meaningless procedures because they make use of the cardinal properties of the valuations.

Due to the monotonicity property of generalized t-norms S_T or S_S and in particular $S_{\text{MIN}}, S_{\text{MAX}}, S_{\Pi^+}, S_{\Pi^-}$ are meaningful procedures for ordinal scaling.

If the valuations are given according to an *interval scale*, the admissible transformations of R are the positive linear transformations $\phi(x) = \alpha x + \beta$ (($\alpha > 0$).

In the case $S_E, S_L, S_{L/E}$ and of course S_T and S_S are meaningful procedures.

If we revisit Example 1 and if we suppose that the weights related to different criteria were measured with a *ratio scale*, i.e. if the admissible transformations are $\phi(x) = \alpha x$ ($\alpha > 0$), we can conclude that all the score functions which were proposed to solve the related ranking problem : $S_E, S_L, S_{L/E}, S_{\text{MIN}}, S_{\text{MAX}}, S_{\Pi^+}, S_{\Pi^-}$ correspond to meaningful procedures.

6.2 Nondominating and nondominated alternatives

Starting from a valued reflexive relation R, we consider (see Section 3.3.2) the degree of strict preference of a over b as

$$P(a,b) = \max\{R(a,b) - R(b,a), 0\}.$$

The valued relation P is irreflexive and antisymmetric. When the relation R is complete, $P = R^d$.

Relation P was first considered by Orlovski (1978) which defined a *nondomination degree* μ_{ND} related to each alternative (a degree to which an element is dominated by no one of the elements of the set A)

$$\begin{aligned}\mu_{ND}(a,R) &= \min_{c\in A}\{1-P(c,a)\}\\ &= \min_{c\in A} P^d(a,c)\\ &= 1-\max_{c\in A} P(c,a)\\ &= 1+\min_{c\in A}\{R(a,c)-R(c,a)\}.\end{aligned}$$

To $\mu_{ND}(a,R)$ corresponds a score function S_{ND} :

$$S_{ND}(a,R) = \inf_{c\in A}\{R(a,c)-R(c,a)\}.$$

In the same spirit, a *nondominance degree* μ_{ND} can be proposed (a degree to which an element is non dominating all the elements of the set A) :

$$\begin{aligned}\mu_{Nd}(a,R) &= \min_{c\in A}(1-P(a,c))\\ &= 1-\max_{c\in A} P(a,c).\end{aligned}$$

The score function related to nondominance corresponds to

$$\begin{aligned}S_{Nd}(a,R) &= \max_{c\in A} P(a,c)\\ &= \max_{c\in A}\{R(a,c)-R(c,a)\}.\end{aligned}$$

The element(s) $a^* \in A$ which correspond to $max_a \mu_{ND}(a,R)$ are considered to be the *nondominated alternative(s)*. This subset constitutes a social choice function.

If $\mu_{ND}(a^*,R) = 1$, a^* is called by Orlovski (1978) an *unfuzzy nondominated alternative* (clearly best alternative). The subset of such alternatives is denoted A^{UND} and

$$A^{UND} : \{a \in A : R(a,c) \geq R(c,a),\ \forall c \in A\}.$$

If $\mu_{Nd}(a^*,R) = 1$, a^* corresponds to an *unfuzzy nondominating alternative* (clearly worst alternative). The subset of undesirable alternatives is denoted A^{UNd} and

$$A^{UNd} : \{a \in A : R(a,c) \leq R(c,a),\ \forall c \in A\}.$$

If we introduce a crisp relation $R_{\geq}$ such that

$$aR_{\geq}b \text{ iff } R(a,b) \geq R(b,a) \ , \quad (\text{i.e. } P(a,b) \geq 0)$$

we immediately obtain :

$$A^{UND} = \{a \in A : aR_{\geq}c \ ,\ \forall c \in A\}.$$

Suppose now that $P_{>}$ corresponds to the antisymmetric part of $R_{\geq}$:

$$aP_{>}b \text{ iff } P(a,b) > 0.$$

If the reflexive relation R is (min) transitive, then $R_{\geq}$ and $P_{>}$ are crisp transitive relations. When A is finite, there exists at least one element a^* with the property : not $cP_{>}a^*$, $\forall c \in A$.

In an equivalent way, $P(c, a^*) = 0$, $\forall c \in A$ and $\mu_{ND}(a^*, R) = 0$.

This corresponds to a result of Orlovski (1978) :

Proposition 6.1 (Orlovski, 1978) *If A is finite and R is reflexive and transitive, A^{UND} is not empty.*

From this result, we immediately obtain the following lemma :

Lemma 6.1 *If A is finite and if R is reflexive and transitive, A^{UNd} is non empty.*

Proof. A^{UND} presents the same definition as A^{UNd} if R is substituted to R^{-1}.

■

Sen (1970) considers a crisp reflexive and complete relation R and defines C as a *social choice function* iff

$$C(X, R) = \{a \in X : aRc\ ,\ \forall c \in X \subseteq A\}$$

is non empty for all $X \subseteq A$.

He proves that the subset C is a social choice function iff the strict preference relation P related to R (aPb iff aRb and not bRa) is acyclic.

It is clear that

$$\begin{aligned} A^{UND} &= \{a \in A : aR_{\geq}c\ ,\ \forall c \in A\} \\ &= C(A, R). \end{aligned}$$

It must be pointed out that a social choice function is not required in our case. We are looking for best alternatives in the fixed set A and we might be able to find non empty $C(X, R)$ for same subset $X \subset A$, despite $A^{UND} \neq \phi$ (Montero and Tejada (1988)).

6.3 Characterization of the net flow score and the min leaving flow score methods

6.3.1 Case of crisp relations

In a paper by Ph. Vincke (1992b), on "exploitation of a crisp relation in a ranking problem", a list of twenty-two properties that a "reasonable" ranking procedure should satisfy is proposed and eleven procedures are considered for the crisp case.

In the case of *crisp relations,*

$$\begin{aligned} R(a, b) &= 1 \text{ iff } a \text{ is not worse than } b \\ &= 0 \text{ otherwise} \end{aligned}$$

Among the procedures which were proposed by Vincke, three of them dominate the others (the set of properties satisfied by the first ones strictly contains the set of properties satisfied by the others).

One of the dominating procedures is based on the transitive closure of R, $\hat{R}$ and one corresponds to the net flow score.

If we consider the equivalence relation E as

$$aEb \text{ iff } a = b \text{ or } a\hat{R}b \text{ and } b\hat{R}a$$

we denote the equivalence class of a, $[a]$ and

$$[a]\, R^r\, [b] \text{ iff } \exists c \in [a] \text{ and } d \in [b] : cRd.$$

Obviously the relation R^r has no circuit and the *rank* of a corresponds to the length of the longest path of R arriving in a.

The ranking procedure corresponding to the ranks of the alternatives is denoted RK. The equivalence classes of $\geq (RK)$ are the elements of A/E. We rank the equivalence classes in the increasing order of their rank in R^r.

If we consider the following tableau :

R	x	y	z	t	ω	Σ^+	$\Sigma^+ - \Sigma^-$
x		1				1	0
y			1			1	0
z	1			1		2	1
t					1	1	0
ω						0	-1
Σ	1	1	1	1	1		

$\hat{R}$ the transitive closure of R corresponds to the following equivalence classes $A\backslash E$: $\{x, y, z\}, \{t\}, \{\omega\}$.

Finally, we obtain, using RK, the following complete procedure

$$(x \sim y \sim z) > (RK)t > (RK)\omega.$$

If we consider the net flow score we obtain

$$z > (S_{L/E})(x \sim y \sim t) > (S_{L/E})\omega.$$

6.3.2 Definition of desirable properties

We first recall that the ranking method $\geq (S)$ based on scores defines a complete ranking on the set of alternatives A in such a way that

$$a \geq (S)b \Leftrightarrow S(a, R) \geq S(b, R)$$

where S is a function (called a score) of the valuation R defined for each alternative :

$$S(a, R) = F\left[R(a, b_1), \ldots, R(a, b_{n-1}), NR(b_1, a), \ldots, NR(b_{n-1}, a)\right]$$

where b_i, $i = 1, \ldots, n-1$, belong to $A \setminus \{a\}$.

It must be mentioned that the ranking $\geq (S)$ *does not satisfy* the Arrow's property of *independence of irrelevant alternatives* : when a is opposed to b with the use of $\geq (S)$, the other alternatives belonging to $A \setminus \{a, b\}$ clearly play a role in the comparison.

It is also evident that the property of *universality* is satisfied : $\geq (S)$ is clearly defined for every binary relation R defined on A and leads always to a transitive complete crisp relation.

Let us now consider some properties that a ranking method based on scores might satisfy :

Neutrality property prevents a built-in favoritism for any alternative. When neutrality is satisfied, the corresponding ranking method treats all alternatives equally; it does not discriminate between alternatives because of their labels. In that case,

$$a \geq (S)b \Leftrightarrow \sigma(a) \geq (S^\sigma)\sigma(b)$$

where σ is a permutation of the alternatives $(a, b, \ldots)$ belonging to A and S^σ applies to valuations R^σ such that $R^\sigma(\sigma(a), \sigma(b)) = R(a, b)$, for all a, b. F is clearly commutative.

Ordinality allows the system to be meaningful when the valuations are defined with the use of an ordinal scale :

$$a \geq (S)b \Leftrightarrow a \geq (S_\phi)b$$

where ϕ is an automorphism of $[0, 1]$ to $[0, 1]$ applied on the valuations R; $\phi(R)$ is a strictly increasing transformation of R such that $\phi(0) = 0$, $\phi(1) = 1$.

Monotonicity says that the ranking method does not respond in the wrong direction to a modification of R. If $R(a, c)$ moves upwards in its valuation or if $R(c, a)$ moves downwards, for some c belonging to $A \setminus \{a\}$ and if the other valuations remain unchanged, then the alternative a should be ranked before b if previously $a \geq (S)b$.

Strong monotonicity says that the method responds in the right direction.
More formally :
Monotonicity means :

$$a \geq (S)b \Rightarrow a \geq (S')b \text{ for all } a, b \text{ belonging to } A.$$

Strong monotonicity means

$$a \geq (S)b \Rightarrow a > (S')b \text{ for all } a, b \text{ belonging to } A$$

where $> (S')$ is the asymmetric part of $\geq (S')$ and where R' is identical to R except that $R(a, c) < R'(a, c)$ or $R(c, a) > R'(c, a)$ for some $c \in A \setminus \{a\}$.

F is respectively a non decreasing (strictly increasing) function of the arguments $R(a, b_i)$ and non increasing (strictly decreasing) function of the arguments $R(b_i, a)$.

Strong row monotonicity (Bouyssou (1992)) says that the position of an alternative should improve in the ranking if its position is improved with regard to all the other alternatives in the valuation :

$$a \geq (S)b \Rightarrow a > (S')b \text{ for all } a, b \text{ belonging to } A$$

where R' is identical to R except that $R'(a,c) > R(a,c)$ for all $c \in A \setminus \{a\}$.

Continuity (Bouyssou, (1992)) says that the small changes in the valuations should not lead to radical changes in the associated ranking : for all R, all sequences R^i, $i = 1, 2, \ldots$ converging to R and all a, b belonging to A

$$\left[a \geq (S^i)b \text{ for all } R^i \text{ in the sequence}\right] \Rightarrow a \geq (S)b.$$

Row egalitarian property (Bouyssou, (1992)) means that averaging the row of the matrix R associated to an alternative cannot decrease its position in the ranking :

$$a \geq (S)b \Rightarrow a \geq (S_a)b$$

where S_a corresponds to R_a which is identical to R except that

$$R_a(a,c) = \left\{ \sum_{d \in A \setminus \{a\}} R(a,d) \right\} / (n-1), \text{ for all } c \in A \setminus \{a\}.$$

Independence of admissible translations on circuits of the valued graph $G(A,R)$ means (Bouyssou, (1991)) : if R' consists in adding a same positive or negative quantity to the valuations of the arcs in an elementary circuit such that R' belongs to $[0,1]^n$,

$$a \geq (S)b \Rightarrow a \geq (S')b.$$

It is clear that *strong column monotonicity* and *column egalitarian properties* are easily defined (columns of the matrix R are substituted to the rows).

In the same spirit, *independence of admissible similarities on circuits* of the valued graph $G(A,R)$ means : if R' consists in multiplying the valuations of the arcs in an elementary circuit by a strict positive value such that R' belongs to $[0,1]^n$,

$$a \geq (S)b \Rightarrow a \geq (S')b.$$

Independence of admissible translations on alternated cycles of the valued graph $G(A,R)$ means (Bouyssou and Perny (1992)) : if R' consists in adding a same positive (in one direction of the cycle) and negative (in the other direction of the cycle) quantity to the valuations of the arcs in a 4 or 6-alternated cycle such that R' belongs to $[0,1]^n$

$$a \geq (S)b \Rightarrow a \geq (S')b.$$

Let us as an example consider a 4-alternated cycle with its associated values

$$(a,b) : .2, \quad (c,b) : .3, \quad (c,d) : .4, \quad (a,d) : .2.$$

If one adds or substracts (.2), we obtain running along the cycle (a,b,c,d,a) :

$$(a,b) : .4, \quad (c,b) : .1, \quad (c,d) : .6, \quad (a,d) : 0.$$

Independence from non discriminating alternatives means (see Vincke (1991) for the crisp case, and Perny (1992) for the valued case) that removing one or several alternatives which do not allow to make a difference between the remaining alternatives does not change the ranking of the remaining alternatives. In formal terms, for all $B \subset A$, for all $b, b' \in B$, $\forall c \in A \setminus B$, such that $R(b,c) = R(b',c)$ and $R(c,b) = R(c,b')$:

$$\geq (S/B) \Leftrightarrow \geq (S)/B$$

where X/B represents the restriction of the relation X to B.

Loyalty means that the ranking procedures $\geq (S)$ will not modify the information in terms of preorders contained in R.

Let us consider the support of the relation R as a crisp relation supp(R) such that

$$\text{supp}(R)(a,b) = 1 \quad \text{iff} \quad R(a,b) > 0.$$

Loyalty says that if supp(R) is a preorder

$$\geq (S) \subseteq \text{supp}(R).$$

6.3.3 Results of net flow and min outflow procedures

The characterization of two classical ranking methods, the net flow score and the min outflow procedures, was obtained by Bouyssou in 1991 and 1992.

Theorem 6.1 *The net flow method is the only ranking method that is neutral, strongly monotonic and independent of admissible translations on circuits.*

Proof

The idea of the proof is the following (see Bouyssou, 1991). It is first proved that for all valued relations A and R', if [R' can be obtained from R through an admissible translation on elementary circuit] then [R' can be obtained from R through a finite number of admissible translations on elementary circuits of length 2 or 3].

As an example, let us consider the circuit $(a,b),(b,c),(c,d),(d,a)$. One wants to add $\delta > 0$ to obtain $R'(a,b) = R(a,b) + \delta \leq 1$, $R'(b,c) = R(b,c) + \delta \leq 1, \ldots$.

To achieve this translation, let us consider the following steps :

(1) add δ to $R(a,b)$, $R(b,d)$ and $R(d,a)$

(2) add $(-\delta)$ to $R(d,b)$ and $R(b,d)$

(3) add δ to $R(d,b)$, $R(b,c)$ and $R(c,d)$.

It is then proved that for all valued relations R and R', if [R' can be obtained from R through an admissible translation on an elementary cycle] thus [R' can be obtained from R through a finite number of admissible translations on elementary circuits].

Let us once more consider an example. One wants to add or substract δ along the cycle $(a,b),(c,b),(c,d),(d,a)$ to obtain $R'(a,b) = R(a,b) + \delta \leq 1$, $R'(c,b) = R(c,b) - \delta \leq 1$, $R'(c,d) = R(c,d) + \delta = 1$ and $R'(d,a) = R(d,a) + \delta \leq 1$. The procedure follows :

(1) add $(-\delta)$ to $R(b,c)$

(2) add δ to $R(a,b)$, $R(b,c)$, $R(c,d)$ and $R(d,a)$.

Finally Bouyssou proved that, for all R, R' on A, [$S_{\text{L/E}}(a,R) = S_{\text{L/E}}(a,R')$ for all a belonging to A] $\Leftrightarrow$ [R' can be obtained from R through a finite number of admissible translations on elementary cycles].

From these properties the theorem can be proved. ■

Theorem 6.2 *The* min *outflow method is the only ranking method that is neutral, ordinal, continuous, row monotonic and row egalitarian.*

The proof of the theorem is given in Bouyssou (1991).
Another characterization of the min outflow procedure was given by Pirlot (1993).

6.3.4 Comparison meaningfulness and scoring procedures

Let us consider here the particular scoring function dealing only with the flow related to $R(a, b_1), \ldots, R(a, b_{n-1})$:

$$S(a, R) = F\left[R(a, b_1), \ldots, R(a, b_{n-1})\right],$$

should be an increasing function of the $(n-1)$ arguments.

$S(a, R)$ can be considered as an aggregator of $(n-1)$ variables related to the same scale of measurement defined by the admissible transformation ϕ, with inverse ϕ^{-1}.

As "a is declared not to be globally worse than b" $(a \geq (S)b)$ iff

$$S(a, R) \geq S(b, R),$$

the aggregator F should be *ϕ-comparison meaningful* (ϕ-c.m.) in the sense that

$$S(a, R) \geq S(b, R) \text{ implies } S(a, \phi R) \geq S(b, \phi R) \tag{6.8}$$

which is equivalent to

$$S(a, R) = S(b, R) \text{ iff } S(a, \phi R) = S(b, \phi R). \tag{6.9}$$

Let us define S as ***ϕ-stable*** iff

$$S(a, \phi R) = \phi S(a, R). \tag{6.10}$$

If one is concerned with an *ordinal scale* (ϕ is a strict monotone increasing transformation), ϕ-comparison meaninfulness corresponds to ordinality (O-c.m.) – see Section 6.2.2 – and ϕ-stability is equivalent to F being ordinally stable (SO), (see Definition 5.9).

When R corresponds to an *interval scale*, ϕ-comparison meaningfulness is called positive linear transformation-comparison-meaningfulness or briefly (PL-c.m.) and ϕ-stability concerns (SPL)-operators F.

The next Proposition relates stability and comparison-meaningfulness.

Proposition 6.2 (i) *If $S(a, R) = F(a, R(a, b_1), \ldots, R(a, b_{n-1}))$ is ordinally stable (SO), then $S(a, R)$ satisfies ordinality (O-c.m.).*

(ii) *If $S(a, R) = F[a, R(a, b_1), \ldots, R(a, b_{n-1})]$ satisfies ordinality (O-c.m.) and idempotency (I), then $S(a, R)$ is ordinally stable (SO).*

Proof. (i) Suppose that ϕ is a strict monotone increasing transformation.

$$\begin{aligned} S(a,R)=S(b,R) &\Leftrightarrow \phi S(a,R)=\phi S(b,R) \\ &\Leftrightarrow S(a,\phi R)=S(b,\phi R) \quad (SO) \end{aligned}$$

which corresponds to Equation (6.9) defining (O-c.m.).

(ii) Consider $S(a,R)=F(a,R_1,\ldots,R_{n-1})=u_0$.

$$u_0=F(a,u_0,\ldots,u_0) \qquad (I).$$

If $F(a,R_1,\ldots,R_{n-1})=F(a,u_0,\ldots,u_0)$, then

$$\begin{aligned} F(a,\phi R_1,\ldots,\phi R_{n-1}) &= F(a,\phi u_0,\ldots,\phi u_0) \quad (O-c.m.) \\ &= \phi u_0 \quad (I) \end{aligned}$$

and

$$F(a,\phi R_1,\ldots,\phi R_{n-1})=S(a,\phi R)=\phi S(a,R).$$

■

Main desirable properties of scores are : continuity (C), monotonicity (M) and neutrality (N) because all alternatives should be treated equally.

Turning back to the results of Section 5.6 related to stability of aggregators, we can conclude that :
(D)&(SO)-CNM scores correspond to min or max and
(D)&(SPL)-CNM scores correspond to min or max or arithmetic mean.

The first result is very close to the characterization obtained by Bouyssou (1991) for min outflow (see Theorem 6.2).

Monotonicity (M), decomposability (D) and ordinal stability (SO) replace respectively row monotonicity, row egalitarian property and ordinality.

These two results explain why special attention should be paid to the following scores :

$$\begin{aligned} S_1(a,R) &= \min\left\{\min_{c\in A\setminus a} R(a,c), \min_{c\in A\setminus a}(1-R(c,a))\right\} \\ &= \min\left\{S_{\text{MIN}}(a,R), 1+S_{\text{MAX}}(a,R)\right\} \end{aligned}$$

$$\begin{aligned} S_2(a,R) &= \max\left\{\max_{c\in A\setminus a} R(a,c), \max_{c\in A\setminus a}(1-R(c,a))\right\} \\ &= \max\left\{\max_{c\in A\setminus a} R(a,c), \max_{c\in A\setminus a} R^d(a,c)\right\} \end{aligned}$$

$$\begin{aligned} S_3(a,R) &= \frac{1}{2n-2}\left\{\sum_{c\in A\setminus a} R(a,c)+(2n-2)-\sum_{c\in A\setminus a} R(c,a)\right\} \\ &= 1+\frac{1}{2n-2}S_{L/E}(a,R). \end{aligned}$$

$S_3(a, R)$ equivalent to the net flow score, has been also characterized by Bouyssou (1991, 1992). In this case too, results are very close (see Theorem 6.1) when strictness is introduced :
$(D)\&(SPL)$ strict CNM scores correspond to the arithmetic mean.

6.4 Aggregation rules and score functions

6.4.1 Should we aggregate first or score first ?

Suppose that a set of actions $A : \{a, b, \ldots\}$ is described and each of the alternatives is characterized by a criterion i, $i = 1, \ldots, m$.

Let R_i be the valued binary relation on A for a given criterion i and ω_i the corresponding "weight".

We want to define a global ranking of the actions. Two different procedures may be used : the pre-ranking methods and the pre-aggregation methods.

The pre-ranking methods : A score $S(a, R_i)$ is first determined for each action a based on R_i,

$$S(a, R_i) = \underset{c \in A \setminus \{a\}}{T} [R_i(a, c)] ,$$

where T represents the extended t-norm T related to a given t-norm. This is done for each criterion i, $i = 1, \ldots, m$.

The global score $S(a; R_1, \ldots, R_m) = M_\omega(S(a, R_1), \ldots, S(a, R_m))$ is related to an *aggregation rule* M_ω, i.e. a non decreasing function $M_\omega : [0,1]^m \to [0,1]$ such that $M(0, \ldots, 0) = 0$ and $M(1, \ldots, 1) = 1$.

Finally,

$$a \geq (S)b \text{ iff } S(a; R_1 \ldots, R_m) \geq S(b; R_1, \ldots, R_m).$$

It must be pointed out that we do not consider here the unanimity condition for the aggregation rule, i.e.

$$M(x, \ldots, x) = x \ , \quad \text{for all } x \in [0, 1] .$$

The pre-aggregation methods : A binary relation R on A is first determined on the basis of the set $\{R_i\}$ and an aggregation rule M_ω is considered

$$R = M_\omega(R_1, \ldots, R_m).$$

The global score $S'(a, R)$ is determined on the basis of a given extended t-norm T

$$S'(a, R) = \underset{c \in A \setminus \{a\}}{T} [R(a, c)] .$$

Finally,

$$a \geq (S')b \text{ iff } S'(a, R) \geq S'(b, R).$$

We will study in Section 6.4.2 conditions under which

$$S'(a, R) = S(a; R_1, \ldots, R_m).$$

6.4.2 Leaving flow, product and min outflow methods

The leaving flow method does not enter into the category of methods which were described in the previous section because

$$S'_L(a,R) = \sum_{c \in A \setminus \{a\}} R(a,c)$$

does not correspond to any t-norm.

However, it is interesting to compare $S'_L(a,R)$ if

$$R(a,c) = \frac{1}{\sum \omega_i} \sum_i \omega_i R_i(a,c)$$

and

$$S_L(a, R_1, \ldots, R_m) = \frac{1}{\sum \omega_i} \sum_i \omega_i S_L(a, R_i)$$

where

$$S_L(a, R_i) = \sum_{c \in A \setminus \{a\}} R_i(a,c).$$

The aggregation rule considered here corresponds to

$$M_\omega(x_1, \ldots, x_m) = \frac{1}{\sum_i \omega_i} \sum_i \omega_i x_i.$$

It is obvious that $S_L(a, R_1, \ldots, R_m) = S'_L(a,R)$ and the pre-ranking and pre-aggregation methods coincide.

We consider g-transforms of the valuations R_i where g is a strictly increasing function from $[0,1]$ to $[0,1]$. A natural extension of the leaving flow score corresponds to

$$gS'^g_L(a,R) = \left[\sum_{c \in A \setminus \{a\}} gR^g(a,c) \right]$$

where

$$R^g(a,c) = g^{-1} \left[\frac{1}{\sum \omega_i} \sum_i \omega_i g R_i(a,c) \right].$$

We can compare $gS'^g_L(a,R)$ to $gS^g_L(a, R_1, \ldots, R_m)$.

$$gS^g_L(a, R_1, \ldots, R_m) = \frac{1}{\sum \omega_i} \left[\sum_i \omega_i g S^g_L(a, R_i) \right]$$

where

$$S^g_L(a, R_i) = g^{-1} \left[\sum_{c \in A \setminus \{a\}} gR_i(a,c) \right].$$

It is once more obvious that

$$S'^g_L(a,R) = S^g_L(a, R_1, \ldots, R_m).$$

Moreover,

$$\geq (S_L^g) \Leftrightarrow \geq (gS_L^g).$$

The leaving flow method is however meaningless except for data related to R_i expressed in an interval or ratio scale, i.e. valuation data having cardinal properties because

$$\geq (S) \text{ is generally different from } \geq (S_\phi).$$

Let us now consider

$$S(a, R_1, \ldots, R_m) = \prod_{i=1}^{m} [S(a, R_i)]^{\omega_i}$$

where

$$S(a, R_i) = \prod_{c \in A \setminus \{a\}} R_i(a, c)$$

and

$$S'(a, R) = \prod_{c \in A \setminus \{a\}} R(a, c)$$

where

$$R(a, c) = \prod_{i=1}^{m} [R_i(a, c)]^{\omega_i}.$$

The aggregation rule corresponds to

$$M_\omega(x_1, \ldots, x_m) = \prod_{i=1}^{m} [x_i]^{\omega_i} \quad , \quad \omega_i \geq 0$$

and the scoring function is based on the extended product t-norm. Once more, $S(a, R_1, \ldots, R_m)$ and $S'(a, R)$ coincide.

Finally, we consider the case where $\omega_1 = \omega_2 = \cdots = \omega_m$ (voting procedure) and

$$S(a, R_1, \ldots, R_m) = \min_{i=1,\ldots,m} S(a, R_i)$$

where

$$S(a, R_i) = \min_{c \in A \setminus \{a\}} R_i(a, c).$$

If

$$S'(a, R) = \min_{c \in A \setminus \{a\}} R(a, c)$$

with

$$R(a, c) = \min_{i=1,\ldots,m} R_i(a, c),$$

$S'(a, R) = S(a, R_1, \ldots, R_m)$.

This time, the scoring function corresponds to the extended min t-norm. One question remains : how to handle problems with unequal weights ?

Answer to this question is given in the next section.

Aggregation, scoring and extended t-norms

Let us reconsider the problem posed in Section 6.4.1, i.e. conditions under which

$$S'(a,R) = S(a, R_1, \dots, R_m).$$

We are now in a position to give an appropriate answer.

Theorem 6.3 *If* $S'(a,R) = \underset{c\in A\setminus\{a\}}{T} [R(a,c)]$, T *: extended t-norm,*

$$R(a,c) = M_\omega \left[R_1(a,c), \dots, R_m(a,c)\right],$$

M *: non decreasing function for* $[0,1]^m \to [0,1]$, $M(0,\dots,0) = 0$, $M(1,\dots,1) = 1$, $\omega_i > 0$, $i = 1,\dots,m$

$$\begin{aligned} S(a, R_1, \dots, R_m) &= M_\omega \left[S(a,R_1), \dots, S(a,R_m)\right] \\ S(a, R_i) &= \underset{c\in A\setminus\{a\}}{T} \left[R_i(a,c)\right], \quad i = 1,\dots,m \end{aligned}$$

the necessary and sufficient condition to obtain

$$S'(a,R) = S(a, R_1, \dots, R_m)$$

is the following :

$$M_\omega(x_1,\dots,x_m) = T\left[f_1(x_1),\dots,f_m(x_m)\right] \tag{6.8}$$

where f_i *are non decreasing functions from* $[0,1]$ *to* $[0,1]$ *such that*

$$f_i(1) = 1 \ \textit{for any} \ i = 1,\dots,m \tag{6.9}$$

$$T(f_1(0),\dots,f_m(0)) = 0 \tag{6.10}$$

$$f_i T\left[x,y\right] = T\left[f_i(x), f_i(y)\right], \quad i = 1,\dots,m \tag{6.11}$$

Proof. Assume that M has the form (6.8). Then

$$\begin{aligned} &M\left[T(x_{11}, x_{12}, \dots, x_{1n}), T(x_{21}, x_{22}, \dots, x_{2n}), \dots, T(x_{m1}, x_{m2}, \dots, x_{mn})\right] \\ &= T\left[f_1\left[T(x_{11},\dots,x_{1n})\right], \dots, f_m\left[T(x_{m1},\dots,x_{mn})\right]\right], \end{aligned}$$

where x_{ij} corresponds to $R_i(a,c_j)$, $i = 1,\dots,m$, $\{c_j\} = A\setminus\{a\}$, $j = 1,\dots,n$.

First, we remark that associativity of T implies that if (6.11) holds, then

$$f_i\left[T(\beta_1,\dots,\beta_m)\right] = T(f_i(\beta_1),\dots,f_i(\beta_m))$$

is also satisfied for each $i = 1,\dots,m$ and $\beta_1,\dots,\beta_m \in [0,1]$. Using this equality, we can continue the previous one as follows :

$$\begin{aligned} &= T\left[T\left[f_1(x_{11}),\dots,f_1(x_{1n})\right],\dots,T\left[f_m(x_{m1}),\dots,f_m(x_{mn})\right]\right] \\ &= T\left[T\left[f_1(x_{11}), f_2(x_{21}),\dots,f_m(x_{m1})\right],\dots,T\left[f_1(x_{1n}),\dots,f_m(x_{mn})\right]\right] \\ &= T\left[M(x_{11}, x_{21},\dots,x_{m1}),\dots,M(x_{1n}, x_{2n},\dots,x_{mn})\right]. \end{aligned}$$

It is obvious that (6.9) implies $M(1,\dots,1) = 1$ and (6.10) implies $M(0,\dots,0) = 0$. If f_i are nondecreasing then so M is. Thus we proved that (6.8) provides us a solution.

To prove the opposite direction, we distinguish two different cases as follows.

CASE 1 : $m \leq n$.

Let $x_{11}, x_{22}, \ldots, x_{mm}$ be arbitrary numbers from $[0,1]$, the rests are 1. Thus

$$\begin{aligned} & M[T(x_{11}, x_{12}, \ldots, x_{1n}), T(x_{21}, x_{22}, \ldots, x_{2n}), \ldots, T(x_{m1}, x_{m2}, \ldots, x_{mn})] \\ &= T[M(x_{11}, x_{21}, \ldots, x_{m1}), M(x_{12}, x_{22}, \ldots, x_{m2}), \ldots, M(x_{1n}, x_{2n}, \ldots, x_{mn})] \end{aligned} \quad (6.12)$$

implies that

$$\begin{aligned} & M(x_{11}, x_{22}, \ldots, x_{mm}) \\ &= T[M(x_{11}, 1, 1, \ldots, 1), M(1, x_{22}, 1, \ldots, 1), \ldots, , M(1, 1, \ldots, 1, x_{mm}), 1, \ldots, 1] \\ &= T[M(x_{11}, 1, 1, \ldots, 1), M(1, x_{22}, 1, \ldots, 1), \ldots, M(1, 1, \ldots, 1, x_{mm})]. \end{aligned} \quad (6.13)$$

Let

$$f_i(x) = M(1, 1, \ldots, 1, x, 1, \ldots, 1),$$

where x is on the ith place of M. Then we have that

$$M(\alpha_1, \alpha_2, \ldots, \alpha_m) = T(f_1(\alpha_1), \ldots, f_m(\alpha_m)).$$

On the other hand, let $x_{ii} = 0$ in (6.13), $i = 1, \ldots, m$. This implies that

$$0 = T(\nu_1, \nu_2, \ldots, \nu_m). \quad (6.14)$$

It is obvious that $f_i(1) = 1$ for every $i = 1, \ldots, m$, because $M(1, 1, \ldots, 1) = 1$.

CASE 2 : $m > n$

We prove the statement by induction on m. Let $m = n + 1$. If $x_{11}, x_{22}, \ldots, x_{nn}$ are arbitrary values from $[0,1]$ and the rests are 1, then (6.12) implies that

$$M(x_{11}, x_{22}, \ldots, x_{nn}, 1) = T[M(x_{11}, 1, \ldots, 1), \ldots, M(1, \ldots, 1, x_{nn}, 1)].$$

Let

$$f_i(x) = M(1, \ldots, 1, x, 1, \ldots, 1)$$

as before, for $i = 1, \ldots, m$. We can see similarly to the Case 1 that

$$M(\beta_1, \ldots, \beta_n, 1) = T(f_1(\beta_1), \ldots, f_n(\beta_n))$$

and

$$M(\beta_1, \ldots, \beta_m) = T[M(\beta_1, \ldots, \beta_{m-1}, 1), M(1, \ldots, 1, \beta_m)]$$

for any $\beta_1, \ldots, \beta_m \in [0,1]$. This means that

$$\begin{aligned} M(\beta_1, \ldots, \beta_m) &= T[f_1(\beta_1), T[f_2(\beta_2), \ldots, f_m(\beta_m)]] \\ &= T[f_1(\beta_1), f_2(\beta_2), \ldots, f_m(\beta_m)]. \end{aligned}$$

We can prove, by induction, that M has the form (6.8) for any $m > n$. The validity of (6.10) is obvious, (6.9) can be proved as in the Case 1. We have only to prove (6.11) now.

Let x_{i1}, x_{i2} be arbitrary numbers from $[0,1]$, the others are 1. Then (6.12) implies the following equalities :

$$\begin{aligned} & M(1,\ldots,1,T\left[x_{i1},x_{i2}\right],1,\ldots,1) \\ & \quad = T\left[(M(1,\ldots,1,x_{i1},1,\ldots,1), M(1,\ldots,1,x_{i2},1,\ldots,1)\right], \end{aligned}$$

i.e.,

$$f_i(T(x_{i1},x_{i2})) = T\left[f_i(x_{i1}), f_i(x_{i2})\right].$$

Thus the theorem is proved. ∎

Let us now consider some particular cases.

6.4.3 The coherent min outflow procedure

We start with m valued relations $R_1,\ldots,R_m$ and the related weights $\omega_1,\ldots,\omega_m$. We are looking at an aggregation procedure M_ω which is coherent with the scoring function $S'(a,R_i) = \min_{c\in A\setminus\{a\}} R_i(a,c)$, i.e. $M(x_1,\ldots,x_m;\omega_1,\ldots,\omega_m)$ such that

$$S(a,R_1,\ldots,R_m) = S'(a,R).$$

Equation (6.11) becomes

$$f_i\left[\min(x,y)\right] = \min\left[f_i(x), f_i(y)\right]. \tag{6.15}$$

A feasible solution of (6.15) satisfying (6.9) and (6.10) is given by

$$f_i(x) = \max(1-\omega_i, x),\ \omega_i \in [0,1],\ \omega_j = \max_i \omega_i = 1 \tag{6.16}$$

Moreover the unanimity condition

$$M(x,\ldots,x) = x$$

is always satisfied because

$$\min_i\left[\max(1-\omega_i,x)\right] = \min\left[x, \min_{i\neq j}\max(1-\omega_i,x)\right] = x.$$

To resume the coherent minimum flow method corresponds to the following pre-ranking procedure :

Calculate $S(a,R_i) = \min_{c\in A\setminus\{a\}} R_i(a,c), \forall i.$
Determine $S(a) = \min_i\left[\max\{S(a,R_i),(1-\omega_i)\}\right]$
where $\omega_i \in [0,1]$, $\max_i \omega_i = 1.$
Rank the actions according to the rule
$a \geq b$ iff $S(a) \geq S(b)$

6.4.4 The coherent product outflow procedure

We consider now the case of an aggregation rule derived from an extended strict continuous Archimedean t-norm. We know from Theorem 1.6 that

$$T(x_1,\ldots,x_m) = \varphi^{-1}\left[\prod_i \varphi(x_i)\right].$$

Equation (6.11) becomes

$$\varphi f_i\left[\varphi^{-1}(\varphi(x).\varphi(y))\right] = \varphi(f_i(x)).\varphi(f_i(y))$$

which is transformed in

$$g_i(u.v) = g_i(u).g_i(v)\ ,\quad 0 \le u,\ v \le 1 \tag{6.17}$$

with the use of

$$g_i(z) = \varphi f_i \varphi^{-1}(z)\ ,\ \varphi(x) = u,\ \varphi(y) = v.$$

A general and unique solution of equation (6.17) can be found in Aczél (1966) as $g_i(u) = u^{\omega_i}$, if g_i is considered to be continuous.

Finally,

$$f_i(x) = \varphi^{-1}\{\varphi(x)^{\omega_i}\}\ ,\quad \omega_i \in [0,1]$$

satisfies conditions (6.9) and (6.10) and

$$M_\omega(x_1,\ldots,x_m) = \varphi^{-1}\left[\prod_i \varphi(x_i)^{\omega_i}\right].$$

If we add the unanimity condition

$$M_\omega(x,\ldots,x) = x,$$

it implies that $\sum \omega_i = 1$.

These results lead us to a pre-ranking procedure :

Calculate $S(a, R_i) = \prod_{c\in A\setminus\{a\}} R_i(a,c)\ ,\ \forall i$
Determine $S(a) = \prod_i [S(a,R_i)^{\omega_i}]$
where $\omega_i \in [0,1]\ ,\ \sum \omega_i = 1.$
Rank the actions according to the rule
$a \geq b$ iff $S(a) \geq S(b)$.

6.4.5 The coherent Lukasiewicz flow procedure is dictatorial

In this section, we assume that the scoring function

$$S(a, R_i) = \underset{c \in A \setminus \{a\}}{T} (R_i(a, c))$$

is based on a continuous Archimedean t-norm with zero divisors.

In that case, we know from Theorem 1.5 that

$$T(x_1, \ldots, x_m) = \varphi^{-1}\left[\max(\sum_{i=1}^{m} \varphi(x_i) - m + 1, 0)\right].$$

Equation (6.11) gives now

$$f_i \varphi^{-1}\{\max(\varphi(x) + \varphi(y) - 1, 0)\} = \varphi^{-1}\left[\max(\varphi f_i(x) + \varphi f_i(y) - 1, 0\right]$$

which can be transformed in

$$1 - \varphi f_i \varphi^{-1}\{\max(1 - (1 - \varphi(x)) - (1 - \varphi(y)), 0)\} = 1 - \max(\varphi f_i(x) + \varphi f_i(y) - 1, 0).$$

If we use

$$g_i(u) = 1 - \varphi f_i \varphi^{-1}(1 - u) \ , \ u = 1 - \varphi(x), \ v = 1 - \varphi(y),$$

we get

$$1 - \varphi f_i \varphi^{-1}\{\max(1 - u - v, 0)\} = 1 - \max\left[1 - g_i(u) - g_i(v), 0\right]$$

and finally

$$g_i\left[\min(u + v, 1)\right] = \min\left[g_i(u) + g_i(v), 1\right] \tag{6.18}$$

where $u, v \in [0, 1]$ (see also Dubois et al. (1994)).

Theorem 6.4 *The general solution of*

$$g(\min(u + v, 1)) = \min(g(u) + g(v), 1) \tag{6.19}$$

under the condition $g(0) = 0$ *is either* $g(x) \equiv 0$ *or* $g(x) = \min\{cx, 1\}$ *with* $c \geq 1$.

Proof. Let $t_0 = \sup\{t; g(t) < 1\}$. Since $g(0) = 0$ and g is continuous, $t_0 > 0$.
Let $t \in [0, t_0]$ and $u = v = \frac{t}{2}$ in (6.19).
Thus (6.19) implies that

$$g(t) = 2 \cdot g(\frac{t}{2}) \tag{6.20}$$

for $t \in [0, t_0]$.

(6.20) implies $g(\frac{t_0}{2}) \leq \frac{1}{2}$. If $x, y \in [0, \frac{t_0}{2}]$ then, by (6.19),

$$g(x + y) = g(x) + g(y). \tag{6.21}$$

Let $t = x + y$ in (6.20). Then, using also (6.21), we get

$$\frac{g(x) + g(y)}{2} = g(\frac{x + y}{2}) \tag{6.22}$$

for $x, y \in [0, \frac{t_0}{2}]$.

The general solution of (6.22) is

$$g(x) = c \cdot x \ , \ c \geq 0 \ \text{ on } \ \left[0, \frac{t_0}{2}\right],$$

see Aczél (1966), Theorem 1 on page 46.

Using (6.20) :

$$g(t) = 2 \cdot g(\frac{t}{2}) = 2 \cdot c \cdot \frac{t}{2} = c \cdot t \quad \text{on} \quad [0, t_0] \, .$$

If $t_0 < 1$ then $g(t) = 1$ on $[t_0, 1]$, thus indeed $g(x) = \min\{cx, 1\}$.

Assume that $c < 1$.

Let $u + v > 1$ and $c(u + v) < 1$ (for example, $u + v = \frac{1}{2}(1 + \frac{1}{c})$). Then by (6.20) :

$$\begin{aligned} c = g(1) &= \min\{\min\{cu, 1\} + \min\{cv, 1\}, 1\} \\ &= \min\{cu + cv, 1\} = c(u + v). \end{aligned}$$

This is impossible when $c > 0$.

Thus $c < 1$ implies $c = 0$.

The last result implies that in (6.18)

$$g_i(u) = \min\{c_i u, 1\} \ \text{ or } \ 0 \ , \ \forall i = 1, \ldots, m, \ \ c_i \geq 1$$

which means that

$$f_i(x) = \varphi^{-1}(\max(0, 1 - c_i + c_i\varphi(x)) \ \text{ or } \ 1 \ , \ \forall i = 1, \ldots, m, c_i \geq 1.$$

In order to satisfy (6.10), at least one j ($j = 1, \ldots, m$) must give $f_j(x) \neq 1$.

Denote $I_k = \{i_1, \ldots, i_k\}$ ($1 \leq k \leq m$) those indices i for which $f_i(x) \neq 1$. Then, according to (6.8) :

$$M_\lambda(x_1, \ldots, x_m) = \varphi^{-1}\left[\max\left\{\sum_{i \in I_k} \max\{1 - c_i + c_i\varphi(x_i), 0\} - k + 1, 0\right\}\right].$$

If we add the unanimity condition $M(x, \ldots, x) = x$, $\forall x \in [0, 1]$, the last formula implies

$$\left(\sum_{i \in I_k} c_i\right) [\varphi(x) - 1] = \varphi(x) - 1, \quad \forall x.$$

This is true only if $\sum_{i \in I_k} c_i = 1$, i.e., only if there exists an index j such that $c_j = 1$ and $c_i = 0$ otherwise. This implies the existence of a dictatorial criterion j.

∎

Chapter 7

Multiple criteria decision making

7.1 Introduction

In this Chapter, we imagine a decision maker (or a group of experts) trying to establish or examine fair procedures to combine opinions about alternatives related to different points of view.

In some situation, it seems more useful to produce a ranking of all the alternatives (ranking procedure) than it is to just produce a clustering of the elements but both directions will be examined (see also Kitainik (1993)).

All procedures are based on pair comparisons, in the sense that processes are linked to some degree of credibility of preference of any alternative over any other alternative even if the input data corresponds to an evaluation $g_j(a)$ related to each alternative $a \in A$ considered for each point of view $j \in J$. We deliberately exclude methods dealing with some combinations of these evaluations. It consists of building a single criterion g by using an aggregation function u by letting :

$$g(a) = u\left[g_1(a), \ldots, g_j(a), \ldots, g_m(a)\right].$$

Any pair of alternatives can thus be compared in a transitive way on the basis of g.

These kinds of methods are studied in the framework of the *multiple attribute utility theory* (MAUT) but also in recent multiple criteria methods based on fuzzy sets aggregations (Dubois and Prade (1982,1990), Dubois, Fodor, Prade and Roubens (1994), Weber (1984)). For more detailed treatment of utility theory, the reader is referred to Fishburn (1970 b), Keeney and Raiffa (1976).

Formally, suppose we have a set of alternatives A, to be labelled $a, b, c, \ldots$ Let J be the set of points of view (usually called criteria) labelled $j = 1, \ldots, m$, each of them having a weight ω_j.

A preference $R_j(a, b)$ reflects the degree (between 0 and 1) to which a is declared to be not worse than b for point of view j. The discredit $D_j(a, b)$ indicates the degree (between 0 and 1) to which a cannot be considered as being preferred to b for point of view j.

The valued binary relations R_j and D_j usually present a structure of fuzzy interval order.

The classical MCDM (Multiple Criteria Decision Making) procedures perform generally in two steps : aggregation and exploitation.

The aggregation phase defines an outranking relation

$$R = M(R_1, \ldots, R_m, D_1, \ldots, D_m; \omega_1, \ldots, \omega_m)$$

which indicates the global preference between every ordered pair of alternatives taking into consideration the weights of the different points of view. R generally does not present any remarkable structure and there is a need to transform R into some other structure for ranking purposes.

The exploitation phase transforms the global information about the alternatives into a global ranking of the elements of A. Usually, three different ways are used :

(W1) : transform R into another valued relation $\hat{R}$ which presents some interesting property needed for ranking purposes, i.e. transitivity.
(W2) : determine a crisp binary relation close to R which presents crisp properties needed for ordering.
(W3) : use a ranking method to obtain a score function.

(W3) is most commonly used in classical procedures like ELECTRE III (Roy (1978)) and PROMETHEE (Brans and Vincke (1985)).

We propose an alternative way : the ranking is given in terms of a fuzzy quasiorder which preserves the valuation until the end of the exploitation phase (Section 7.4).

7.2 Monocriterion binary preference relation

In this section, we are considering a valued binary relation R_j where $R_j(a,b)$ represents the intensity or degree of preference of alternative a over alternative b for point of view $j : 0 \leq R_j \leq 1$.

$R_j(a,b) = 0$ means that there is no preference of a over b.

$R_j(a,b) = 1$ means that a is definitely preferred to b.

$R_j(a,b) > R_j(c,d)$ means that the preference of a over b is more credible than the preference of c over d.

R_j is a probabilistic relation if $R_j + R_j^{-1} = 1$.

R_j is a crisp relation of R_j equals 0 or 1.

We will see how to build such a relation if an evaluation (together with an indifference and a preference threshold) or a fuzzy number is given for each alternative and a given point of view.

In the first part (Section 7.2.1), the valuation refers to the intensity of the preference of a over b which increases when the difference between crisp evaluations of a and b is increasing. In this context, the scales of measurement involved are precise.

In the second part (Section 7.2.2), fuzzy sets are considered to cope with the uncertainty of the evaluation. The scales of measurement correspond to imprecise measures or linguistic labels. We are dealing with a fuzzy performance which is interpreted as the degree of possibility that the evaluation takes a given value.

7.2.1 An evaluation corresponds to each alternative for a given point of view (measurement is precise)

Suppose that $g_j(a)$ represents an evaluation (physical or monetary value) of alternative a for point of view j, i.e. a function that associates each alternative a of the set A with a real number indicating the performance of that alternative according to point of view j. The function g_j is called a criterion. Without loss of generality, we will suppose that the higher the evaluation, the better the alternative satisfies the expert.

The *intensity of preference* (shortly the *preference*) of a over b for point of view j is given by

$$R_j(a,b) = f_j\left[g_j(a), Ng_j(b)\right] \tag{7.1}$$

where f_j is a non decreasing function of both arguments and N is a strong negation.

The preference of a over a should be equal to one. This implies that R_j is reflexive and

$$R_j(a,a) = f_j\left[g_j(a), Ng_j(a)\right] = 1.$$

The following proposition has been obtained by Perny (1992)

Proposition 7.1 *The preference $R_j(a,b)$ defined with (7.1) is a fuzzy semiorder (R_j is a reflexive, complete, semitransitive and Ferrers valued relation).*

Proof. R_j is obviously complete.

Suppose that $g_j(a) \geq g_j(b)$. Then

$$R_j(a,b) = f_j\left[g_j(a), Ng_j(b)\right] \geq f_j\left[g_j(b), Ng_j(b)\right] = 1.$$

Suppose now that $g_j(a) \leq g_j(b)$:

$$R_j(b,a) = f_j\left[g_j(b), Ng_j(a)\right] \geq f_j\left[g_j(b), Ng_j(b)\right] = 1.$$

Finally, $\max\left[R_j(a,b), R_j(b,a)\right] = 1$.

R_j is semitransitive.

Suppose that the property does not hold. For at least one quadruple, a, b, c, d :

$$\min\left[R_j(a,b), R_j(b,d)\right] > \max\left[R_j(a,c), R_j(c,d)\right]$$

and $R_j(a,b) > R_j(a,c)$, $R_j(b,d) > R_j(c,d)$. Consequently,

$$f_j\left[g_j(a), Ng_j(b)\right] > f_j\left[g_j(a), Ng_j(c)\right]$$

and

$$f_j\left[g_j(b), Ng_j(d)\right] > f_j\left[g_j(c), Ng_j(d)\right].$$

We then have :

$$Ng_j(b) > Ng_j(c) \text{ or } g_j(b) < g_j(c) \text{ and } g_j(b) > g_j(c), \text{ a contradiction.}$$

Using the same arguments,it is easily proved that R_j is a Ferrers relation. ■

Corollary 7.1 *Every λ-cut of R_j defined by (7.1) is a crisp complete semiorder.*

Proof. The proof is obvious.

Let us now consider a particular case of (7.1) where f_j is a non decreasing function of $g_j(a)$ and $(-g_j(b))$.

We define after Roy (1978) the *concordance index*

$$R_j(a,b) = \frac{PT\,[g_j(a)] - \min(g_j(b) - g_j(a),\ PT\,[g_j(a)])}{PT\,[g_j(a)] - \min((g_j(b) - g_j(a),\ IT\,[g_j(a)])} \tag{7.2}$$

where PT and IT are non decreasing functions of the argument $g_j(a)$ and correspond respectively to a *preference threshold* and an *indifference threshold* which may vary along the scale of the criterion. The concordance between a and b depends on the *difference* between the evaluations $g_j(a)$ and $g_j(b)$, expressing the performance of the alternatives a and b under j.

The shape of $R_j(a,b)$ and $R_j(b,a)$ is represented in Figure 7.1 as a function of $g_j(b)$.

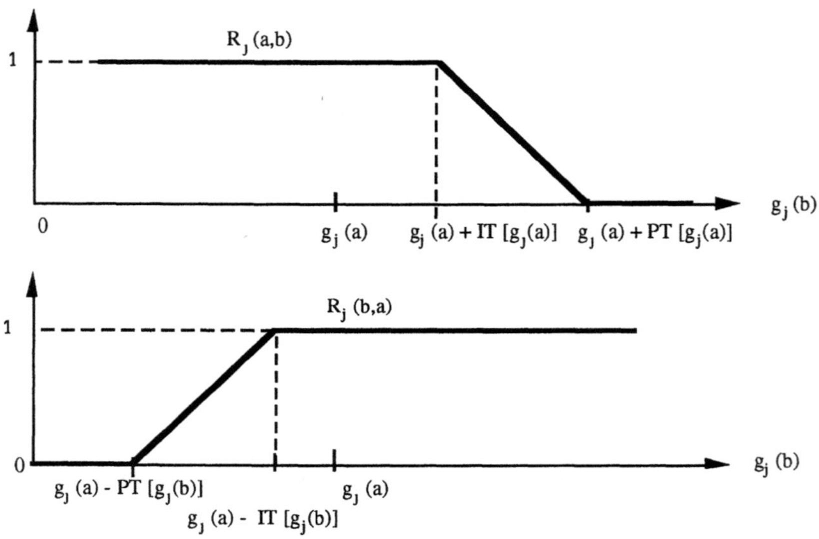

Fig. 7.1

It is important to control the *meaningfulness* of the concordance index. Briefly, following the ideas of Suppes (1959) and Roberts (1985), an index involving scales of measurement is called meaningful if its value is unchanged whenever all scales used in the index are transformed by admissible transformations. Looking at the index of concordance, we obtain the next result.

Proposition 7.2 *R is meaningful if g_j is an interval scale with range R, if the indifference threshold corresponds to a constant or a proportion of g_j and if the preference threshold is expressed as a proportion of g_j.*

Proof. Admissible transformations for an interval scale correspond to $g_j' = rg_j + t$, $r > 0$. Suppose now that

$$IT\,[g_j(a)] = C \text{ (or } \alpha g_j(a)) \text{ and } PT\,[g_j(a)] = \beta g_j(a), \quad 0 \leq \alpha < \beta.$$

We have

$$
\begin{aligned}
R_j(a,b) &= 1 \text{ if } g_j(b) \leq g_j(a) + IT\left[g_j(a)\right], \\
&= 0 \text{ if } g_j(b) \geq g_j(a) + PT\left[g_j(a)\right], \\
&= \frac{PT\left[g_j(a)\right] - \left[g_j(b) - g_j(a)\right]}{PT\left[g_j(a)\right] - IT\left[g_j(a)\right]}, \quad \text{otherwise.}
\end{aligned}
$$

Using the admissible transformations, it comes

$$IT\left[g'(a)\right] = rC \text{ (or } \alpha rg(a)) \ , \ PT\left[g'(a)\right] = \beta rg(a).$$

Clearly,

$$
\begin{aligned}
g_j(b) \leq g_j(a) + IT\left[g_j(a)\right] &\Rightarrow g_j'(b) \leq g_j'(a) + IT\left[g_j'(a)\right], \\
g_j(b) \geq g_j(a) + PT\left[g_j(a)\right] &\Rightarrow g_j'(b) \geq g_j'(a) + PT\left[g_j'(a)\right],
\end{aligned}
$$

$$\frac{PT\left[g_j(a)\right] - \left[g_j(b) - g_j(a)\right]}{PT\left[g_j(a)\right] - IT\left[g_j(a)\right]} = \frac{PT\left[g_j'(a)\right] - \left[g_j'(b) - g_j'(a)\right]}{PT\left[g_j'(a)\right] - IT\left[g_j'(a)\right]},$$

and R_j is invariant. ■

Example 7.1 (Perny, 1992 and Skalka and al., 1992) Let us consider in Table 7.1 the maximum speed (ratio scale) of eight cars with their thresholds.

Maximum speed and their thresholds.

Type	max. speed : g_s in km/hour	IT = 5 in km/h	PT = .10 g_s in km/h
1. VW Golf C	140	5	14
2. Renault R9 GTL	150	5	15
3. Citroën GSA X1	160	5	16
4. Peugeot P305 GR	153	5	15.3
5. Talbot HOR.GLS	164	5	16.4
6. Audi 80 CL	148	5	14.8
7. Renault R18 GTL	155	5	15.5
8. Alfa SUD TI-NR	170	5	17

Table 7.1

For the Renault R9 (car 2), we obtain easily $R_s(2,k)$, $k = 1, \ldots, 8$, with the diagram of Figure 7.2

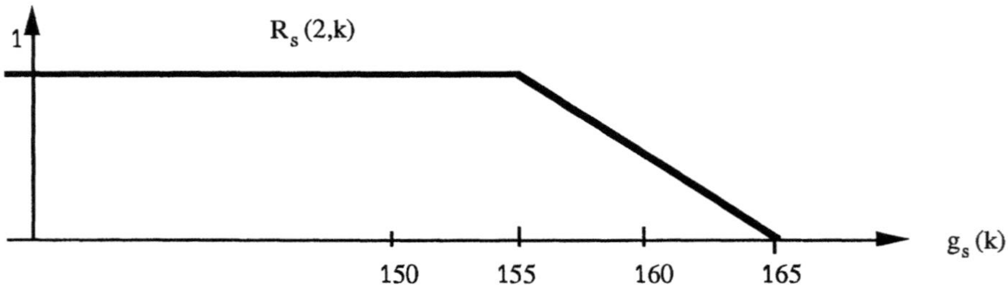

Fig. 7.2

and $R_s(2,1) = 1$, $R_s(2,2) = 1$, $R_s(2,3) = .5$, $R_s(2,4) = 1$, $R_s(2,5) = .1$, $R_s(2,6) = 1$, $R_s(2,7) = 1$, $R_s(2,8) = 0$.

The complete tableau indicating the concordance indices for the maximum speed is shown in Table 7.2.

Concordance index for maximum speed of 8 cars

	1	2	3	4	5	6	7	8
1	1	.44	0	.11	0	.67	0	0
2	1	1	.50	1	.10	1	1	0
3	1	1	1	1	1	1	1	.55
4	1	1	.81	1	.42	1	1	0
5	1	1	1	1	1	1	1	.91
6	1	1	.29	1	0	1	.80	0
7	1	1	1	1	.62	1	1	.05
8	1	1	1	1	1	1	1	1

Table 7.2

From the valued complete relation R_j we now define the intensity of strict preference of a over b, $P_j(a,b)$ and the intensity of indifference of a and b, $I_j(a,b)$. The intensity of incomparability is always zero because R_j is complete.

Two major axiomatics were proposed in Chapter 3.

$$\begin{cases} P_j(a,b) = \max(R_j(a,b) - R_j^{-1}(a,b), 0) \\ I_j(a,b) = \min(R_j(a,b)\ ,\ R_j^{-1}(a,b)) \end{cases}$$

$$\begin{cases} P_j(a,b) = \min(R_j(a,b)\ ,\ 1 - R_j^{-1}(a,b)) \\ I_j(a,b) = \max(R_j(a,b) + R_j(b,a) - 1, 0) \end{cases}$$

In this case, both give the same result

$$\begin{aligned} P_j(a,b) &= R_j^d(a,b) = 1 - R_j(b,a) \\ I_j(a,b) &= \min(R_j(a,b), R_j(b,a)). \end{aligned}$$

Using Figure 7.1, we easily obtain the graphical representation of (P_j, I_j) as in Figure 7.3.

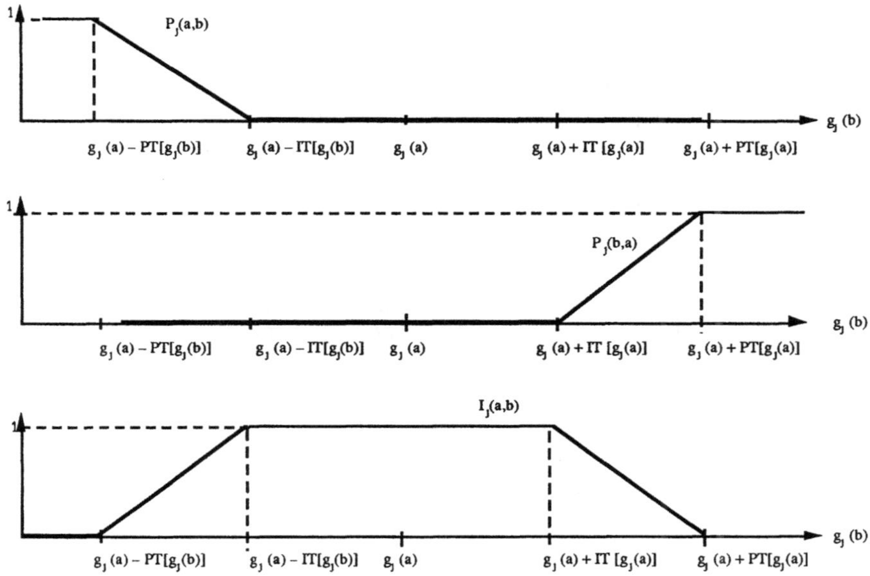

Fig. 7.3

Corollary 7.2 *The intensity of strict preference as defined with the dual relation of R_j is a fuzzy partial order (irreflexive, antisymmetric and transitive valued relation).*

Proof. Immediate.

Considering the example 7.1, we obtain in Table 7.3 the intensity of strict preference P for maximum speed of eight cars and its related support (λ-cut of P_j with $\lambda = \min\{P_j > 0\}$).

Intensity of strict preference for
maximum speed of eight cars

	1	2	3	4	5	6	7	8
1	0	0	0	0	0	0	0	0
2	.56	0	0	0	0	0	0	0
3	1	.5	0	.19	0	.71	0	0
4	.89	0	0	0	0	0	0	0
5	1	.9	0	.58	0	1	.38	0
6	.33	0	0	0	0	0	0	0
7	1	0	0	0	0	.20	0	0
8	1	1	.45	1	.09	1	.95	0

Supp(P_s)

	1	2	3	4	5	6	7	8
1	0	0	0	0	0	0	0	0
2	1	0	0	0	0	0	0	0
3	1	1	0	1	0	1	0	0
4	1	0	0	0	0	0	0	0
5	1	1	0	1	0	1	1	0
6	1	0	0	0	0	0	0	0
7	1	0	0	0	0	1	0	0
8	1	1	1	1	1	1	1	0

Table 7.3

The support of P_s is a partial order (every λ-cut of a fuzzy partial order is a crisp partial order) as it can be seen with the the use of the Hasse diagram presented in Figure 7.4

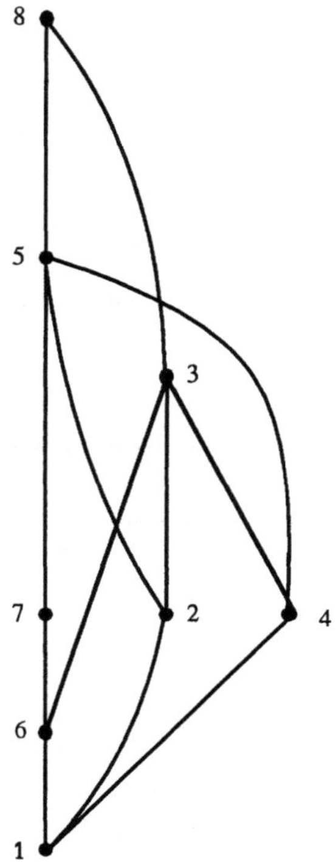

Fig. 7.4

The range of variation of $g_j(b)$ can be divided in five zones :

(1) $g_j(b) \geq g_j(a) + PT\,[g_j(a)]$, where

$$\underline{P_j(b,a) = 1} \text{ and } P_j(a,b) = I_j(a,b) = 0.$$

(2) $g_j(a) + IT\,[g_j(a)] < g_j(b) < g_j(a) + PT\,[g_j(a)]$, where

$$\underline{0 < P_j(b,a) < 1},\ \underline{0 < I_j(a,b) < 1},\ P_j(a,b) = 0.$$

(3) $g_j(a) - IT\,[g_j(b)] \leq g_j(b) \leq g_j(a) + IT\,[g_j(a)]$, where

$$\underline{I_j(a,b) = 1} \text{ and } P_j(a,b) = P_j(b,a) = 0.$$

(4) $g_j(a) - PT\,[g_j(b)] < g_j(b) < (a) - IT\,[g_j(b)]$, where

$$\underline{0 < P_j(a,b) < 1}\ ,\ \underline{0 < I_j(a,b) < 1}\ ,\ P_j(b,a) = 0.$$

(5) $g_j(b) \leq g_j(a) - PT\,[g_j(b)]$, where

$$\underline{P_j(a,b) = 1} \text{ and } P_j(b,a) = I_j(a,b) = 0.$$

To the valued pair (P_j, I_j) is associated a crisp triple (P, Q, I) with the following assignment :

bPa (b is strongly preferred to a) in zone (1)
bQa (b is weakly preferred to a) in zone (2)
aIb (a is indifferent to b) in zone (3)
aQb (a is weakly preferred to b) in zone (4)
aPb (b is strongly preferred to a) in zone (5)

which corresponds to the decomposition in Figure 7.5.

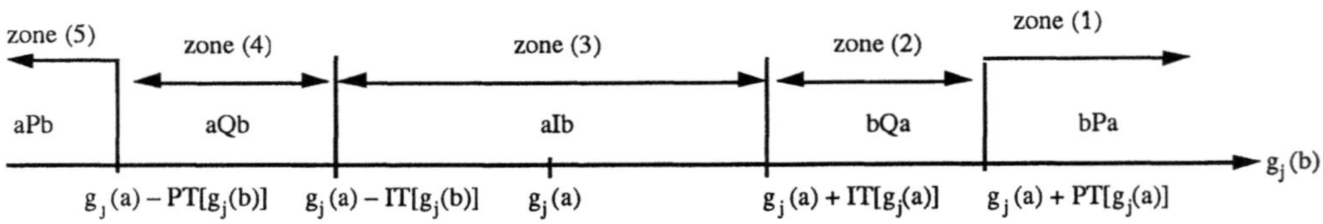

Fig. 7.5

The crisp preference structure (P, Q, I) is called a crisp *pseudo-order* and has been extensively studied by Roy and Vincke (1984,1987) and Vincke (1988). It explains why the criterion g_j linked to the (P, Q, I) structure is called a *pseudo-criterion*.

If $IT\,[g_j(a)]$ and $PT\,[g_j(a)]$ are constant, IT_j and PT_j, we can express the functions of Fig. 7.3 in terms of $g_j(a) - g_j(b)$ to obtain Figure 7.6.

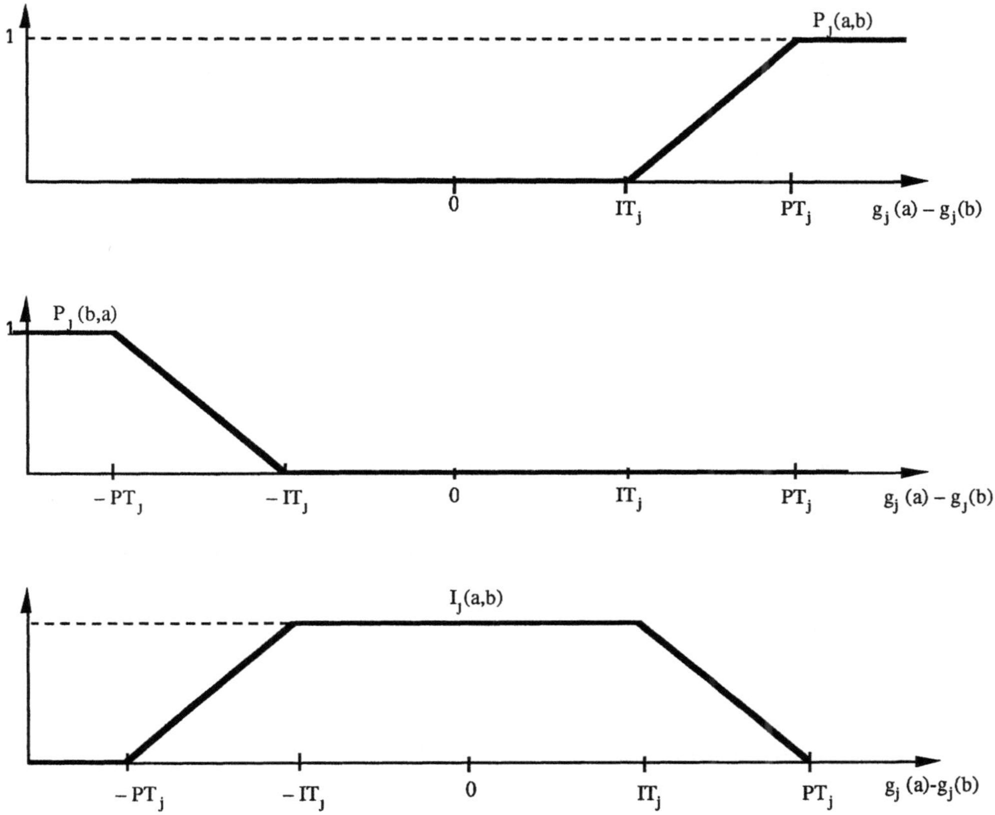

Fig. 7.6

$P_j(a,b)$ and $P_j(b,a)$ are used in this form in the method PROMETHEE (Brans and Vincke, 1985).

Particular cases of a pseudo-criterion are called : interval-criterion, semi-criterion and real criterion.

An *interval-criterion* is associated to a concordance index such that $IT\,[g_j(a)] = PT\,[g_j(a)] = T\,[g_j(a)]$. In that particular case, $R_j(a,b)$ corresponds to a crisp relation R_j :

$$\begin{aligned} R_j(a,b) &= 1\ (aR_jb)\ \text{ iff }\ g_j(b) \leq (a) + T\,[g_j(a)] \\ &= 0\ (\text{not } aR_jb)\ \text{otherwise.} \end{aligned}$$

As an immediate consequence, we have

$$\begin{array}{lll} aP_jb(aR_jb \text{ and not } bR_ja) & \text{iff} & g_j(a) > g_j(b) + T\,[g_j(b)] \\ bP_ja(bR_jb \text{ and not } aR_jb) & \text{iff} & g_j(b) > g_j(a) + T\,[g_j(a)] \end{array}$$

$$aI_jb(aR_jb \text{ and not } bR_ja) \quad \text{iff} \quad g_j(a) \leq g_j(b) + T\,[g_j(b)]$$
$$\text{and} \quad g_j(b) \leq g_j(a) + T\,[g_j(a)]\,.$$

These inequalities characterize an interval order (Fishburn (1970a), Roberts (1976,1979), Roubens and Vincke (1985)).

A *semi-criterion* corresponds to a concordance index defined on the basis of a constant threshold T and

$$IT\,[g_j(a)] = PT\,[g_j(a)] = T.$$

In that case : $R_j(a,b) = 1$ iff $g_j(b) \leq g_j(a) + T$, and

$$\begin{aligned} aP_jb &\text{ iff } g_j(a) > g_j(b) + T \\ bP_ja &\text{ iff } g_j(b) > g_j(a) + T \\ aI_jb &\text{ iff } |g_j(b) - g_j(a)| \leq T. \end{aligned}$$

These equations characterize a semiorder (Scott and Suppes (1958), Roberts (1976, 1979), Roubens and Vincke (1985)).

We may think of a criterion g_j with no threshold to define a *true criterion* :

$$R_j(a,b) = 1 \text{ iff } g_j(a) \geq g_j(b).$$

Then indifference corresponds to the crisp binary relation I_j defined by

$$aI_jb \text{ iff } g_j(a) = g_j(b)$$

and preference corresponds to the strict inequality

$$aP_jb \text{ iff } g_j(a) > g_j(b).$$

In order to emphasize that a very important value of an evaluation (related to b) might bring into disrepute a low value of another evaluation (related to a), it seems interesting to add to the information $R_j(a,b) = 0$, a *degree of discredit* (shortly a *discredit*) of a over b for point of view j :

$$D_j(a,b) = h_j\,[g_j(b), Ng_j(a)] \tag{7.3}$$

where h_j is a non decreasing function of both arguments and N is a strong negation, with $D_j(a,a) = 0$ and $\min\,[R_j(a,b), D_j(a,b)] = 0$.

This implies that

$$R_j(a,b) > 0 \Rightarrow D_j(a,b) = 0 \ \ (\text{or } ND_j(a,b) = 1).$$

A non zero preference and a non zero discredit cannot coexist.

The discredit of a over b for point of view j is called after Roy (1978) in ELECTRE III, the valued *discordance index* if h_j is a linear function with parameters PT and VT where VT corresponds to a *veto threshold* :

$$D_j(a,b) = \min\left\{1, \max\left(0, \frac{g_j(b) - g_j(a) - PT\,[g_j(a)]}{VT(g_j(a)) - PT(g_j(a))}\right)\right\}. \tag{7.4}$$

The veto threshold expresses also that the existence of a discordant point of view ($D_j(a,b) = 1$) prohibits accepting the idea that a is globally preferred to b, as it will be seen in the next section concerning aggregation.

The shapes of $R_j(a,b)$ and $D_j(a,b)$ are represented in Figure 7.7 :

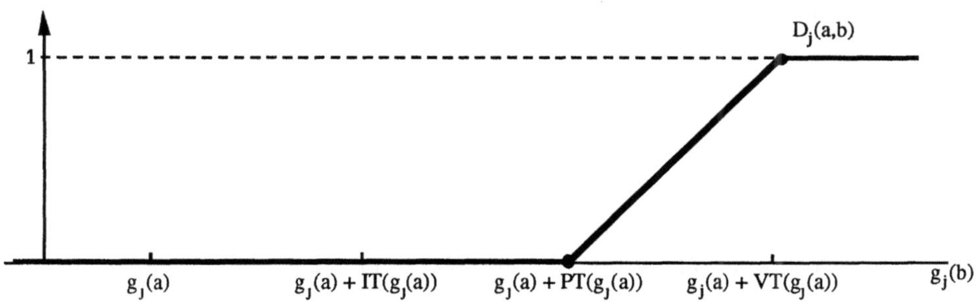

Fig. 7.7

A *crisp* discordance index was introduced by Roy (1968) in ELECTRE I as

$$D_j(a,b) = \begin{cases} 1 & \text{if } g_j(b) \geq g_j(a) + VT_j \\ 0 & \text{otherwise} \end{cases}$$

The discredit $D_j(a,b)$ presents some nice properties summarized in the following proposition.

Proposition 7.3 *The discredit $D_j(a,b)$ defined with (7.3) is a fuzzy partial order (irreflexive, antisymmetric and transitive valued relation).*

Proof. Suppose that $g_j(a) \geq g_j(b)$,

$$D_j(a,b) = h_j\left[g_j(b), Ng_j(a)\right] \leq h_j\left[g_j(a), Ng_j(a)\right] = 0.$$

Suppose now that $g_j(a) \leq g_j(b)$,

$$D_j(b,a) = h_j\left[g_j(a), Ng_j(b)\right] \leq h_j\left[g_j(b), Ng_j(b)\right] = 0.$$

D_j is antisymmetric : $\min\left[D_j(a,b), D_j(b,a)\right] = 0$.

D_j is a semitransitive and Ferrers relation. The proof follows the same lines as in Proposition 7.1

D_j being semitransitive and irreflexive is transitive.

Semitransitivity implies : $\min\left[R_j(a,b), R_j(b,c)\right] \leq \max\left[R_j(a,c), R_j(c,c)\right]$ and $R_j(a,c) \geq \min\left[R_j(a,b), R_j(b,c)\right]$, $\forall a,b,c \in A$.

■

Corollary 7.3 *Every λ-cut of D_j defined with (7.3) is a crisp partial order.*

Proof. Immediate.

Corollary 7.4 *The non-discordance index (ND_j) as defined with (7.4) is a fuzzy semiorder (ND_j is a reflexive, complete, semitransitive and Ferrers relation).*

Proof.

$$D_j(a,a) = 0 \Rightarrow ND_j(a,a) = 1;$$

ND_j is reflexive.

$$\min\left[D_j(a,b), D_j(b,a)\right] = 0 \Rightarrow N\max(ND_j(a,b), ND_j(b,a)) = 0,$$

or

$$\max\{ND_j(a,b), ND_j(b,a)\} = 1;$$

ND_j is complete.

$$\begin{aligned}
\min\{D_j(a,b), D_j(b,c)\} &\leq \max\{D_j(a,d), D_j(d,c)\} \Rightarrow \\
N\max\{ND_j(a,b), ND_j(b,c)\} &\leq N\min\{ND_j(a,d), ND_j(d,c)\} \Rightarrow \\
\min\{ND_j(a,d), ND_j(d,c)\} &\leq \max\{ND_j(a,b), ND_j(b,c)\}, \forall a,b,c,d \in A;
\end{aligned}$$

ND_j is semitransitive.

The proof is the same for Ferrers property.

■

Example 7.2 (Perny, 1992, and Skalka and al., 1992) Let us come back to the example 7.1 completed with a veto threshold

$$VT(g_s(a)) = 10 + .10g_s(a).$$

We successively obtain Table 7.4 giving preference and veto thresholds and Table 8.5 showing the discordance indices :

Maximum speed and their
preference and veto thresholds

Type	max speed : g_s in km/h	PT $= .10g_s$ in km/h	VT $= 10 + .10g_s$ in km/h
1. VW Golf C	140	14	24
2. Renault R9 GTL	150	15	25
3. Citroën GSA X1	160	16	26
4. Peugeot P305 GLS	153	15.3	25.3
5. Talbot HOR.GLS	164	16.4	26.4
6. Audi 80 CL	148	14.8	24.8
7. Renault R18 GTL	155	15.5	25.5
8. Alfa SUD TI-NR	170	17	27

Table 7.4

Discordance index for maximum speed of 8 cars

	1	2	3	4	5	6	7	8
1	0	0	.60	0	1	0	.10	1
2	0	0	0	0	0	0	0	.50
3	0	0	0	0	0	0	0	0
4	0	0	0	0	0	0	0	.17
5	0	0	0	0	0	0	0	0
6	0	0	0	0	.12	0	0	.72
7	0	0	0	0	0	0	0	0
8	0	0	0	0	0	0	0	0

Table 7.5

7.2.2 A fuzzy interval corresponds to each alternative for a given point of view (measurement is imprecise)

We may decide to translate imprecision or ambiguity in the measurement of an alternative for a given point of view with the use of a *fuzzy set* in R which is a set of ordered pairs $\{x, \mu_j^a(x)\}$ where $\mu_j^a(x)$ is termed "the grade of membership of x" for alternative a related to j.

The grade of membership can be interpreted as the degree of compatibility (degree of possibility) of the measurement of a with x. The fuzzy set will be supposed to be *normal* ($\sup_x \mu_j^a(x) = 1$) and *convex* ($\forall x, y, z \in R,\ y \in [x, z],\ \mu_j^a(y) \leq \min\{\mu_j^a(x), \mu_j^a(z)\}$).

In the case of a *fuzzy interval* (normal, convex fuzzy set), every (strong) λ-cut of $\mu_j^a(x)$ is a closed interval $I_{j,\lambda}^a : \{x : \mu_j^a(x) \geq \lambda\}$ and the information about the alternatives, given in a λ level, corresponds to an interval order on A.

A particular example of a fuzzy interval corresponds to a trapezoidal fuzzy number (see Figure 7.8) defined with parameters $(a_j^-, a_j^+, \sigma_j^-(a), \sigma_j^+(a))$ and the grade of membership

$$\begin{aligned}
\mu_j^a(x) &= 1 - \frac{a_j^- - x}{\sigma_j^-(a)} && \text{if } a_j^- - \sigma_j^-(a) \leq x \leq a_j^- \\
&= 1 && \text{if } a_j^- \leq x \leq a_j^+ \\
&= 1 - \frac{x - a_j^+}{\sigma_j^+(a)} && \text{if } a_j^+ \leq x \leq a_j^+ + \sigma_j^+(a) \\
&= 0 \text{ otherwise.}
\end{aligned}$$

The *kernel* of μ_j^a corresponds to the interval $\left[a_j^-, a_j^+\right]$.

The *support* of μ_j^a corresponds to the interval $\left[a_j^- - \sigma_j^-(a), a_j^+ + \sigma_j^+(a)\right]$.

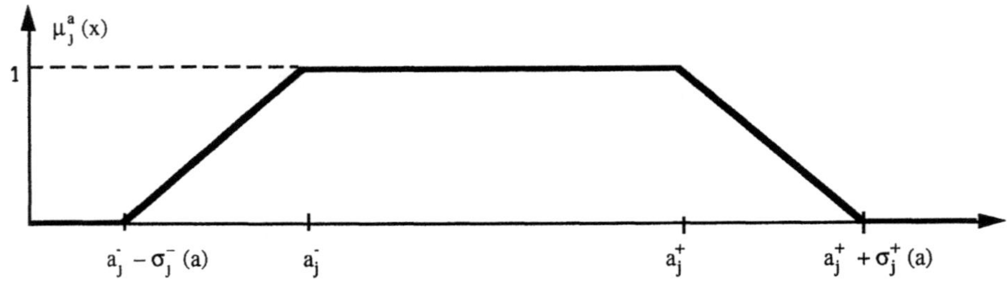

Fig. 7.8

The *degree of credibility* (shortly *credibility*) of the preference of a over b that refers to uncertainty linked to the measurement is obtained from the comparison of the fuzzy intervals corresponding to a and b.

A comparison method should satisfy the following requirements :

- it should be sensitive to the specific range and shape of the grades of membership
- it should be independant of the irrelevant alternatives (the order relation between a and b should not depend on the existence of another fuzzy interval c)
- it should satisfy transitivity, as a natural extension of the crisp property : "if a is preferred to b and if b is preferred to c, then a is preferred to c".

Two procedures will be proposed. The first one deals with possibility measure and the second one takes into account the shapes of the membership functions and satisfies the three desirable conditions.

The *credibility of the preference* of a over b for point of view j can be defined as the possibility that $a \geq b$:

$$\Pi_j(a \geq b) = \bigvee_{x \geq y} \left[\mu_j^a(x) \wedge \mu_j^b(y)\right] = \sup_{x \geq y} \left[\min(\mu_j^a(x), \mu_j^b(y))\right] \tag{7.5}$$

The possibility that $a \geq_j b$ has been considered by many scholars to compare fuzzy numbers (Tong and Bonissone (1980), Dubois and Prade (1983), Buckley (1985), Roubens and Vincke (1988)). The following properties has been obtained by Roubens and Vincke (1988) and in a revisited way by Dubois and Prade (1991a).

Proposition 7.4 *The credibility as defined with (7.5) is a fuzzy interval order (Π_j is a reflexive, complete and Ferrers valued relation) and*

$$\min(\Pi_j(a,b), \Pi_j(b,a)) = \sup_x \min(\mu_j^a(x), \mu_j^b(x)).$$

Proof. Omitted

Corollary 7.5 *Every λ-cut of Π_j defined by (7.5) is a crisp complete interval order.*

Proof. Immediate.

If μ^a is a symmetrical fuzzy interval ($\sigma_j^+(a) = \sigma_j^-(a)$), it is always possible to introduce in a formal way a valuation $g_j(a)$ and thresholds $IT\,[g_j(a)]$ and $PT\,[g_j(a)]$ such that

$$\begin{aligned} a_j^+ &= g_j(a) + \frac{IT\,[g_j(a)]}{2} \\ a_j^+ + \sigma_j^+(a) &= g_j(a) + \frac{PT\,[g_j(a)]}{2} \\ a_j^- &= g_j(a) - \frac{IT\,[g_j(a)]}{2} \\ a_j^- - \sigma_j^-(a) &= g_j(a) - \frac{PT\,[g_j(a)]}{2}. \end{aligned}$$

The lengths of the kernel and the support of the fuzzy interval μ_j^a are respectively equal to $IT\,[g_j(a)]$ and $PT\,[g_j(a)]$. We then obtain (see Figure 7.9) :

$$g_j(b) - g_j(a) \leq \frac{IT\,[g_j(a)] + IT\,[g_j(b)]}{2}$$

implies

$$\Pi_j(a \geq b) = 1;$$

$$IT\,[g_j(a)] + IT\,[g_j(b)] < g_j(b) - g_j(a) < \frac{PT\,[g_j(a)] + PT\,[g_j(b)]}{2}$$

implies

$$\Pi_j(a \geq b) = h_j(a \cap b) = \sup_x \min\left[\mu_j^a(x), \mu_j^b(x)\right];$$

$$g_j(b) - g_j(a) \geq \frac{PT\,[g_j(a)] + PT\,[g_j(b)]}{2}$$

implies

$$\Pi_j(a \geq b) = 0.$$

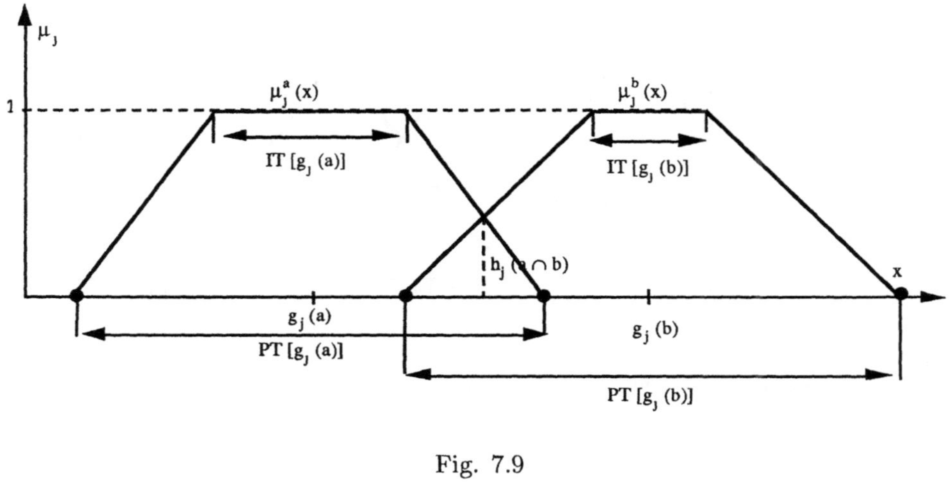

Fig. 7.9

Finally,

$$\Pi_j(a \geq b) = \frac{\frac{PT[g_j(a)]+PT[g_j(b)]}{2} - \min\left\{g_j(b) - g_j(a), \frac{PT[g_j(a)]+PT[g_j(b)]}{2}\right\}}{\frac{PT[g_j(a)]+PT[g_j(b)]}{2} - \min\left\{g_j(b) - g_j(a), \frac{IT[g_j(a)]+IT[g_j(b)]}{2}\right\}}, \tag{7.6}$$

Comparing formula (7.2) and (7.6), we immediately see that the concordance index of Roy related to $a \geq_j b$ and the possibility that $a \geq_j b$ are equivalent if $IT\,[g_j(a)]$ and $PT\,[g_j(a)]$ are independent from $g_j(a)$ and may be denoted IT_j and PT_j. In that case,

$$R_j(a,b) = \Pi_j(a \geq b) = \frac{PT_j - \min\{g_j(b) - g_j(a), PT_j\}}{PT_j - \min\{g_j(b) - g_j(a), IT_j\}}.$$

If we reconsider Example 7.1, we build the eight fuzzy intervals corresponding to the eight cars as in Figure 7.10

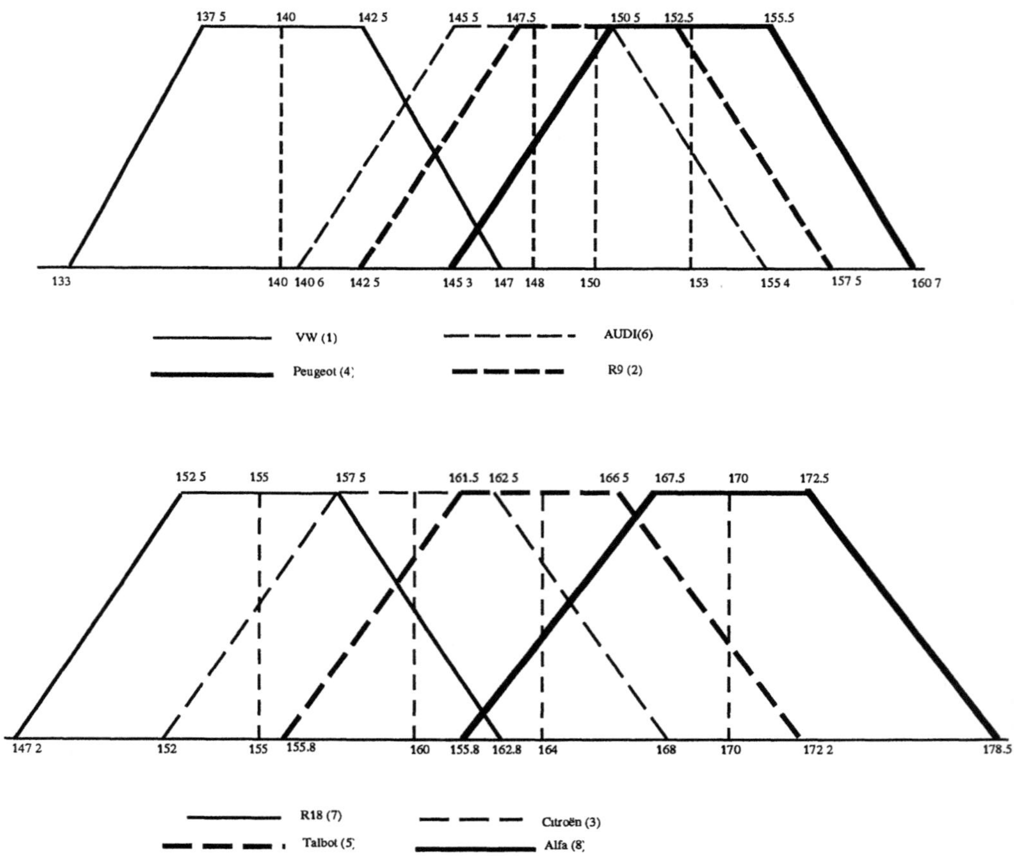

Fig. 7.10

The possibilities are summarized in Table 7.6 (compare with Table 7.2).

π	1	2	3	4	5	6	7	8
1	1	.47	0	.17	0	.68	0	0
2	1	1	.52	1	.16	1	1	0
3	1	1	1	1	1	1	1	.57
4	1	1	.81	1	.45	1	1	0
5	1	1	1	1	1	1	1	.91
6	1	1	.33	1	0	1	.80	0
7	1	1	1	1	.63	1	1	.11
8	1	1	1	1	1	1	1	1

Table 7.6

If the indifference and preference thresholds $IT\,[g_j(a)]$, $PT\,[g_j(a)]$ are linear functions of $g_j(a)$:

$$\begin{aligned} IT\,[g_j(a)] &= \alpha_I + \beta_I g_j(a) \\ PT\,[g_j(a)] &= \alpha_P + \beta_P g_j(a) \end{aligned}$$

we obtain that :

$$g_j(b) \leq \left(\frac{2\alpha_I}{2-\beta_I}\right) + \left(\frac{2+\beta_I}{2-\beta_I}\right) g_j(a) \text{ implies } \Pi_j(a \geq b) = 1,$$

$$\left(\frac{2\alpha_I}{2-\beta_I}\right) + \left(\frac{2+\beta_I}{2-\beta_I}\right) g_j(a) < g_j(b) \leq \left(\frac{2\alpha_P}{2-\beta_P}\right) + \left(\frac{2+\beta_P}{2-\beta_P}\right) g_j(a)$$

implies $0 < \Pi_j(a \geq b) < 1$ and $\Pi_j(a,b)$ is linearly decreasing from 1 to 0 when $g_j(b)$ is encreasing,

$$g_j(b) \geq \left(\frac{2\alpha_P}{2-\beta_P}\right) + \left(\frac{2+\beta_P}{2-\beta_P}\right) g_j(a) \text{ implies } \Pi_j(a \geq b) = 0,$$

or

$$\Pi_j(a \geq b) = \frac{\alpha_P + \beta_P\left(\frac{g_j(a)+g_j(b)}{2}\right) - \min\left(g_j(b) - g_j(a),\ \alpha_P + \beta_P\left(\frac{g_j(a)+g_j(b)}{2}\right)\right)}{\alpha_P + \beta_P\left(\frac{g_j(a)+g_j(b)}{2}\right) - \min\left(g_j(b) - g_j(a),\ \alpha_I + \beta_I\left(\frac{g_j(a)+g_j(b)}{2}\right)\right)}.$$

$\Pi_j(a \geq b)$ is a function of $g_j(a)$ and $Ng_j(b)$, linearly decreasing with both arguments. It follows that :

Proposition 7.5 *If $\mu_j^a(x)$ and $\mu_j^b(x)$ are symmetric fuzzy intervals such that the supports PT_j and the kernels IT_j are linear functions of the centers of gravity g_j,*

$\Pi_j(a \geq b)$ is a fuzzy semiorder.

Proof. Consequence of Proposition 7.1.

We now turn to a third example.

Example 7.3 (Roubens and Vincke, 1988) Consider five alternatives a, b, c, d and e which were rated on a scale going from 0 to 10 :

a lies between 8.5 and 9.5
b lies possibly between 3 and 10 and certainly between 6 and 8
c lies possibly between 2 and 6 and certainly between 3 and 4
d lies possibly between 3.5 and 6.5 and certainly between 4.5 and 5.5
e lies possibly between 2.5 and 8.5 and certainly between 3.5 and 7

Using trapezoidal fuzzy numbers, we obtain the graphical representation in Figure 7.11.

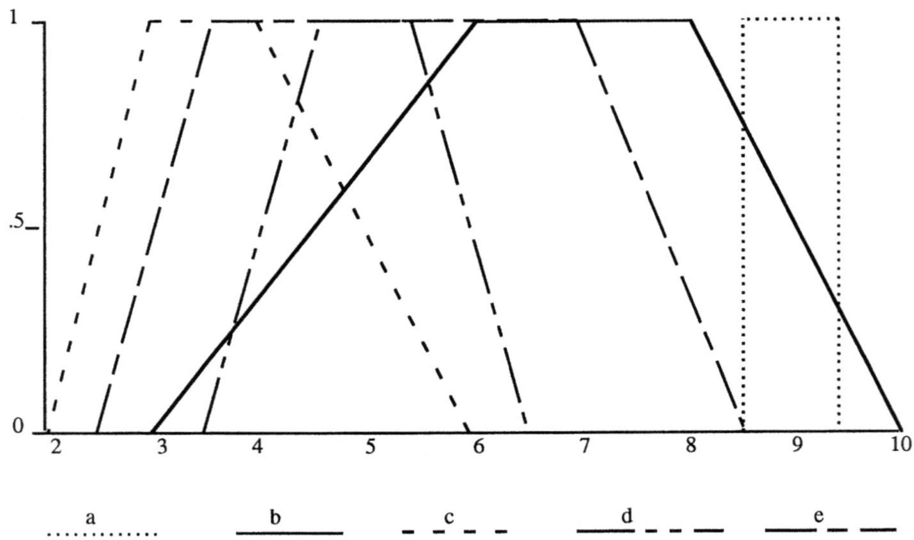

Fig. 7.11

The possibility degrees $\Pi(a \geq b)$ are presented in Table 7.7.

Possibility $\Pi(a \geq b)$ for five alternatives

	a	b	c	d	e
a	1	1	1	1	1
b	.75	1	1	1	1
c	0	.6	1	.833	1
d	0	.875	1	1	1
e	0	1	1	1	1

Table 7.7

Π is clearly not a fuzzy semiorder because

$$1 = \min\{\Pi(c \geq e),\ \Pi(e \geq b)\} > \max\{\Pi(c \geq d),\ \Pi(d \geq b)\} = .833.$$

Considering the levels $\lambda = .6,\ .75,\ .833,\ .875$ and 1, we obtain the crisp λ-cut structures summarized in Table 7.8.

λ-cuts for Π.

λ	nested crisp λ-cuts	type of crisp structure	indifference graphs
.6	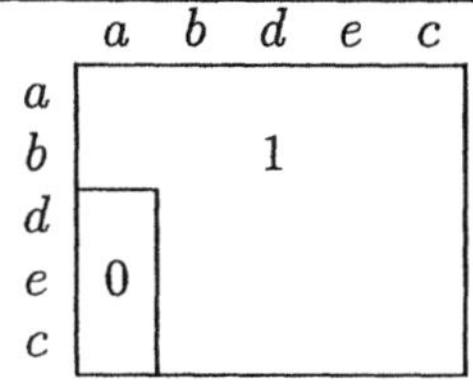	Supp_{Π} : semiorder	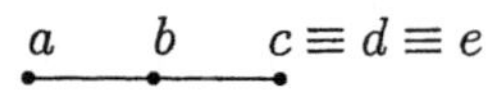
.75	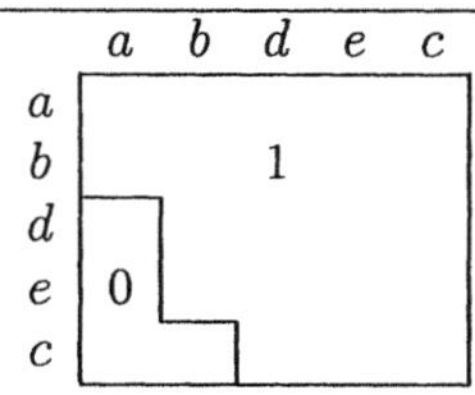	semiorder	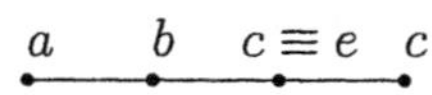
.833	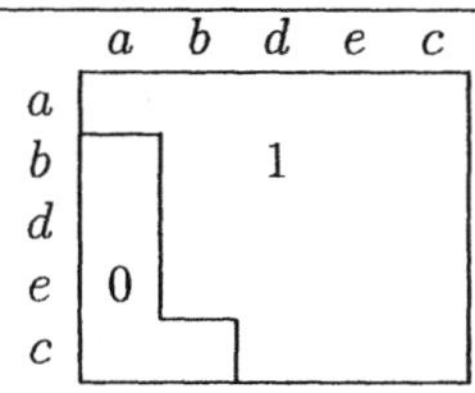	semiorder	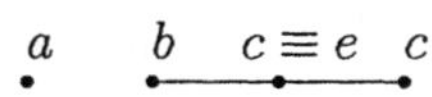
.875	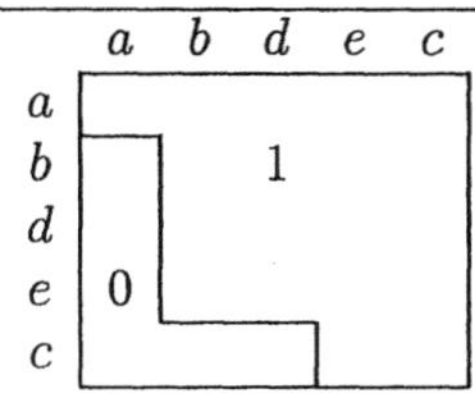	semiorder	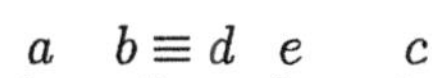
1	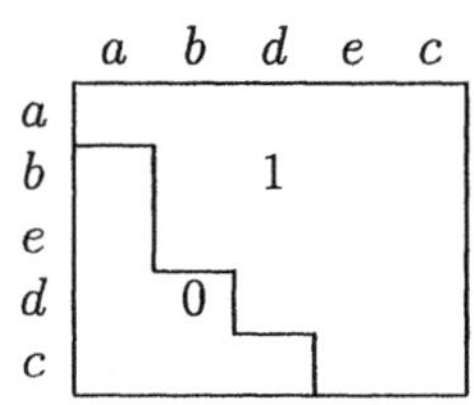	interval order	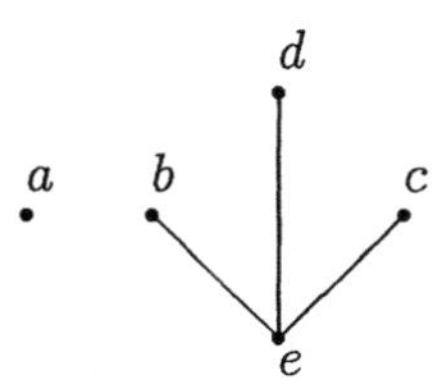

Table 7.8

As already mentioned in Section 7.2.1, the axiomatics developed in Chapter 3 gives the *credibility of strict preference* and *indifference* :

$$P_j(a,b) \quad = \quad R_j^d(a,b) = 1 - \Pi_j(b \geq a) = \mathcal{N}_j(a > b) \tag{7.7}$$

$$I_j(a,b) = \min\left[\Pi_j(a \geq b), \Pi_j(b \geq a)\right] \quad (7.8)$$

The credibility of strict preference of a over b for point of view j corresponds to the necessity that a is strictly better than b for j. This is an immediate consequence of the following relation between possibility and necessity degrees : $\Pi(A) = 1 - \mathcal{N}(\bar{A})$, where A and $\bar{A}$ are complementary events (see Dubois and Prade (1987) for an axiomatic definition of possibility and necessity).

Corollary 7.6 *The credibility of strict preference as defined with (7.7) is a fuzzy partial order and every λ-cut is a crisp partial order.*

Proof. Immediate.

Coming back to the example 7.3, we obtain in Table 7.9 the credibility of strict preference and its related support which produces a partial order shown in Figure 7.12.

$\mathcal{N}(a > b)$ for five alternatives

	a	b	e	d	c
a	0	.25	1	1	1
b	0	0	0	.125	.4
d	0	0	0	0	.167
e	0	0	0	0	0
c	0	0	0	0	0

Supp $\mathcal{N}$

	a	b	e	d	c
a	0	1	1	1	1
b	0	0	0	1	1
d	0	0	0	0	1
e	0	0	0	0	0
c	0	0	0	0	0

Table 7.9

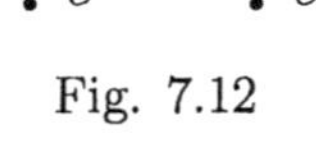

Fig. 7.12

$P_j(a,b)$ as defined by (7.8) is not the unique credibility of strict preference to deal with imprecise information.

If we consider two fuzzy intervals $\{x, \mu_j^a(x)\}$ and $\{x, \mu_j^b(x)\}$ with finite supports, we may introduce (see Roubens (1990))

$$\begin{aligned}
\inf I_j^a(\lambda) &= \inf_x \{x : \mu_j^a(x) > \lambda\} \\
\sup I_j^a(\lambda) &= \sup_x \{x : \mu_j^a(x) > \lambda\} \\
U_j(a,b) &: \{\lambda : \inf I_j^a(\lambda) \geq \inf I_j^b(\lambda)\} \\
V_j(a,b) &: \{\lambda : \sup I_j^a(\lambda) \geq \sup I_j^b(\lambda)\}
\end{aligned}$$

We define a transitive compensatory $>_j$ in the sense that values of λ for which $\inf I_j^a(\lambda) + \sup I_j^a(\lambda) \geq \inf I_j^b(\lambda) + \sup I_j^b(\lambda)$ are compensating those values for which the inequality does not hold.

The degree to which $(a >_j b)$ can be considered as (before normalization) :

$$\begin{aligned}\Delta_j(a,b) &= \int_{U_1(a,b)} \left[\inf I_j^a(\lambda) - \inf I_j^b(\lambda)\right] d\lambda \\ &+ \int_{V_1(a,b)} \left[\sup I_j^a(\lambda) - \sup I_j^b(\lambda)\right] d\lambda \\ &- \int_{[0,1]\setminus U_1(a,b)} \left[\inf I_j^b(\lambda) - \inf I_j^a(\lambda)\right] d\lambda \\ &- \int_{[0,1]\setminus V_1(a,b)} \left[\sup I_j^b(\lambda) - \sup I_j^a(\lambda)\right] d\lambda, \text{ if positive}\end{aligned}$$

$\Delta_j(a,b) = 0$, otherwise.

If we call $\inf I_j^a(\lambda) + \sup I_j^a(\lambda) = sum\ I_j^a(\lambda)$, then

$$\begin{aligned}\Delta_j(a,b) &= \left[\int_0^1 \left[sum\ I_j^a(\lambda) - sum\ I_j^b(\lambda)\right] d\lambda\right] \\ &= 0, \text{ otherwise.}\end{aligned}$$

For a trapezoidal fuzzy numbers, with parameters $(a_j^-, a_j^+, \sigma_j^-(a), \sigma_j^+(a))$,

$$\int_0^1 \left[\inf I_j^a(\lambda) + \sup I_j^a(\lambda)\right] d\lambda = a_j^- - \frac{1}{2}\sigma_j^-(a) + a_j^+ + \frac{1}{2}\sigma_j^+(a),$$

and

$$\begin{aligned}\Delta_j(a,b) &= a_j^- - \frac{1}{2}\sigma_j^-(a) + a_j^+ + \frac{1}{2}\sigma_j^+(a) - \left(b_j^- - \frac{1}{2}\sigma_j^-(b)\right) - \left(b_j^+ + \frac{1}{2}\sigma_j^+(b)\right) \\ &= \inf I_j^a(.5) + \sup I_j^a(.5) - \inf I_j^b(.5) - \sup I_j^b(.5),\end{aligned}$$

if positive.

In order to obtain a valued binary relation, we define the degree to which $(a >_j b)$ as

$$P_j'(a,b) = \Delta_j(a,b) / \max_{a,b} \Delta_j(a,b)$$

because $\Delta_j(a,b)$ is finite for every pair (a,b).

Proposition 7.6 *The credibility of strict preference $P_j'(a,b)$ is a fuzzy partial order (irreflexive, antisymmetric and transitive valued binary relation) and every λ-cut is a crisp partial order.*

Proof. Irreflexivity and antisymmetry are obvious. Let us prove that

$$P_j'(a,c) \geq \min\left[P_j'(a,b), P_j'(b,c)\right], \text{ for any given } a,b,c \text{ in } A$$

If $P_j'(a,b) = P_j'(b,c) = 0$, the transitivity condition is satisfied. We now consider that $P_j'(a,b)$ and $P_j'(b,c)$ are positive and we suppose without restriction that $P_j'(a,b) \geq P_j'(b,c) > 0$. In that case, we have to prove that $P_j'(a,c) \geq P_j'(b,c)$.

$$\Delta_j(a,b) = \int_0^1 \left[sum\ I_j^a(\lambda) - sum\ I_j^b(\lambda)\right] d\lambda > 0,$$

and

$$\int_0^1 sum\ I_j^a(\lambda)d\lambda > \int sum\ I_j^b(\lambda)d\lambda.$$

From the last inequality, we may conclude that

$$\int_0^1 \left[sum\ I_j^a(\lambda) - sum\ I_j^c(\lambda)\right] d\lambda > \int_0^1 \left[sum\ I_j^b(\lambda) - sum\ I_j^c(\lambda)\right] d\lambda,$$

or $P_j'(a,c) > P_j'(b,c) > 0$, which prove the transitiveness.

Every λ-cut of a fuzzy partial order is a crisp partial order. ■

Coming back to example 8.3, we obtain Table 7.10

a	$a^- - \frac{1}{2}\sigma^-(a)$	$a^+ + \frac{1}{2}\sigma^+(a)$	sum
a	8.50	9.50	18
b	4.50	8	12.50
c	2.50	5	7.50
d	4	6	10
e	3	7.75	10.75

Table 7.10

From Table 7.10, the $P'(a,b)$ can be easily derived as show in Table 7.11.

	a	b	e	d	c
a	0	.52	.69	.76	1
b	0	0	.17	.24	.48
e	0	0	0	.07	.31
d	0	0	0	0	.24
c	0	0	0	0	0

Table 7.11

From Table 7.11, we can easily build a nested family (a chain) of crisp partial orders

$$R_{.07} \supset R_{.17} \supset R_{.24} \supset R_{.48} \supset R_{.52} \supset R_{.69} \supset R_{.76} \supset R_1$$

7.3 Aggregation of monocriterion preference relations

Preference relations between every ordered pair of alternatives have been modelled in the previous section. We are turning now to the aggregation phase to produce a global system of *outranking relations*.

The problem facing us is the following : consider the profile of m fuzzy interval orders $(R_1, \ldots, R_m)$, each R_j corresponding to a point of view weighted by ω_j defined on a ratio scale.

We are first looking for an operator $M(R_1, \ldots, R_m; \omega_1, \ldots, \omega_m)$, called outranking relation, which should preserve two properties of R_j : completeness and negative transitivity (in the max-min sense).

If $R = M(R_1, \ldots, R_m; \omega_1, \ldots, \omega_m)$ is complete, we know from results obtained in Chapter 3 that in an axiomatical way, the strict preference P corresponds to R^d. R being complete and negatively transitive, R^d is antisymmetric and (max-min) transitive and every λ-cut of the strict preference relation P is a crisp partial order. In fact, max-min transitivity of P is a necessary and sufficient condition to obtain a crisp partial order for every λ-cut of P. These parametrized partial orders might be considered as an appropriate answer to the combination (aggregation + ranking).

7.3.1 How to preserve transitivity in the aggregation procedure ?

Our problem is now to find an operator M such that $R = M(R_1, \ldots, R_m; \omega_1, \ldots, \omega_m)$ is complete and negatively transitive if R_j present the same properties.

One answer is given by

$$M(R_1, \ldots, R_m) = \max_{j \in J} h_j(R_j),$$

where h_j are non decreasing functions from $[0, 1]$ to $[0, 1]$, J is the index set $\{1, \ldots, m\}$ and $h_j(1) = 1$ for some $j \in J$, $h_j(0) = 0$ for all $j \in J$.

The completeness of M directly follows from the property that $h_j(1) = 1$ for some j. We also have that M is monotonic and $M(0, \ldots, 0) = 0$, $M(1, \ldots, 1) = 1$.

We now turn to the negative transitivity of M and we consider a lemma.

Lemma 7.1 *The following inequality holds :*

$$\max_{j \in J} h_j(\max(x_j, y_j)) \le \max\left\{\max_{j \in J} h_j(x_j), \max_{j \in J} h_j(y_j)\right\}.$$

Proof.

$$\begin{aligned} \max_{j \in J} h_j(\max(x_j, y_j)) &= \max\left\{\max_{\substack{j \\ x_j > y_j}} h_j(x_j), \max_{\substack{j \\ x_j \le y_j}} h_j(y_j)\right\} \\ &\le \max\left\{\max_{j \in J} h_j(x_j), \max_{j \in J} h_j(y_j)\right\}. \end{aligned}$$

■

With the previous lemma, we obtain :

Theorem 7.1 *Suppose that $R_1, \ldots, R_m$ are complete and negatively transitive valued binary relations. Then the aggregated valued relation defined as*

$$R = M(R_1, \ldots, R_m)$$

is a complete and negatively transtive valued binary relation if and only if

$$R = \max_{j \in J} h_j(R_j),$$

where h_j are non decreasing functions from $[0,1]$ to $[0,1]$ with $h_j(1) = 1$, for some $j \in J$ and $h_j(0) = 0$, for all $j \in J$.

Proof. (Sufficiency).

R is negatively transitive. Indeed, we have the following equalities-inequalities chain for all $a, b, c \in A$:

$$\begin{aligned} R(a,b) &= \max_{j\in J} h_j(R_j(a,b)) \\ &\leq \max_{j\in J} h_j\left[\max(R_j(a,c),R_j(c,b))\right], \text{ } R_j \text{ are negatively transitive relations} \\ &\leq \max\left[\max_{j\in J} h_j(R_j(a,c)), \max_{j\in J}(R_j(c,b))\right], \text{ lemma 7.1} \\ &\leq \max\left[R(a,c),R(c,b)\right]. \end{aligned}$$

(Necessity).

Let $x_j = R_j(a,b)$, $y_j = R_j(a,c)$, $z_j = R_j(c,b)$, $j \in J$.

If R is negatively transitive, then we have the following inequality :

$$M(x_1,\ldots,x_m) \leq \max\left[M(y_1,\ldots,y_m), M(z_1,\ldots,z_m)\right].$$

We generally have that : $x_j \leq \max(y_j, z_j)$. In the particular case where $x_j = \max(y_j, z_j)$, since M is non decreasing in each place, we obtain that :

$$\begin{aligned} M(x_1,\ldots,x_m) &= M(\max(y_1,z_1),\ldots,\max(y_m,z_m)) \\ &\geq M(y_1,\ldots,y_m) \\ &\geq M(z_1,\ldots,z_m) \\ &\geq \max\left[M(y_1,\ldots,y_m), M(z_1,\ldots,z_m)\right]. \end{aligned}$$

Finally, when $x_j = \max(y_j, z_j)$, we have that :

$$M(x_1,\ldots,x_m) = \max\left[M(y_1,\ldots,y_m), M(z_1,\ldots,z_m)\right].$$

Suppose first that $x_1 = y_1$, $x_2 = z_2$, $z_1 = y_2 = x_3 = \cdots = x_m = 0$. In that case,

$$M(x_1,x_2,0,\ldots,0) = \max(M(x_1,0,\ldots,0), M(0,x_2,0,\ldots,0)).$$

Using this inequality and mathematical induction, we conclude that :

$$M(x_1,\ldots,x_m) = \max\left[M(x_1,0,\ldots,0), M(0,x_2,0,\ldots,0),\ldots,M(0,\ldots,0,x_m)\right].$$

That is, introducing $h_j(x_j) = M(0,\ldots,0,x_j,0,\ldots,0)$, we have that

$$M(x_1,\ldots,x_m) = \max_{j\in J} h_j(x_j).$$

$h_j(0) = 0$, for all $j \in J$, since $M(0,\ldots,0) = 0$. There exists $j \in J$ such that $h_j(1) = 1$ because $M(1,\ldots,1) = 1$. Functions h_j are non decreasing by non decreasingness of M.

∎

This theorem and the related proof are very close to those obtained by Leclerc (1984) in the framework of consensus functions on transitively valued relations.

The dual theorem for antisymmetric and transitive relations can be immediately deduced.

Theorem 7.2 *Suppose that $R_1, \ldots, R_m$ are antisymmetric and transitive valued binary relations. The aggregated valued relation defined as*

$$R = M'(R_1, \ldots, R_m)$$

is an antisymmetric and transitive valued binary relation if and only if

$$R = \min_{j \in J} f_j(R_j)$$

where f_j are non decreasing function from $[0,1]$ to $[0,1]$ wit $f_j(0) = 0$, for some $j \in J$ and $h_j(1) = 1$, for all $j \in J$.

Proof. Consider $R'_j = 1 - R_j$, $j \in J$. These binary relations are complete and negatively transitive. From Theorem 7.1 we deduce that

$$R' = \max_{j \in J} h_j(R'_j)$$

is complete and negatively transitive. It immediately comes out that

$$R = 1 - R' = 1 - \max_{j \in J} h_j(1 - R_j) = \min_{j \in J} [1 - h_j(1 - R_j)]$$

is antisymmetric and transitive.

■

Coimplications and Sugeno integrals can be used to represent M.
If

$$M(x_1, \ldots, x_m) = \bigvee_{j \in J} I_c^{\rightarrow}(1 - \omega_j, x_j)$$

$h_j(x_j) = I_c^{\rightarrow}(1 - \omega_j, x_j)$ is non decreasing in argument x_j and $h(0) = I_c^{\rightarrow}(1 - \omega_j, 0) = 0$, for all $j \in J$. Moreover there exists at least one $j^* \in J$ such that $\omega_{j^*} = 1$ and $h_{j^*}(1) = I_c^{\rightarrow}(0, 1) = 1$.

An important particular case corresponds to the weighted maximum (Dubois and Prade, 1986) :

$$h_j(R_j) = \min(\omega_j, R_j) \text{ if } \max_{j \in J} \omega_j = 1.$$

It is the valued translation of the linguistic aggregation procedure : "a is globally not worse than b if a is not worse than b for at least one weighted point of view" :

$$\begin{aligned} R(a,b) &= M(R_1(a,b), \ldots, R_m(a,b); \omega_1, \ldots, \omega_m) \\ &= \bigvee_{j \in J} [\omega_j \wedge R_j(a,b)] \end{aligned}$$

where $\wedge$ corresponds to the t-norm min and $\vee_j$ to the associated generalized t-conorm max.

If we want to determine the global credibility of the proposition "a is strictly better than b", two ways might be used to achieve this goal, each having two steps :

1a. Aggregate the monocriterion relations R_j to obtain a global outranking relation $R = M(R_1, \ldots, R_m; \omega_1, \ldots, \omega_m)$.

1b. Define $P = p(R, R^{-1})$.

2a. For each point of view j, define the strict preference $P_j = p'(R_j, R_j^{-1})$.

2b. Define $P' = M'(P_1, \ldots, P_m; \omega_1, \ldots, \omega_m)$.

We can illustrate these two procedures on Figure 7.13.

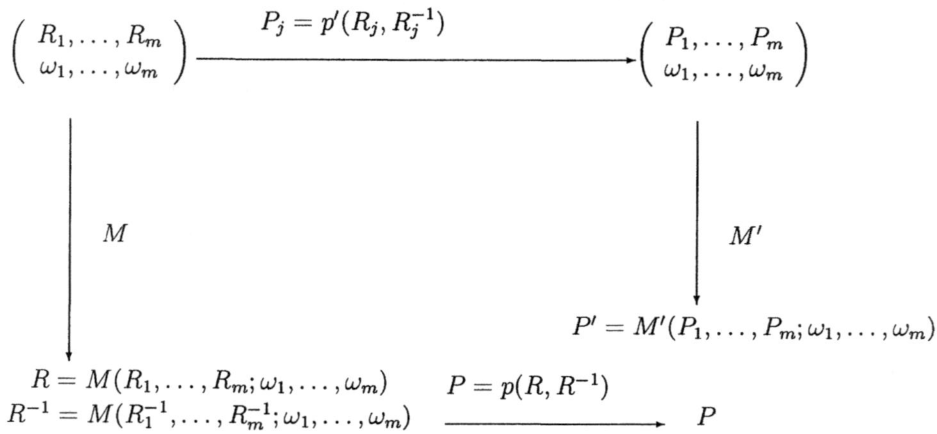

Fig. 7.13

We now examine conditions under which both procedures (1a,1b) and (2a, 2b) coincide.

Consider

$$R = M(R_1, \ldots, R_m; \omega_1, \ldots, \omega_m) = \max_{j \in J} h_j(R_j).$$

Due to the completeness of R, the axiomatical analysis of strict preference (see Chapter 3) gives :

$$P = R^d = 1 - \max_{j \in J} h_j(R_j^{-1})$$

which is an antisymmetric and transitive relation :

$$\min(P(a,b), P(b,a)) = 0$$

$$P(a,b) \geq \min(P(a,c), P(c,b)) \;, \quad \forall a, b, c \in A.$$

Using the same arguments, we define $P_j = R_j^d$ and we consider

$$P' = M'(P_1, \ldots, P_m; \omega_1, \ldots, \omega_m) = M'(R_1^d, \ldots, R_m^d; \omega_1, \ldots, \omega_m) = P.$$

We then obtain,

$$M'(1 - R_1^{-1}, \ldots, 1 - R_m^{-1}; \omega_1, \ldots, \omega_m) = 1 - \max_{j \in J} h_j(R_j^{-1})$$

and finally,

$$M'(x_1, \ldots, x_m) = \min_{j\in J} [1 - h_j(1 - x_j)] .$$

Implications and Sugeno integrals can be once more used to represent M'.
If $h_j(x_j) = I_c^{\rightarrow}(1 - \omega_j, x_j)$,

$$\begin{aligned} M'(x_1, \ldots, x_m) &= \bigwedge_j [1 - h_j(1 - x_j)] = \bigwedge_j [1 - I_c^{\rightarrow}(1 - \omega_j, 1 - x_j)] \\ &= \bigwedge_j I^{\rightarrow}(\omega_j, x_j). \end{aligned}$$

A convenient particular pair (M, M') to obtain $P = P'$ is thus given by the pair of weighted maximum and minimum (see Dubois and Prade, 1986)

$$\begin{aligned} M(R_1, \ldots, R_m; \omega_1, \ldots, \omega_m) &= \max_{j\in J} \min(\omega_j, R_j) \\ M'(P_1, \ldots, P_m; \omega_1, \ldots, \omega_m) &= \min_{j\in J} \max(1 - \omega_j, P_j). \end{aligned}$$

We finally obtain a synthesis of the previous result in Figure 7.14 :

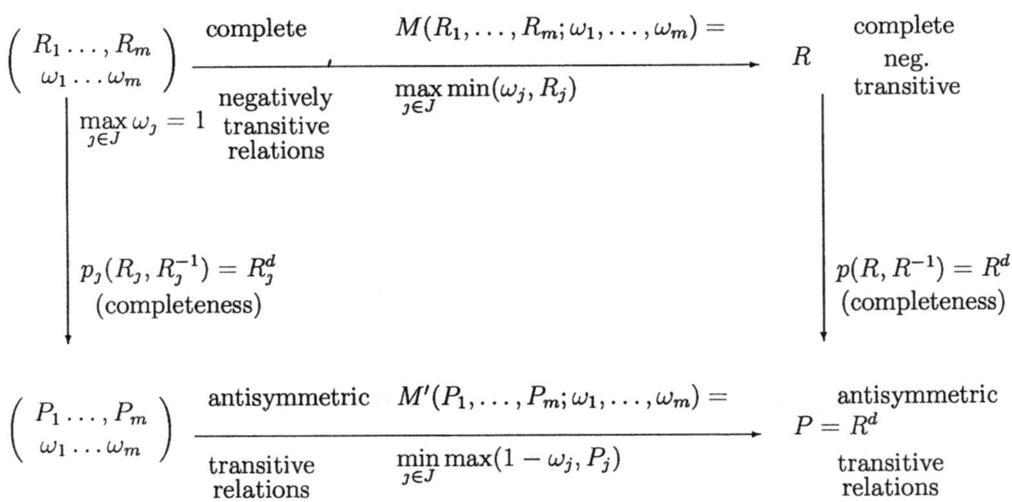

Fig. 7.14

Every λ-cut of the valued relation P is a crisp antisymmetric and transitive relation, i.e. a crisp partial order.

The answer seems however rather unsatisfactory. It gives for most practical situations, non discriminant results (most of the elements of the matrix R are equal to one). This situation is the consequence of the choice of the aggregation rule : "... for at least one weighted point of view ..." as we will see in the Section 7.3.3 (Example 7.6)

7.3.2 Crisp case and a possibility theorem related to a voting procedure

We consider here the case where all criteria (voters) are of the same importance. $R_j \in \{0,1\}$ represent crisp complete preorders. R_j are complete ($R_j(a,b)$ or $R_j(b,a) = 1$) and transitive. These conditions imply negative transitivity : not $R(a,b)$ and not $R(b,c)$ implies $P(b,a)$ and $P(c,b)$, transitivity implies $P(c,a)$ which corresponds to not $R(a,c)$.

The utility functions f_j must be the same for every voter and finally $f_j(0) = 0$ and $f_j(1) = 1$. The only possible answer is

$$R(a,b) = \max_j R_j(a,b).$$

This procedure is non dictatorial, independent from irrelevant alternatives and satisfies the unanimity principle (see Fishbrun (1970 b)). It is however non universal giving non transitive global answers. Let us, for example, consider the following votes :

$$aP_jbP_jcP_jd \quad , \quad j = 1,2,3$$
$$dP_4aP_4bP_4c.$$

$R(a,b) = 1$, $R(b,a) = 0$, $R(b,c) = 1$, $R(c,b) = 0$, $R(a,c) = 1$, $R(c,a) = 0$, $R(a,d) = R(d,a) = R(d,c) = R(c,d) = R(d,b) = R(b,d) = 1$.

R is complete and negatively transitive but the global response is not transitive : $R(c,d) = 1$ and $R(d,a) = 1$, but $R(c,a) = 0$.

If we consider the dual situation :

$$R(a,b) = \min_j R_j(a,b),$$

we obtain an antisymmetric and transitive crisp relation (partial order). Completeness cannot be assumed.

Coming back to the example :
$R(a,b) = 1$, $R(b,a) = 0$, $R(b,c) = 1$, $R(c,b) = 0$, $R(a,c) = 1$, $R(c,a) = 0$, $R(a,d) = R(d,a) = R(d,c) = R(c,d) = R(d,b) = R(b,d) = 0$.

7.3.3 Aggregation operators based on empiric rules

We are now concerned with some aggregation rules which translate in valued terms some natural qualitative connection between the points of view (Bellman and Zadeh (1970), Yager (1977), Dubois and Prade (1980), Kitainik (1993), Kitainik, Orlovsky and Roubens (1990) among others) :

"a outranks b" (or equivalently, a is globally not worse than b) if

"a is not worse than b for at least one point of view" (7.9)

"a is not worse than b for all points of view" (7.10)

"a is not worse than b for at least one weighted point of view" (7.11)

"a is not worse than b for all weighted points of view" (7.12)

"a is not worse than b for at least one significant point of view" (7.13)

"a is not worse than b for all significant points of view" (7.14)

In terms of fuzzy logical connectives ($\wedge, \vee, \neg, I^{\rightarrow}$) as presented in Chapter 1, (7.9)–(7.14) may be rewritten as

$$M_{\vee}(R_1, \ldots, R_m) = \bigvee_j R_j \tag{7.15}$$

$$M_{\wedge}(R_1, \ldots, R_m) = \bigwedge_j R_j \tag{7.16}$$

$$M_{\vee,\omega}(R_1, \ldots, R_m) = \bigvee_j [R_j \wedge \omega_j] \tag{7.17}$$

$$M_{\wedge,\omega}(R_1, \ldots, R_m) = \bigwedge_j [R_j \vee (1 - \omega_j)] \tag{7.18}$$

$$M_{I_c^{\rightarrow},\omega}(R_1, \ldots, R_m) = \bigvee_j [I_c^{\rightarrow}(1 - \omega_j, R_j)] \tag{7.19}$$

$$M_{I^{\rightarrow},\omega}(R_1, \ldots, R_m) = \bigwedge_j [I^{\rightarrow}(\omega_j, R_j)] \tag{7.20}$$

$I^{\rightarrow}$ and $I_c^{\rightarrow}$ represent respectively implications and coimplications satisfying (I-6) – (I-6') and (I-10) – (I-10').

If T = min, π or L, using implications and coimplications of type R and S (for notations, see Sections 1.8 and 1.9), we obtain the results of Table 7.12.

Some examples of aggregation operators
based on empiric rules

	$M_{I_c^{\rightarrow}}$	$M_{I^{\rightarrow}}$
$I_c^{\rightarrow}$ of type T $I^{\rightarrow}$ of type S	"... for at least one weigthed weighted point of view..."	"... for all weighted points of view ..."
Kleene-Dienes	$\max_j \min(\omega_j, R_j)$	$\min_j \max(1-\omega_j, R_j)$
Rechenbach	$\max_j \omega_j R_j$	$\min_j (1-\omega_j+\omega_j R_j)$
Lukasiewicz	$\max_j (\omega_j + R_j - 1)$	$\min_j (1-\omega_j+R_j)$
$I_c^{\rightarrow}, I^{\rightarrow}$ of type R	"... for at least one significant weighted point of view..."	"... for all significant points of view ..."
Gödel	$\min_{j \atop R_j > 1-\omega_j} R_j$ or 0 if $R_j \leq 1-\omega_j$ for all j	$\min_{j \atop R_j < \omega_j} R_j$ or 1 if $R_j \geq \omega_j$
Goguen	$\max_j \left(\frac{R_j+\omega_j-1}{\omega_j}\right)$	$\min \left(\frac{R_j}{\omega_j}\right)$
Lukasiewicz	$\max_j (\omega_j + R_j - 1)$	$\min(1-\omega_j+R_j)$

Table 7.12

All the operators of Table 7.12 preserve negative transitivity (resp. transitivity) and are idempotent and compensative.

More sophisticated combinations might be introduced if we consider at the same time, preference degrees R_j and non discredit degrees ND_j.

Let us go into the approach of Roy in ELECTRE III (1978) : "a outranks b if a is globally in concordance with b *and* if a is not discordant vis à vis b for all discordant points of view".

The degree of significance of a discordant point of view corresponds to $I_T^{\rightarrow}(NC, ND_j)$, where $I_T^{\rightarrow}$ represents an R-implication associated with the t-norm (see Definition 1.16). Finally :

$$R(a,b) = T\{M(R_1, \ldots, R_m; \omega_1', \ldots, \omega_m'), T\left[I_T^{\rightarrow}(NC, ND_1), \ldots, I_T^{\rightarrow}(NC, ND_m)\right]\} \tag{7.21}$$

If one wants to introduce the condition $I^{\rightarrow}(x,0) = 0$, $x \neq 0$, then if $ND_j(a,b) = 0$ for some j (a is totally discordant vis a vis b for j)

$$R(a,b) = T\{M, T\left[\ldots, I_T^{\rightarrow}(NC, 0), \ldots,\right]\} = 0$$

From Proposition 1.11, we know that $I_T^{\rightarrow}(x,0) = 0$ if and only if I is an R-implication with the corresponding conjunction being a positive t-norm ($T = \Pi$ and Goguen's I or $T = \min$ and Gödel's I are convenient).

Roy considered the weighted mean aggregation operator as M, $Nx = 1 - x$, $T = \prod$ and $I_{\vec{T}}$, the Goguen's implication :

$$I_{\vec{T}}(1 - C,\ 1 - D_j) = \min\left(1, \frac{1 - D_j}{1 - C}\right)$$

where

$$C = \frac{1}{\sum \omega_j} \sum_j \omega_j R_j = \sum_j \omega'_j R_j$$

to obtain

$$\begin{aligned} R(a,b) &= C(a,b) \cdot \prod_{D_j(a,b) > C(a,b)} \frac{1 - D_j(a,b)}{1 - C(a,b)}. \\ &= C(a,b) \text{ if all } D_j(a,b) \leq C(a,b) \end{aligned} \qquad (7.22)$$

Other combinations might have been used, e.g. $M = M_4$, $Nx = 1 - x$, $T = \min$ and $I_{\vec{T}}$, the Gödel's implication, to obtain

$$\begin{aligned} R(a,b) &= \min\{C^{\min}(a,b), \min_{D_j(a,b) > C(a,b)} (1 - D_j(a,b))\} \\ &= C^{\min}(a,b) \text{ if all } 1 - D_j(a,b) \leq C(a,b) \end{aligned}$$

where

$$C^{\min} = \min_j \max(1 - \omega_j, R_j).$$

Example 7.4 (based on Baas and Kwakernaak, 1977) Consider three alternatives a, b, c which are rated in linguistic way as being "poor, fair, fair to good, good, very good and not clear" for our criteria 1,2,3,4. The linguistic terms are transformed into fuzzy sets which membership functions are presented in Figure 7.15 ($\mu_j(x) = 1$, $x \in [0, 1]$ for "not clear") :

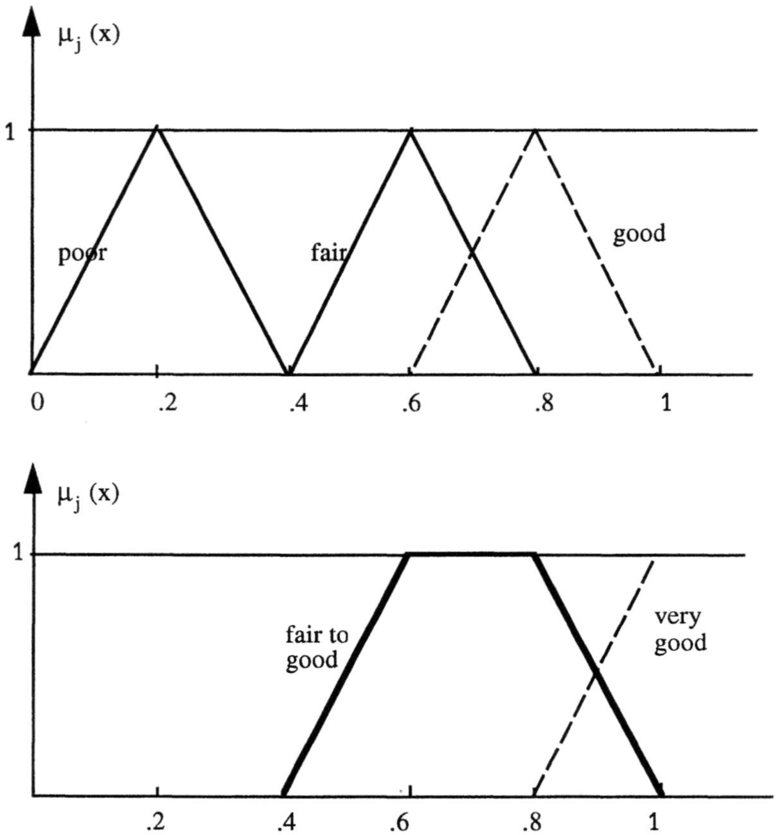

Fig. 7.15

The result of the ratings delivered by the decision maker is given in Table 8.13.

Rating of 3 alternatives a, b, c
with respect to 4 criteria 1,2,3,4

$\mu_j^i(x)$	$i = a$	b	c	ω_j
$j = 1$	good	very good	fair	1
2	poor	poor	poor	.625
3	poor	fair to good	fair	.625
4	good	not clear	fair	.25

Table 7.13

The credibility $R_j(a, b)$, as the "possibility degree that a is not worse than b" (see

7.2.2), is obtained in the Table 7.14

Π_1	a	b	c
a	1	.5	1
b	1	1	1
c	.5	0	1

$\omega_1 = 1$

Π_2	a	b	c
a	1	1	1
b	1	1	1
c	1	1	1

$\omega_2 = .625$

Π_3	a	b	c
a	1	0	0
b	1	1	1
c	1	1	1

$\omega_3 = .625$

Π_4	a	b	c
a	1	1	1
b	1	1	1
c	.5	1	1

$\omega_4 = .25$

Table 7.14

If we consider the supports of these valued relations, $\Pi_{j,\lambda>0}$, it is easily seen that these crisp relations (complete interval orders, see Corollary 7.5) can be represented in terms of intervals of the real line (see Roubens and Vincke, 1985) as in Figure 7.16 :

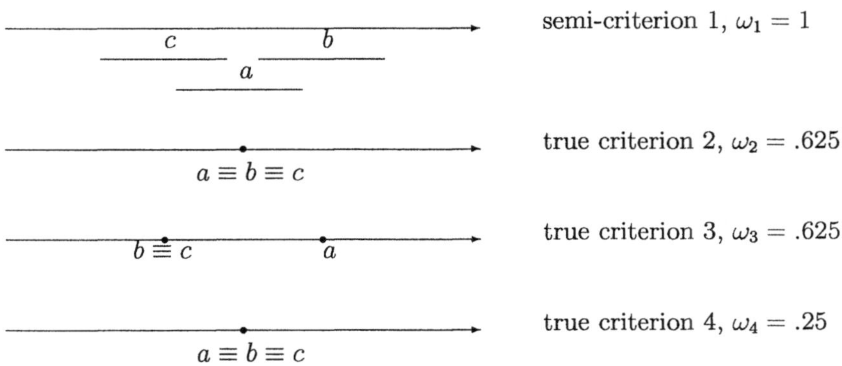

Fig. 7.16

The following empiric rules are now applied to the input data given by Table 8.14 to produce the outranking relations :

$R^I(a,b)$: a is not worse than b if a is not worse than b for at least one point of view ($M_\vee$).

$R^{II}(a,b)$: a is not worse than b if a is not worse than b for all points of view ($M_\wedge$).

$R^{III}(a,b)$: a is not worse than b if a is not worse than b for at least one weighted point of view (Kleene-Dienes)

$R^{IV}(a,b)$, : a is not worse than b if a is not worse than b for all weighted points of view (Kleene-Dienes).

$R^V(a,b)$, : a is not worse than b if a is not worse than b for at least one significant point of view (Gödel).

$R^{VI}(a,b)$, : a is not worse than b if a is not worse than b for all significant points of view (Gödel).

These outranking relations are summarized in Table 7.15

.. at least one point of view

R^I	a	b	c
a	1	1	1
b	1	1	1
c	1	1	1

$$R^I = \max_j R_j$$

.. all points of view

R^{II}	a	b	c
a	1	0	0
b	1	1	1
c	.5	0	1

$$R^{II} = \min_j R_j$$

.. at least one weighted point of view

R^{III}	a	b	c
a	1	.375	1
b	1	1	1
c	.375	.375	1

$$R^{III} = \max_j \min(\omega_j, R_j)$$

.. all weighted points of view

R^{IV}	a	b	c
a	1	.375	.375
b	1	1	1
c	.5	.0	1

$$R^{IV} = \min_j \max(1 - \omega_j, R_j)$$

.. at least one significant point of view

R^V	a	b	c
a	1	1	1
b	1	1	1
c	1	1	1

$$R^V = \max_{R_j > 1 - \omega_j} R_j (\text{or } 0)$$

.. all significant points of view

R^{VI}	a	b	c
a	1	0	0
b	1	1	1
c	.5	.0	1

$$R^{VI} = \min_{R_j < \omega_j} R_j (\text{or } 1)$$

P^{III}	a	b	c
a	0	0	.625
b	.625	0	.625
c	0	0	0

$$P^{III} = 1 - (R^{III})^{-1}$$

Table 7.15

We know from Section 7.3.1 that P^{III}, the strict preference relation related to R^{III} (as given in Table 7.15) is antisymmetric and transitive. One can easily observe that the strict outranking graph represented in Figure 7.17 corresponds to P^{III}.

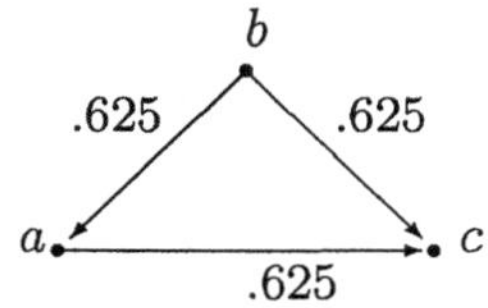

p^{III}

Fig. 7.17

Example 7.5 (Orlovski, 1983) A regional government plans to choose a location for the construction of liquefied gas terminal. The presence of such a terminal at any of the locations considered involves certain degrees of risk associated with great environmental damages that might occur in cases of some catastrophic events at the terminal site. Thus, the government desires to choose the location where such risk is as minimal as possible.

We assume that there are four possible locations in the region in question : $L1, L2$, $L3$ and $L4$. We also assume that the government invited four experts in risk analysis : $E1, E2, E3$ and $E4$ and relies on their joint opinion. However, the government values the experts' opinions differently : opinions of one expert are respected to some degree more than the opinion of another. We assume that the government describes its attitude (or respect) to the experts' opinions with the following matrix of a fuzzy relation "not less important" (Table 7.16) :

	$E1$	$E2$	$E3$	$E4$
$E1$	1	.4	.6	0
$E2$	1	1	.8	1
$E3$	.2	1	1	1
$E4$	.8	0	1	1

Table 7.16

For instance, the element $(E1, E3)$ of this matrix is equal to 0.6. This means that the government considers the opinion of expert $E1$ to be not less important than that of expert $E3$ to a degree 0.6. The elements $(E3, E4)$ and $(E4, E3)$ are both equal to 1, and this means that experts $E3$ and $E4$ are definitely (to a degree 1) equivalent from the government's viewpoint, and so on.

We might also assume that the government describes its attitude to the experts' opinion with the following weights

$$\omega_1 = 1, \quad \omega_2 = .8, \quad \omega_3 = .4, \quad \omega_4 = .4$$

Each of the experts compares the alternative locations with each other in terms of potential risks associated with the construction of the terminal at these locations. The results of these comparisons are represented by matrices. If for example, in such a matrix an element $(L2, L3)$ is equal to 1, then to the corresponding expert's opinion, the risk of constructing the terminal at $L2$ is not greater than at $L3$. If an expert is not definite about this comparison, he may characterize its degree with a number smaller than 1.

In our case, the expert's matrices (or preferences between the alternative locations) are as follows (Table 7.17) :

R_1	$L1$	$L2$	$L3$	$L4$
$L1$	1	.8	1	0
$L2$	0	1	.2	1
$L3$	0	.8	1	0
$L4$	0	0	0	1

R_2	$L1$	$L2$	$L3$	$L4$
$L1$	1	.1	.5	.3
$L2$	.8	1	.8	.8
$L3$	.5	.3	1	0
$L4$	.8	0	0	1

R_3	$L1$	$L2$	$L3$	$L4$
$L1$	1	0	.8	0
$L2$	0	1	0	0
$L3$	.1	0	1	.4
$L4$	1	1	1	1

R_4	$L1$	$L2$	$L3$	$L4$
$L1$	1	1	.9	0
$L2$	0	1	1	1
$L3$	.4	0	1	0
$L4$	0	0	0	1

Table 7.17

Let us reconsider the same empiric rules as in the previous example. We obtain R^I to R^{VI} in Table 7.18 :

	$L1$	$L2$	$L3$	$L4$
$L1$	1	1	1	.3
$L2$	.8	1	1	1
$L3$	.5	.8	1	.4
$L4$	1	1	1	1

$$R^I = \max_j R_j$$

	$L1$	$L2$	$L3$	$L4$
$L1$	1	0	.5	0
$L2$	0	1	0	0
$L3$	0	0	1	0
$L4$	0	0	0	1

$$R^{II} = \min_j R_j$$

	$L1$	$L2$	$L3$	$L4$
$L1$	1	.8	1	.3
$L2$	.8	1	.8	1
$L3$	.5	.8	1	.4
$L4$	.8	.4	.4	1

$$R^{III} = \max_j \min(\omega_j, R_j)$$

	$L1$	$L2$	$L3$	$L4$
$L1$	1	.2	.5	0
$L2$	0	1	.2	.6
$L3$	0	.3	1	0
$L4$	0	0	0	1

$$R^{IV} = \min_j \max(1 - \omega_j, R_j)$$

	$L1$	$L2$	$L3$	$L4$
$L1$	1	1	1	.3
$L2$	.8	1	1	1
$L3$	.5	.8	1	0
$L4$	1	1	1	1

$$R^V = \max_{R_j > 1 - \omega_j} R_j \text{ (or 0)}$$

	$L1$	$L2$	$L3$	$L4$
$L1$	1	0	.5	0
$L2$	0	1	0	0
$L3$	0	0	1	0
$L4$	0	0	0	1

$$R^{VI} = \min_{R_j < \omega_j} R_j \text{ (or 1)}$$

Table 7.18

R_1 to R_4 are not complete, negatively transitive relations and we cannot apply the results of Section 7.3.1.

Example 7.6 (Perny, 1992 and Skalka and al., 1992) Let us come back to example 7.1. We want to compare eight cars on the basis of the following points of view : maximum speed (s), in km/h; volume (v) in cm^3; price (p) in FF; fuel consumption per 100 km at 120 km/hour (c) in liters.

The input data is summarized in Table 7.19.

Input data for comparison of eight cars

points of view	maximum speed	volume	price	consumption
weights tresholds	$\omega_s = 2/3$	$\omega_V = 2/3$	$\omega_p = 1$	$\omega_c = 1$
indifference $IT(g)$	5	$.05g_v$	$500 + .02g_p$	$.10 + .05g_c$
preference $PT(g)$	$.10g_s$	$.10g_v$	$1000 + .10g_p$	$.10 + .10g_c$
veto $VT(g)$	$10 + .10g_s$	$.40 + .10g_v$	$2000 + .15g_p$	$.20 + .15g_c$
brands	g_s : in km/h	g_v : in cm^3	g_p : in FF	g_c : in liters
1. VW GOLF C	140	6.13	−41.360	−7.8
2. Renault R9 GTL	150	6.70	−45.700	−7.5
3. Citroën GSA XI	160	6.63	−46.450	−8.2
4. Peugeot P305 GR	153	6.91	−48.200	−8.4
5. Talbot HOR. GLS	164	6.65	−48.800	−8.5
6. Audi 80 CL	148	7.36	−50.830	−7.0
7. Renault R18 GTL	155	7.40	−51.700	−8.1
8. Alfa SUD TI-NR	170	6.19	−52.500	−7.8

Table 7.19

The concordance $R_j(a,b)$ and discordance indices $D_j(a,b)$ are given in Table 7.20 :

R_c	1	2	3	4	5	6	7	8
1	1	1	1	1	1	.21	1	1
2	1	1	1	1	1	.93	1	1
3	1	.54	1	1	1	0	1	1
4	.81	.1	1	1	1	0	1	.81
5	.59	0	1	1	1	0	1	.59
6	1	1	1	1	1	1	1	1
7	1	.77	1	1	1	0	1	1
8	1	1	1	1	1	.21	1	1

R_p	1	2	3	4	5	6	7	8
1	1	1	1	1	1	1	1	1
2	.2	1	1	1	1	1	1	1
3	.01	1	1	1	1	1	1	1
4	0	.72	.92	1	1	1	1	1
5	0	.58	.77	1	1	1	1	1
6	0	.10	.28	.72	.87	1	1	1
7	0	0	.09	.51	.66	1	1	1
8	0	0	0	.33	.48	.96	1	1

R_s	1	2	3	4	5	6	7	8
1	1	.44	0	.11	0	.67	0	0
2	1	1	.50	1	.10	1	1	0
3	1	1	1	1	1	1	1	.55
4	1	1	.81	1	.42	1	1	0
5	1	1	1	1	1	1	1	.91
6	1	1	.29	1	0	1	.8	0
7	1	1	1	1	.62	1	1	.09
8	1	1	1	1	1	1	1	1

R_v	1	2	3	4	5	6	7	8
1	1	.31	.51	0	.45	0	0	1
2	1	1	1	1	1	.22	.11	1
3	1	1	1	1	1	.02	0	1
4	1	1	1	1	1	.79	.69	1
5	1	1	1	1	1	.07	0	1
6	1	1	1	1	1	1	1	1
7	1	1	1	1	1	1	1	1
8	1	.49	.68	0	.63	0	0	1

D_c	1	2	3	4	5	6	7	8
1	0	0	0	0	0	0	0	0
2	0	0	0	0	0	0	0	0
3	0	0	0	0	0	.55	0	0
4	0	0	0	0	0	.88	0	0
5	0	.10	0	0	0	1	0	0
6	0	0	0	0	0	0	0	0
7	0	0	0	0	0	.38	0	0
8	0	0	0	0	0	0	0	0

D_p	1	2	3	4	5	6	7	8
1	0	0	0	0	0	0	0	0
2	0	0	0	0	0	0	0	0
3	0	0	0	0	0	0	0	0
4	.57	0	0	0	0	0	0	0
5	.76	0	0	0	0	0	0	0
6	1	0	0	0	0	0	0	0
7	1	.14	0	0	0	0	0	0
8	1	.38	.13	0	0	0	0	0

D_s	1	2	3	4	5	6	7	8
1	0	0	.60	0	1	0	.10	1
2	0	0	0	0	0	0	0	.50
3	0	0	0	0	0	0	0	0
4	0	0	0	0	0	0	0	.17
5	0	0	0	0	0	0	0	0
6	0	0	0	0	.12	0	0	.72
7	0	0	0	0	0	0	0	0
8	0	0	0	0	0	0	0	0

D_v	1	2	3	4	5	6	7	8
1	0	0	0	.22	0	1	1	0
2	0	0	0	0	0	0	0	0
3	0	0	0	0	0	0	.07	0
4	0	0	0	0	0	0	0	0
5	0	0	0	0	0	0	.03	0
6	0	0	0	0	0	0	0	0
7	0	0	0	0	0	0	0	0
8	0	.0	0	.07	0	1	1	0

Table 7.20

We are first using the empiric rule (7.22) defined by Roy in ELECTRE III (1978). If

$$\begin{aligned} C(a,b) &= \frac{1}{\sum \omega_j} \sum_j \omega_j R_j(a,b) \\ ND(a,b) &= \prod_{\substack{j \\ D_j(a,b) > C(a,b)}} \frac{1 - D_j(a,b)}{1 - C(a,b)} \\ R(a,b) &= C(a,b) \cdot ND(a,b) \end{aligned}$$

we obtain C and R according to the elements of Table 7.21 :

C	1	2	3	4	5	6	7	8
1	1	.75	.70	.62	.69	.49	.60	.80
2	.76	1	.90	1	.82	.82	.82	.80
3	.70	.86	1	1	1	.50	.80	.91
4	.64	.65	.94	1	.88	.66	.94	.74
5	.58	.57	.93	1	1	.51	.80	.86
6	.70	.73	.64	.92	.76	1	.96	.80
7	.70	.63	.73	.85	.82	.70	1	.81
8	.70	.60	.64	.60	.77	.55	.80	1

R	1	2	3	4	5	6	7	8
1	1	.75	.70	.62	0	0	0	0
2	.76	1	.90	1	.82	.82	.82	.80
3	.70	.86	1	1	1	.46	.80	.91
4	.64	.65	.94	1	.88	.22	.94	.74
5	.33	.57	.93	1	1	0	.80	.86
6	0	.73	.64	.92	.76	1	.96	.80
7	0	.63	.73	.85	.82	.70	1	.81
8	0	.60	.64	.60	.77	0	0	1

Table 7.21

Let us now turn to another empiric rule proposed with relation (7.22). If

$$\begin{aligned} C^{\min}(a,b) &= \min_j \max(1 - \omega_j, R_j(a,b)) \\ ND^{\min}(a,b) &= \min_{D_j(a,b) > C(a,b)} (1 - D_j(a,b)) \text{ or } 1 \text{ if } D_j(a,b) \le C(a,b),\ \forall j \\ R(a,b) &= \min(C^{\min}(a,b), ND^{\min}(a,b)) \end{aligned}$$

we obtain $C^{\min}$ and R according to the elements of Table 7.22 :

C^{min}	1	2	3	4	5	6	7	8
1	1	.67	.67	.67	.67	.21	.67	.67
2	.20	1	.67	1	.67	.67	.67	.67
3	.01	.54	1	1	1	0	.67	.67
4	0	.10	.81	1	.67	0	.69	.67
5	0	0	.77	1	1	0	.67	.59
6	0	.10	.28	.72	.67	1	.67	.67
7	0	0	.09	.51	.66	0	1	.67
8	0	0	0	.33	.48	.21	.67	1

R	1	2	3	4	5	6	7	8
1	1	.67	.67	.67	.67	.21	.67	.67
2	.20	1	.67	1	.67	.67	.67	.67
3	.01	.54	1	1	1	0	.67	.67
4	0	.10	.81	1	.67	0	.69	.67
5	0	0	.77	1	1	0	.67	.59
6	0	.10	.28	.72	.67	1	.67	.28
7	0	0	.09	.51	.66	0	1	.67
8	0	0	0	.33	.48	0	0	1

Table 7.22

Finally, if we apply the non-discriminant but transitive rule "... at least one weighted point of view ..." (formula 7.17), we obtain the following concordance matrix C^{max} (Table 7.23) :

C^{max}	1	2	3	4	5	6	7	8
1	1	1	1	1	1	1	1	1
2	1	1	1	1	1	1	1	1
3	1	1	1	1	1	1	1	1
4	.67	.67	1	1	1	1	1	1
5	.67	.67	1	1	1	1	1	1
6	1	1	1	1	1	1	1	1
7	1	.77	1	1	1	1	1	1
8	1	1	1	1	1	.96	1	1

Table 7.23

7.4 Ranking and choice procedures

Aggregation of monocriterion preference relations has been realized in the previous section. We are now dealing with an outranking relation R which generally presents no other property than to be reflexive.

Starting from R, we can determine a ranking of the alternatives on the basis of scores. In that case, we obtain a total preorder (crisp linear quasiorder) as suggested in Chapter 6.

However, in order to preserve the valuation that was kept in the aggregation procedure, we might prefer a valued structure to approach R.

We investigate in this section the problem of ranking the alternatives in terms of a fuzzy or a crisp partial preorder (quasiorder). Every cut relation associated with a fuzzy quasiorder corresponds to a crisp quasiorder (Proposition 2.29). The reflexive outranking relation must then be transformed into a reflexive and transitive valued relation. In order to reach this goal three solutions are proposed.

The *first solution* corresponds to the transitive closure $\hat{R}$ which is the upper nearest reflexive and transitive relation to R (Proposition 2.23). It is also the minimal transitive relation *including* R.

The indifference relation $I = R \wedge R^{-1}$ (see Chapter 3) being reflexive and symmetric is a *proximity relation.*

$\hat{I}$, the transitive closure of I, preserves reflexivity and symmetry and offers transitivity. It is a *similarity relation.*

Following Proposition 2.23, $\hat{I}$ is also the minimal (and the nearest) similarity relation including R.

There exists an indirect way to compute the transitive closure $\hat{I}$. Consider *ultrametric* distance $d = 1 - I$ (irreflexive, symmetric and negatively transitive relation). Use the Kruskal algorithm (see Christofides (1975)) to define the shortest spanning tree related to the indirected graph whose nodes correspond to the alternatives and whose vertices are valued by d. The vertices of the spanning tree correspond to the $(1 - \lambda)$-cuts related to the equivalence classes of the partition tree. (for details, see Zahn (1971)).

The *second solution* is linked to the traces of R. We know that R^ℓ and R^r are fuzzy quasiorders *included* in R (Propositions 2.7 and 2.19). In the case of min-transitivity,

$$R^r(a,b) = \inf_{\substack{c \\ R(a,c) < R(b,c)}} R(a,c) \text{ or } 1 \text{ if } R(a,c) \geq R(b,c), \text{ for all } c.$$

$R^r(a,b)$ is a minimum leaving flow related to a restricted by the presence of b.

$$R^\ell(b,a) = \inf_{\substack{c \\ R(c,a) < R(c,b)}} R(c,a) \text{ or } 1 \text{ if } R(c,a) \geq R(c,b), \text{ for all } c.$$

$R^\ell(b,a)$ corresponds to the minimum entering flow related to a restricted by the presence of b.

Similar rankings will be given by the quasiorders R^r and $(R^{-1})^r$.

Finally, $Tr = R^r \wedge (R^{-1})^r$ is a valued quasiorder (see Proposition 2.28) included in R.

In order to express the indifferences between the alternatives, we consider $\hat{\hat{I}} = Tr \wedge Tr^{-1}$ which is a similarity relation.

A *third solution* corresponds to crisp partial or complete preorders obtained from scoring functions as defined in Chapter 6.

Example 7.6 is now used to compare the three approaches and to obtain best and worst choices.

7.4.1 Transitive closure

Let us first consider the transitive closure $\hat{R}$. This reflexive and transitive relation is given in Table 7.24, for Example 7.6.

$\hat{R}$, transitive closure of R

	1	2	3	4	5	6	7	8
1	1.	.75	.75	.75	.75	.75	.75	.75
2	.76	1	.94	1	.94	.82	.94	.91
3	.76	.86	1	1	1	.82	.94	.91
4	.76	.86	.94	1	.94	.82	.94	.91
5	.76	.86	.94	1	1	.82	.94	.91
6	.76	.86	.92	.92	.92	1	.96	.91
7	.76	.85	.85	.85	.85	.82	1	.85
8	.76	.77	.77	.77	.77	.77	.77	1

Table 7.24

We can consider the cut relations of $\hat{R}$ in terms of Hasse diagrams. These crisp partial preorders (quasiorders) appear in Figure 7.18.

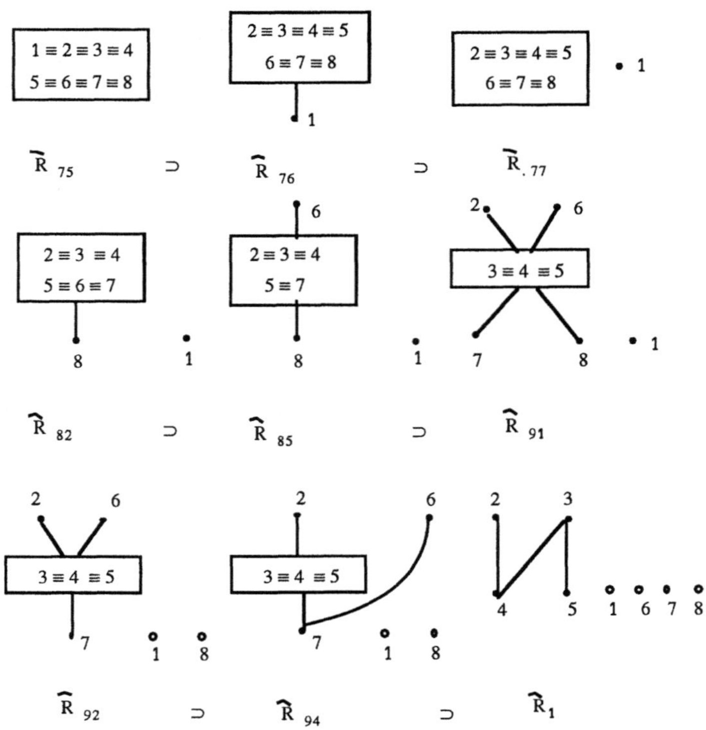

Fig. 7.18

The proximity relation $I = R \wedge R^{-1}$ and its transitive closure, the similarity relation $\hat{I}$ appear respectively in Tables 7.25 and 7.26.

Proximity Relation I

	1	2	3	4	5	6	7	8
1	1	.75	.70	.62	0	0	0	0
2		1	.86	.65	.57	.73	.63	.80
3			1	.94	.93	.46	.73	.64
4				1	.88	.22	.85	.60
5					1	0	.80	.77
6						1	.70	0
7							1	0
8								1

Table 7.25

Similarity relation $\hat{I}$

	1	2	3	4	5	6	7	8
1	1	.75	.75	.75	.75	.73	.75	.75
2		1	.86	.86	.86	.73	.85	.80
2			1	.94	.93	.73	.85	.80
4				1	.93	.73	.85	.80
5					1	.73	.85	.80
6						1	.73	.73
7							1	.80
8								1

Table 7.26

The similarity relation $\hat{I}$ can be viewed as a partition tree (see Section 4.2) and corresponds to Figure 7.19

Partition tree related to $\hat{I}$

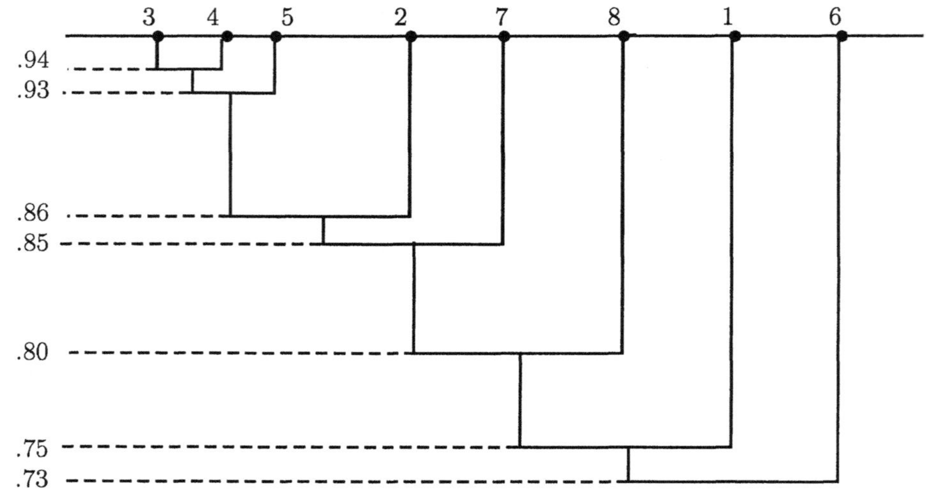

Fig. 7.19

This explains the presence of clusters (indifferent alternatives) at a rather high level of cutting levels in the crisp quasiorders.

As already mentioned, the partition tree can be easily obtained using the d-matrix and the related shortest spanning tree. The d-matrix for Example 7.6 correspond to Table 7.27 and the related shortest spanning tree to Fig. 7.20

Ultrametric $d = 1 - I$

	1	2	3	4	5	6	7	8
1	0	.25	.30	.38	1	1	1	1
2		0	.14	.35	.43	.27	.37	.20
3			0	.06	.07	.34	.27	.36
4				0	.12	.78	.15	.40
5					0	1	.20	.23
6						0	.30	1
7							0	1
8								0

Table 7.27

Shortest spanning tree

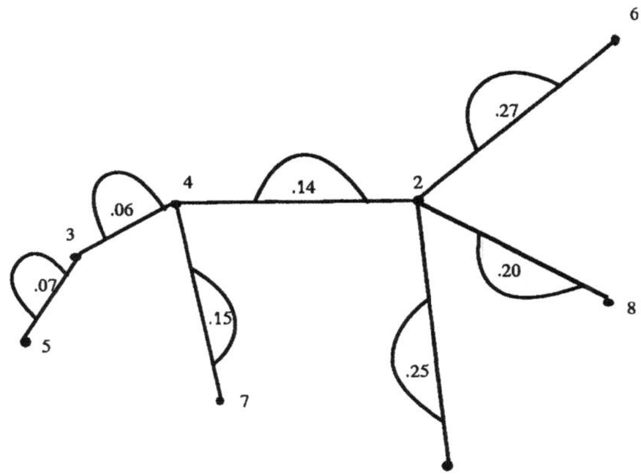

Fig. 7.20

7.4.2 Intersection of traces

We first consider the right traces related to R and R^{-1}. They are presented in Table 7.28.

Right traces R^r and $(R^{-1})^r$

R^r	1	2	3	4	5	6	7	8
1	1	0	0	0	0	0	0	0
2	.76	1	.80	.82	.80	.82	.80	.80
3	.70	.46	1	.80	1	.46	.46	.91
4	.64	.22	.22	1	.74	.22	.22	.74
5	.33	0	0	0	1	0	0	.57
6	0	0	0	0	0	1	.64	.76
7	0	0	0	0	0	.63	1	.81
8	0	0	0	0	0	0	0	1

$(R^{-1})^r$	1	2	3	4	5	6	7	8
1	1	0	0	0	0	0	0	0
2	.75	1	.57	.57	.57	.63	.57	.57
3	.70	.64	1	.64	.64	.64	.64	.64
4	.62	.62	.60	1	.60	.92	.85	.60
5	0	0	0	0	1	.76	.76	.76
6	0	0	0	0	0	1	0	0
7	0	0	0	0	0	.96	1	0
8	0	0	0	0	.74	.80	.74	1

Table 7.28

The intersection $R^r \wedge (R^{-1})^r$ is a fuzzy quasiorder included in R. This intersection appears in Table 7.29 and the related cut relations in Figure 7.21.

$$Tr = R^r \wedge (R^{-1})^r$$

	1	2	3	4	5	6	7	8
1	1	0	0	0	0	0	0	0
2	.75	1	.57	.57	.57	.63	.57	.57
3	.70	.46	1	.64	.64	.46	.46	.64
4	.62	.22	.22	1	.60	.22	.22	.60
5	0	0	0	0	1	0	0	.57
6	0	0	0	0	0	1	0	0
7	0	0	0	0	0	.63	1	0
8	0	0	0	0	0	0	0	1

Table 7.29

Cut relations in Tr

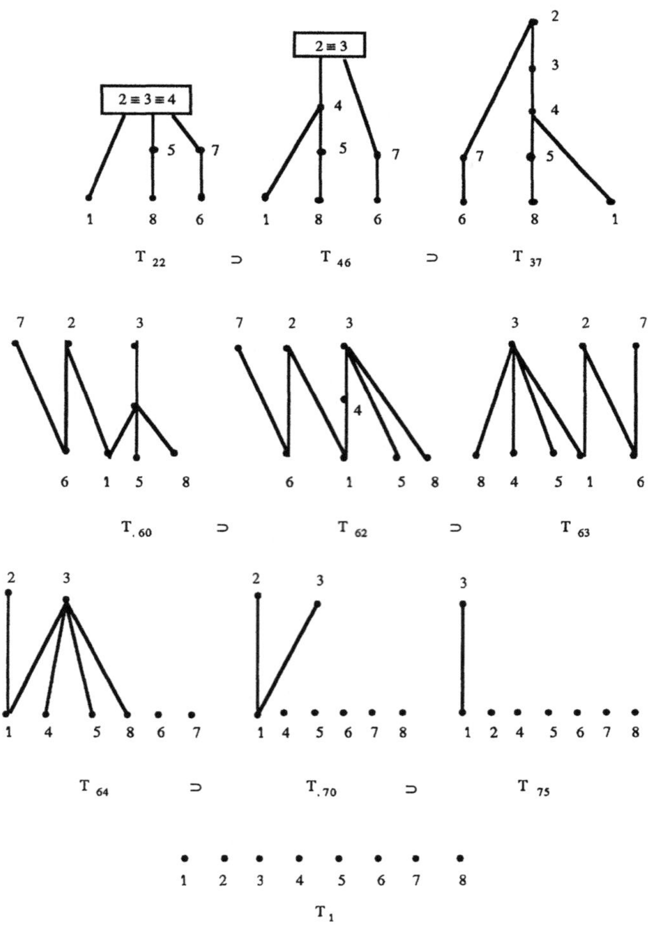

Fig. 7.21

The similary relation $\hat{\hat{I}} = Tr \wedge Tr^{-1}$ is shown in Table 7.30 and the related partition tree appears in Figure 7.22.

$$\hat{\hat{I}} = Tr \wedge Tr^{-1}$$

	1	2	3	4	5	6	7	8
1	1	0	0	0	0	0	0	0
2		1	.46	.22	0	0	0	0
3			1	.22	0	0	0	0
4				1	0	0	0	0
5					1	0	0	0
6						1	0	0
7							1	0
8								1

Table 7.30

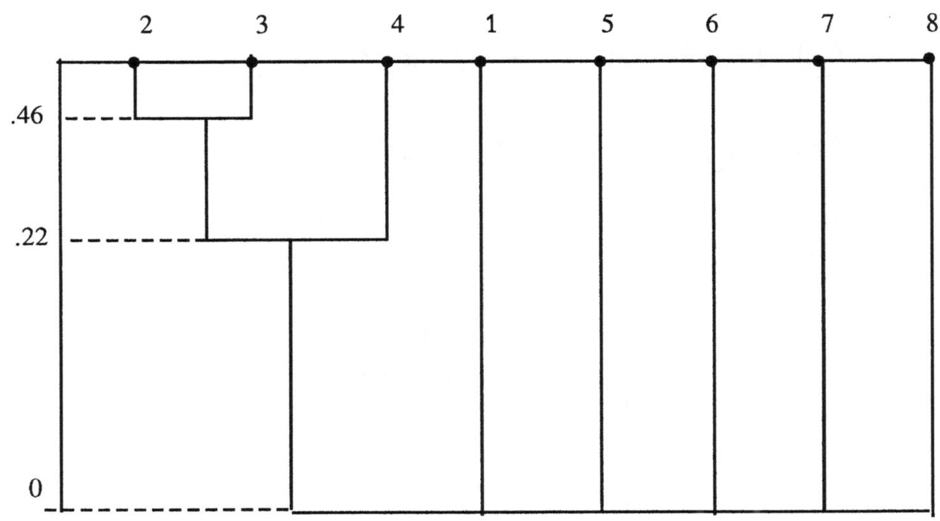

Fig. 7.22

The last picture shows that only a few pairs of alternatives have significant indifference levels and explains that nearly no indifference classes appear in Fig. 7.21.

7.4.3 Scoring functions and undominated(ing) alternatives

We are now considering the rankings (see Section 6.1.2) obtained with the use of the leaving flow S_L, the entering flow S_E, the net flow $(S_{L/E})$ and the intersection of S_L and S_E $(S_L \wedge S_E)$.

For the example related to Table 7.21, we obtain Table 7.31 :

a	S_L	S_E	$S_{L/E}$
1	2.07	−2.43	−.36
2	5.92	−4.79	1.13
3	5.73	−5.48	.25
4	5.01	−5.99	−.98
5	4.49	−5.05	−.56
6	4.81	−2.20	2.61
7	4.55	−4.32	.23
8	2.61	−4.92	−2.31

Table 7.31

These score functions give the following rankings :

S_L ⇒ 

S_E ⇒ 

$S_{L/E}$ ⇒ 

The intersection of S_L and S_E ($S_L \wedge S_E$) gives a partial ranking (quasi-order) shown in Figure 7.23 using a Hasse diagram :

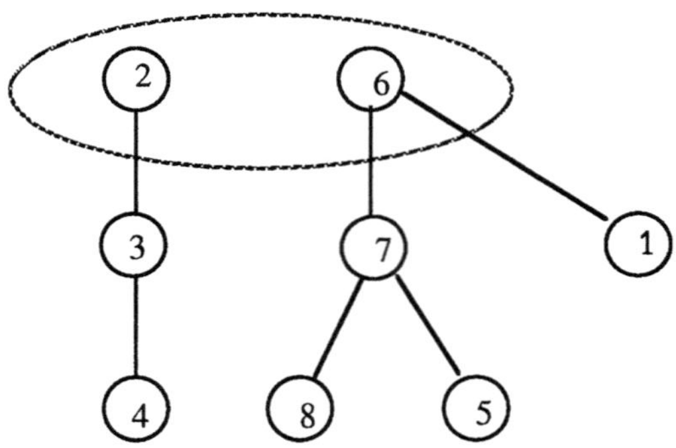

Fig. 7.23

The kernel of this quasi-order (elements of the kernel are incomparable and elements outside the kernel are outranked by at least one element of the kernel) represents the "best" alternatives and corresponds in the example to alternatives 2 and 6 (subset surrounded by dotted lines in Fig. 7.23).

Let us finally consider nondominance and nondomination concepts as introduced in Section 6.

For the same example corresponding to Table 7.21, we immediately obtain Tables 7.32 and 7.33 representing P and P^d.

P	1	2	3	4	5	6	7	8	$\max_{c \neq a} P(a,c)$
1	0	0	0	0	0	0	0	0	0
2	.01	0	.04	.35	.25	.09	.19	.20	.35
3	0	0	0	.06	.07	0	.07	.27	.27
4	.02	0	0	0	0	0	.09	.14	.14
5	.33	0	0	.12	0	0	0	.09	.33
6	0	0	.18	.70	.76	0	.26	.80	.76
7	0	0	0	0	.02	0	0	.81	.81
8	0	0	0	0	0	0	0	0	0

Table 7.32

P^d	1	2	3	4	5	6	7	8	$\min_{c \neq a} P^d(a,c)$
1	1	.99	1	.98	.67	1	1	1	.67
2	1	1	1	1	1	1	1	1	1
3	1	.96	1	1	1	.82	1	1	.82
4	1	.65	.94	1	.88	.30	1	1	.30
5	1	.75	.93	1	1	.24	.98	1	.24
6	1	.91	1	1	1	1	1	1	.91
7	1	.81	.93	.91	1	.74	1	1	.74
8	1	.80	.73	.86	.91	.20	.19	1	.19

Table 7.33

The scoring functions (S_{ND} et S_{Nd}) give the following rankings :

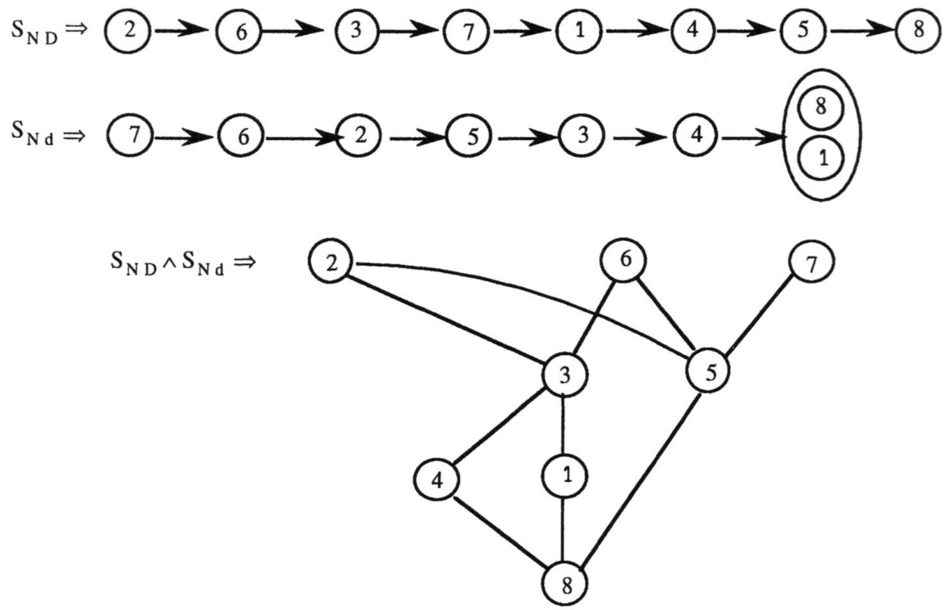

Fig. 7.24

Alternative 2 is an unfuzzy undominated alternative. It is clearly a good choice.

In the same spirit, alternatives 1 and 8 are unfuzzy undominating alternatives. They are clearly a bad choice.

All this information can be summarized in Figure 7.25 indicating second best and second worst choices.

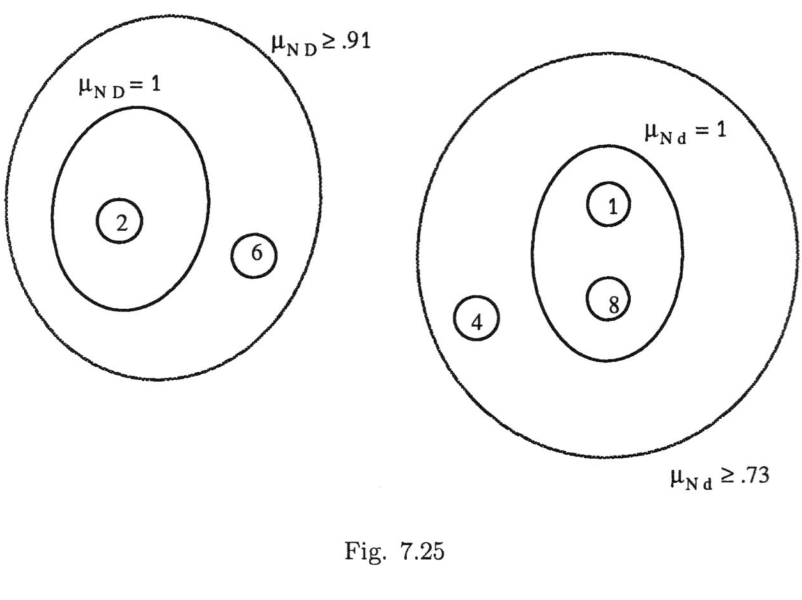

Fig. 7.25

It is clear that the crisp relation $P_>$ ($aP_>b$ iff $P(a,c) > 0$) defined (only one's are indicated) in Table 7.34 is not acyclic ($5P4, 4P7, 7P5$) despite $A^{UND} \neq \phi$.

$P_>$	1	2	3	4	5	6	7	8
1								
2	1		1	1	1	1	1	1
3				1	1		1	1
4	1							1
5	1			1				1
6			1	1	1		1	1
7					1			1
8								

Table 7.34

Chapter 8

Summary, perspectives and open problems

In this chapter we reformulate some results included in the book from a more practical point of view. This may help non-specialists to catch the main points and go into details (if they need any) in the corresponding Sections. Some unsolved theoretical problems are also presented.

8.1 Operators, transitivity and axiomatics

The first inevitable step of any kind of application of fuzzy logic consists in the proper choice of logical operators such as negation, conjunction, disjunction, implication and others. Although t-norms are well-accepted and appropriate models for **AND**, they form a too broad class from a practical point of view. However, theoretical results presented in Chapter 1 can help the user in his/her choice.

Let us recall that, basically, two main classes of t-norms can be distinguished (Section 1.3.1) :

- t-norms having zero divisors
- positive t-norms.

This distinction is very natural, as one could recognize during the development of this book.

In Chapter 2, some properties (e.g. T-asymmetry, see Proposition 2.13) of valued binary relations mainly depend on the underlying class of t-norms (positive or having zero divisors).

The functional equations considered in Chapter 3 immediately imply that the t-conorm should be nilpotent (i.e. the corresponding t-norm must have zero divisors, see Section 1.4.2 and Section 3.3.1).

Requiring further properties (such as left-continuity, Archimedean property), the picture becomes clearer (see Figures 8.1 and 8.2).

Each of these t-norms can be considered as *prototypes* of the corresponding class illustrated by the Table 8.1 (φ is an automorphism of the unit interval).

Notice that $\min(x,y) = \varphi^{-1}(\min(\varphi(x),\varphi(y)))$ and $Z(x,y) = \varphi^{-1}(Z(\varphi(x),\varphi(y)))$ for any automorphism φ, thus in these cases the corresponding classes consist of the prototype only.

Therefore, at the first stage, we propose to choose one of these classes on the basis of the available information.

Prototypes of t-norms having zero divisors

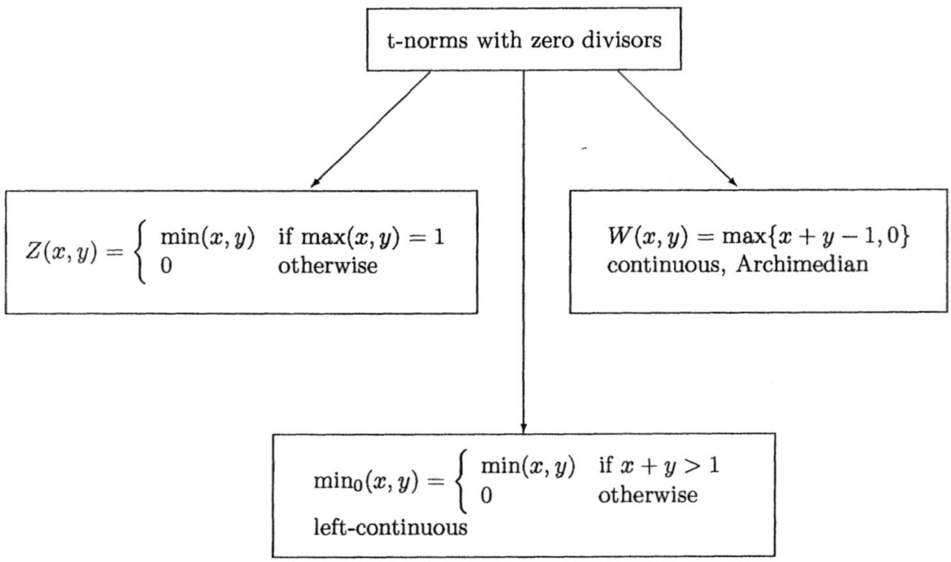

Fig. 8.1

Prototypes of positive t-norms

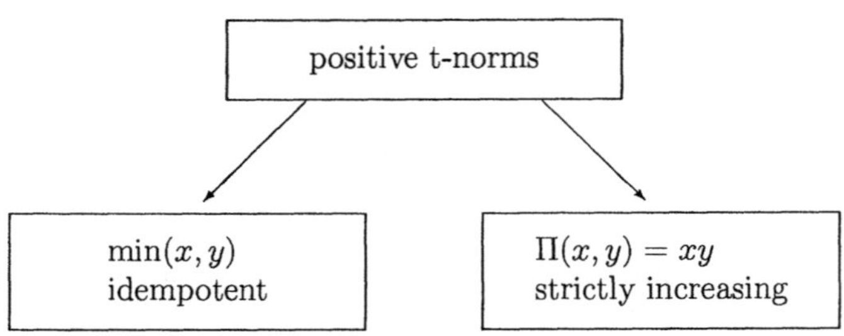

Fig. 8.2

Some important subclasses of t-norms

Prototype	General member of the corresponding class
$\min(x, y)$	$\min(x, y)$
xy	$\varphi^{-1}(\varphi(x)\varphi(y))$
$\min_0(x, y)$	$\begin{cases} \min(x, y) & \text{if } \varphi(x) + \varphi(y) > 1 \\ 0 & \text{otherwise} \end{cases}$
$\max(x + y - 1, 0)$	$\varphi^{-1}(\max\{\varphi(x) + \varphi(y) - 1, 0\})$
$Z(x, y)$	$Z(x, y)$

Table 8.1

For example, this is done in Chapter 3. The selected axioms and the functional equations represent pieces of information and, as a consequence, only the class of Lukasiewicz-like t-norms can appear in the De Morgan triple (Lemma 3.2). Moreover, after fixing the automorphism φ, the operations are unique.

This result of Chapter 3 points out another important issue of the proper choice for the logical operators: different kinds of connectives (negation, conjunction, disjunction, implication, etc) should be in a 'harmony' with each other. This harmony can be expressed by some interrelations among the operators. The most basic and necessary connection is represented by the De Morgan law. This supports that, after choosing one class of t-norms in Table 8.1 and fixing the automorphism φ, we have to use the strong negation defined by φ: $N_\varphi(x) = \varphi^{-1}(1 - \varphi(x))$ (see Lemma 3.3). Finally, the appropriate t-conorm is obtained by the De Morgan law.

The situation is a little bit more complicated when we need also an implication. Should it be an R-implication or an S-implication? Perhaps a QL-implication? (Section 1.8)

In the problems investigated in the present book, R-implications play an almost exclusive role especially in Chapters 2 and 4 (transitivity type conditions, see e.g. Proposition 2.20). R-implications appear very naturally, by formulating some properties related to a t-norm in an equivalent way. Then, obviously, we must use the R-implication defined by the related t-norm in question. The connection between equivalences and implications also supports the use of R-implications (see Proposition 1.13 and the remark at the end of Section 1.9.2).

Sometimes it happens that R- and S-implications are identical for a De Morgan triple. Among the prototypes mentioned before, only the nilpotent minimum and the Lukasiewicz t-norm are such that we have (with $N(x) = 1 - x$)

$$I_T^{\rightarrow}(x, y) = N(T(x, N(y))).$$

As a conclusion, the most attractive operations (in the sense that they satisfy the largest number of important theoretical conditions, see Table 1.1) are given in Table 8.2.

Operations with the most attractive properties

T	$\varphi^{-1}(\max\{\varphi(x)+\varphi(y)-1,0\})$	$\begin{cases}\min(x,y) & \text{if } \varphi(x)+\varphi(y)>1\\ 0 & \text{otherwise}\end{cases}$
S	$\varphi^{-1}(\min\{\varphi(x)+\varphi(y),1\})$	$\begin{cases}1 & \text{if } \varphi(x)+\varphi(y)\geq 1\\ \max(x,y) & \text{otherwise}\end{cases}$
N	$\varphi^{-1}(1-\varphi(x))$	$\varphi^{-1}(1-\varphi(x))$
$I^{\rightarrow}$	$\varphi^{-1}(\min\{1-\varphi(x)+\varphi(y),1\})$	$\begin{cases}1 & \text{if } x\leq y\\ \varphi^{-1}(\max\{1-\varphi(x),\varphi(y)\}) & \text{otherwise}\end{cases}$

Table 8.2

In spite of these facts, the role of the minimum and related connectives is also extremely important, due to the behaviour of α-cuts in accordance with their related properties (see Propositions 2.4 and 2.12). This is summarized (together with strict operations) in Table 8.3.

T	$\min(x,y)$	$\varphi^{-1}(\varphi(x)\varphi(y))$
S	$\max(x,y)$	$\varphi^{-1}(\varphi(x)+\varphi(y)-\varphi(x)\varphi(y))$
N	$\varphi^{-1}(1-\varphi(x))$	$\varphi^{-1}(1-\varphi(x))$
R-implication	$\begin{cases}1 & \text{if } x\leq y\\ y & \text{otherwise}\end{cases}$	$\varphi^{-1}\left(\min\left\{1,\frac{\varphi(y)}{\varphi(x)}\right\}\right)$
S-implication	$\varphi^{-1}(\max(1-\varphi(x),\varphi(y)))$	$\varphi^{-1}(1-\varphi(x)+\varphi(x)\varphi(y))$

Table 8.3

The characterization of implications defined by Łukasiewicz-like t-norms is an important point in Chapter 1 (Theorem 1.15), which gives an equivalent characterization of the family of Łukasiewicz-like t-norms. We think that a similar result can be obtained for implications defined by the nilpotent minimum and in an equivalent way, for the nilpotent minimum family, too.

Open problem 1. *Characterize the nilpotent minimum family of t-norms.*

We have shown the important role of R-implications in representation of T-transitive valued binary relations (Theorems 2.1 and 2.2). For T-preorders, relations having the form

$$R(a,b) = I_T^{\rightarrow}(h(b),h(a))$$

have also been characterized (Theorem 4.10). It would be nice to have a similar result for not necessarily reflexive valued binary relations, too.

Open problem 2. *Characterize those valued binary relations R on A such that there exist two functions $h: A \rightarrow [0,1]$, $k: A \rightarrow [0,1]$, $h(a) \geq k(a)$:*

$$R(a,b) = I_T^{\rightarrow}(h(b),k(a)).$$

Turning back to the T-transitivity property, its preservation is an interesting problem in several contexts. On one hand, when T is a continuous Archimedean t-norm with additive genartor f and we have T-transitive preference relations $R_1, \ldots, R_m$ representing m criteria then the use of the aggregation

$$M(R_1, \ldots, R_m) = f^{-1}\left(\frac{1}{m}\sum_{i=1}^{m} f(R_i)\right)$$

implies that $R = M(R_1, \ldots, R_m)$ is also T-transitive (Theorem 5.11). Moreover,

$$M(R_1, \ldots, R_m) = \min_{i=1,m} f_i(R_i)$$

is the most general form of an aggregation that ensures the transitivity of $R = M(R_1, \ldots, R_m)$ when $R_1, \ldots, R_m$ are transitive valued binary relations (see Theorem 7.2).

On the other hand, it was proved that if R is transitive then $P = p(R, R^{-1})$ is also transitive by using a function $p : [0,1]^2 \to [0,1]$ which is nondecreasing in the first place and nonincreasing in the second argument with $p(x,x) = 0$ for all $x \in [0,1]$ (Theorem 3.7). In particular, $P = W_\varphi(R, N_\varphi R^{-1})$ preserves transitivity among the solutions of the functional equation systems described in Section 3.3.

If a valued binary relation R is not T-transitive then we can approach it both from the top and from the bottom by the T-transitive closure and by a maximal transitive relation contained by R, respectively (Propositions 2.22 and 2.23 and Theorems 2.3 and 2.4).

It would be desirable to establish other constructions of a maximal transitive relation, by the exploitation of the traces (these can be obtained easily in one step). Moreover, further connections between the representation of a T-transitive relation and that of its traces (which are always reflexive and T-transitive valued binary relations) should be proved.

Open problem 3. *Under which condition does the following statement hold? If R is a T-transitive valued binary relation on A*

$$R(a,b) = \inf_{\gamma \in \Gamma} I_T^{\rightarrow}(h_\gamma(b), k_\gamma(a))$$

($h_\gamma \geq k_\gamma$) then its traces R^ℓ and R^r have the following representation:

$$\begin{aligned} R^\ell(a,b) &= \inf_{\gamma \in \Gamma} I_T^{\rightarrow}(k_\gamma(b), k_\gamma(a)), \\ R^r(a,b) &= \inf_{\gamma \in \Gamma} I_T^{\rightarrow}(h_\gamma(b), h_\gamma(a)). \end{aligned}$$

We would like to point out a dangerous 'procedure' often used in the literature. It corresponds to the 'automatic fuzzification' of crisp notions (see e.g. Proposition 3.1): without any careful axiomatic study we might loose some important properties of the defined notion which were satisfied in the crisp case, or we can even obtain contraditory results.

Therefore, we would like to urge the reader to predetermine what he/she wants to preserve from the crisp properties (in other words, axiomatize the problem), then investigate if it is possible to satisfy the requirements.

It might happen (as in Chapter 3) that we obtain a wide class of solutions to the problem under consideration (like the functions p, i, j, see Theorems 3.1). Then some additional conditions may help to get a unique solution (Theorems 3.3, 3.4 and 3.5).

At the same time, axiomatic approaches can support some intuitively acceptable proposals. For example, the strict preference relation

$$P(a, b) = \max(R(a, b) - R(b, a), 0)$$

extensively used by people having theoretical or practical interest is well-supported by the axiomatics developed in Section 3.3.2 (Theorem 3.4).

This is also the case with similarity classes of a T-similarity relation. Definition of Zadeh completely fits into the axiomatic framework of T-classes: these two notions are the same for T-similarity relations (Theorem 4.4).

8.2 Modelling and aggregation in MCDM

Let us consider a set $A : \{a, b, c, \ldots\}$ of possible courses of action (alternatives) and $j \in J$, the attributes (criteria, points of view) with which alternative performances are observed.

The performance evaluation (score, rating, ...) of alternative a with respect to attribute j can be precise (crisp) or imprecise (fuzzy or linguistic).

We denote the crisp performance rating $g_j(a)$ and we introduce a fuzzy number to deal with imprecise evaluation $\mu_j^a(x)$, $x \in R$, and to capture the subjectivity of human behavior.

We now define (modelling phase) for every pair of alternatives $(a, b) \in A^2$ with respect to a single attribute $j \in J$ an *intensity of the preference* of a over b on the basis of $g(a), g(b)$ and indifference and preference thresholds $IT[g_j(a)]$, $IT[g_j(b)]$, $PT[g_j(a)]$, $PT[g_j(b)]$ (Section 7.2.1) or a *degree of credibility* of the preference of a over b using $\mu_j^a(x)$, $\mu_j^b(x)$ (Section 7.2.2).

The *concordance index* (intensity of preference)

$$R_j(a, b) = \frac{PT[g_j(a)] - \min(g_j(b) - g_j(a), PT[g_j(a)])}{PT[g_j(a)] - \min(g_j(b) - g_j(a), IT[g_j(a)])}$$

and *the possibility degree* (degree of credibility)

$$\pi_j(a, b) = \sup_{x \geq y}[\min(\mu_j^a(x), \mu_j^b(y))]$$

both present the poperties of reflexivity, completeness and negative transitivity.

In the case of precise measurement, when a veto thresthold $VT(g_j)$ is introduced, the *non discordance index*

$$ND_j(a, b) = \max\left\{0, \min\left(1, \frac{VT(g_j(a)) - (g_j(b) - g_j(a))}{VT(g_j(a)) - PT(g_j(a))}\right)\right\}$$

also presents the properties of reflexivity, completeness and negative transitivity.

Concordance and nondiscordance indices as well as possibility degrees have been defined for every pair of alternatives and with respect to a single attribute.

The next step (aggregation phase) consists in aggregating the single attribute valued binary relation into a unified (global) relation with the use of *weights* ω_j, $j \in J$.

The choice of an aggregation operator is linked to some "desirable" properties (Sections 5.3, 5.7 and 5.8).

The most valuable property to preserve is the negative transitivity (or the transitivity, by duality). The answer is however somehow narrow (Section 7.3.1) :
"the class of weighted aggregators that preserve the negative transitivity of a valued binary relation which is negatively transitive (as R_j, ND_j, π_j) corresponds to

$$R(a,b) = \max_j h_j[R_j(a,b)]$$

where h_j are non decreasing functions from $[0,1]$ to $[0,1]$ with $h_j(1) = 1$ for some $j \in J$ and $h_j(0) = 0$, for all $j \in J$."

This family contains the *weighted maximum*

$$R(a,b) = \max_j\{\min(\omega_j, R_j(a,b))\} \ , \ \max_{j \in J} \omega_j = 1,$$

which corresponds to the qualitative connection :
"a is globally not worse than b for at least one weighted criterion".

Another class of popular weighted aggregators corresponds to the quasi-linear means (Section 5.8) :

$$R(a,b) = f^{-1}\left[\sum_j \omega'_j f R_j(a,b)\right] \ , \ \sum_{j \in J} \omega'_j = 1,$$

where f is a strictly monotonic continuous function mapping [0,1] onto [0,1].

This last class of aggregators is characterized by the following "desirable" properties : continuity (which avoids chaotic reactions of the aggregator), strict monotonicity (positive responsiveness), idempotency (unanimity), bisymmetry. Negative transitivity is however lost.

Some particular quasi-linear means are often used :

(1) the weighted mean :

$$R(a,b) = \sum_j \omega'_j R_j(a,b)$$

which perserves the stability property related to interval scales with independent zeroes and same unit ($SPLU$).

(2) the weighted geometric mean

$$R(a,b) = \prod_j [R_j(a,b)]^{\omega'_j}$$

which preserves stability property (SSI) related to independent ratio scales.

The last step (exploitation phase) consists in defining global choices (best and worst actions) among the set of alternatives A or a partial or complete ranking of the elements of A.

8.3 Choice problem

Choice is usually based on the following scoring functions (Section 6.2)

- the non domination degree

$$\mu_{ND}(a,R) = \min_{c\in A}(1 - P(c,a)) = \min_{c} P^d(a,c)$$

- the non dominance degree

$$\mu_{Nd}(a,R) = \min_{c\in A}(1 - P(a,c)) = 1 - \max_{c\in A} P(a,c)$$

where P corresponds to the strict preference

$$\begin{aligned} P(a,b) &= \max(R(a,b) - R(b,a), 0) \\ &= W(R(a,b), 1 - R(b,a)) \end{aligned}$$

(W represents the Lukasiewicz t-norm).

Clearly best actions correspond to alternatives belonging to A^{UND}, the subset of unfuzzy nondominated elements :

$$A^{UND} : \{a \in A : \mu_{ND}(a,R) = 1\}.$$

Clearly worst choices correspond to alternatives belonging to A^{UNd}, the subset of unfuzzy nondominating elements :

$$A^{UNd} : \{a \in A : \mu_{Nd}(a,R) = 1\}.$$

A sufficient condition to obtain non empty A^{UND} and A^{UNd} corresponds to the transitivity of R.

Open problem 4. *Determine the necessary and sufficient conditions to obtain A^{UND} or $A^{UNd} \neq \phi$ (condition defined by Montero and Tejada (1988) is too strong and concerns social choice functions; see Sections 6.2 and 7.4.3).*

8.4 Ranking problem

Ranking of alternatives can be obtained with the use of scoring functions or with the help of transitive relations which are "close" to the outranking relation R.

The scoring function

$$S_{L/E}(a,R) = \sum_{c}(R(a,c) - R(c,a)),$$

called net flow score, presents some "desirable" properties as strict monotonicity, continuity and stablity for interval scales with the same unit (Section 6.2.4).

We know that the general expression of a reflexive and transitive relation corresponds to (Section 2.5.5) :

$$Tr(a,b) = \inf_{\gamma \in \Gamma} I^{\rightarrow}_{\min} [h_\gamma(b), h_\gamma(a)]$$

where $I^{\rightarrow}_{\min}$ represents the Gödel implication, and h_γ is a family of functions from A to $[0,1]$ and $\gamma \in \Gamma$.

$$Tr(a,b) = \inf_{\substack{\gamma \\ h_\gamma(b) > h_\gamma(a)}} h_\gamma(a) \text{ or } 1 \text{ if } h_\gamma(b) \leq h_\gamma(q),\ \forall \gamma \in \Gamma.$$

It seems natural to consider the particular family

$$h_c(a) = R(a,c) \text{ or } R(c,a) \quad , \quad c \in A$$

to obtain

$$\begin{aligned} Tr_1(a,b) &= \inf_{\substack{c \\ R(b,c) > R(a,c)}} R(a,c) \text{ or } 1 \text{ if } R(b,c) \leq R(a,c),\ \forall c \in A \\ &= R^r(a,b), \text{ the right trace of } R \\ Tr_2(a,b) &= \inf_{\substack{c \\ R(c,b) > R(c,a)}} R(c,a) \text{ or } 1 \text{ if } R(c,b) \leq R(c,a),\ \forall c \in A \\ &= R^\ell(b,a) = (R^{-1})^r(a,b), \text{ the right trace of } R^{-1}. \end{aligned}$$

Tr_1 included in R and Tr_2 are transitive relations (Section 2.5.4).

$Tr_1 \wedge Tr_2$ is also a transitive relation included in R (Section 2.5.8).

R^r, $(R^{-1})^r$, $R^r \wedge (R^{-1})^r$ correspond to fuzzy quasiorders (fuzzy partial preorders) (Section 7.4.2).

Their λ-cuts give crisp partial preorders; it seems to be a readable answer to the problematics of ranking preserving the indifference and incomparability that naturally appear in MCDM exploitation phase.

None of these three binary relations are however maximal transitivity relation contained in R.

The construction of such maximal transitive relation $\check{R}$ has been produced in a recursive way :

$$\check{R}(a_1, a) = R(a_1, a) \quad , \quad \forall a \in A : \ \{a_1, \ldots, a_n\}.$$

Assuming that $\check{R}(a_j, a)$ is defined, let

$$\check{R}(a_{j+1}, a_k) = \begin{cases} \min\{R(a_{j+1}, a_k), U(a_{j+1}, a_k), V(a_{j+1}, a_k)\} \text{ if } k < j+1 \\ \min\{R(a_{j+1}, a_k), V(a_{j+1}, a_k)\} \text{ if } k \geq j+1 \end{cases}$$

where

$$\begin{aligned} U(a_{j+1}, a_k) &= \min_{c \in A} I^{\rightarrow}_{\min}(\check{R}(a_k, c), R(a_{j+1}, c)) \\ V(a_{j+1}, a_k) &= \min_{i \leq j} I^{\rightarrow}_{\min}(\check{R}(a_i, a_{j+1}), R(a_i, a_k)), \end{aligned}$$

it is proved that $\check{R}$ is transitive and $\check{R} \subseteq R$. Moreover, if Q is a transitive relation such that $\check{R} \subseteq Q \subseteq R$, then $Q = \check{R}$.

This last result is the counterpart of the classical result related to the transitive closure $\hat{R}$:

$$\hat{R}(a,b) = \sup_m R^m(a,b)$$

where $R^m = R^{m-1} \circ R$ and $R^2(a,b) = \sup_c \min\{R(a,c), R(c,b)\}$.

$\hat{R}$ is known to be the closest and minimal transitive relation including R ($\hat{R} \supseteq R$).

References

[1] J. Aczél (1948), On mean values, *Bulletin of the American Math. Society*, **54** : 392–400.

[2] J. Aczél (1949), Sur les opérations définies pour des nombres réels, *Bull. Soc. Math. de France*, **76** : 59–64.

[3] J. Aczél (1966), *Lectures on Functional Equations and Applications*, (Academic Press, New York).

[4] J. Aczél (1984), On weighted synthesis of judgements, *Aequationes Mathematicae* **27** : 288–307.

[5] J. Aczél and C. Alsina (1987), Synthesizing judgements : a functional equations approach, *Math. Modelling*, **9** : 311–320.

[6] J. Aczél and J. Dhombres (1989), *Functional Equations in Several Variables with applications to Mathematics, Information Theory and to the Natural and Social Sciences*, (Cambridge Univ. Press, Cambridge).

[7] J. Aczél and F.S. Roberts (1989), On the possible merging functions, *Math. Social Sciences*, **17** : 205–243.

[8] J. Aczél, F.S. Roberts and Z. Rosenbaum (1986), On scientific laws without dimensional constants, *Journal of Math. Analysis and Appl.*, **119** : 389–416.

[9] C. Alsina (1985), On a family of connectives for fuzzy sets, *Fuzzy Sets and Systems,* **16** : 231–235.

[10] C. Alsina, E. Trillas and L. Valverde (1983), On some logical connectives for fuzzy sets theory, *J. Math. Anal. Appl.,* **93** : 15–26.

[11] K.J. Arrow (1963), *Social Choice and Individual Values,* (Wiley, New York).

[12] J.F. Baldwin, and N.C.F. Guild (1979), Comparison of fuzzy sets on the same decision space, *Fuzzy Sets and Systems*, **2** : 213-231.

[13] J.P. Barthelemy and B. Monjardet (1981), The median procedure in cluster analysis and social choice theory, *Mathematical Social Sciences,* **1** : 235-267.

[14] R. E. Bellman and M. Giertz (1973), On the analytic formalism of the theory of fuzzy sets, *Inform. Sci.,* **5** : 149–156.

[15] R.E. Bellman and J.A. Zadeh (1970), Decision-making in a fuzzy environment, *Management Science,* **17** : 141-164.

[16] R. Bellman and L. Zadeh (1977), Local and fuzzy logic, in: J.C. Dunn and G. Epstein (Eds.), *Modern Uses of Multiple-valued Logic*, (D. Reidel, Dordrecht) : 103–156.

[17] G. Birkhoff (1948), *Lattice Theory* (2nd edition, New York).

[18] B. Black (1958), *The Theory of Committees and Elections,* (Cambridge University Press, Cambridge).

[19] D. Bouyssou (1991), A note on the "min in favour" ranking method for valued preference relations, in: M. Cerny, D. Gluckaufova and D. Loula (Eds.), *Proc. International Workshop on Multicriteria Decision Makings*, (Liblice, Czechoslovakia, Prague).

[20] D. Bouyssou (1992), Ranking methods based on valued preference relations : a characterization of the net flow method, *European Journal of Operational Research*, **60** : 61—68.

[21] D. Bouyssou and P. Perny (1992), Ranking methods for valued preference relations : a characterization of a method based on leaving and entering flows, *European Journal of Operational Reserach*, **61** : 186-194.

[22] J.P. Brans and Ph. Vincke (1985), A preference ranking organization method, *Management Science*, **31** : 647-656.

[23] J.J. Buckley (1985), Ranking alternatives using fuzzy numbers, *Fuzzy Sets and Systems*, **15** : 21-31.

[24] M.K. Chakraborty and M. Das (1983), Studies in fuzzy relations over fuzzy subsets, *Fuzzy Sets and Systems,* **9** : 79–89.

[25] G. Choquet (1953-1954), Theory of capacities, *Annales de l'Institut Fourier*, **5** : 131-295.

[26] N. Christofides (1975), *Graph Theory : An Algorithmic Approach,* (Academic Press, London).

[27] C.H. Coombs (1952), A Theory of Psychological Scaling, *Eng. Res. Bull.* **34**, University of Michigan Press, Ann Arbor, Michigan.

[28] M. Defays (1978), Analyse hiérarchique des préférences et généralisations de la transitivité, *Math. Sci. Hum.*, **16** : 5–27.

[29] A. Di Nola, S. Sessa, W. Pedrycz and E. Sanchez (1989), *Fuzzy Relational Equations and Their Applications to Knowledge Engineering,* (Kluwer, Dordrecht).

[30] J.-P. Doignon, B. Monjardet, M. Roubens and Ph. Vincke (1986), Biorders families, valued relations and preference modelling, *J. of Math. Psychology,* **30** : 435–480.

[31] J. Dombi (1982a), A general class of fuzzy operators, the De Morgan class of fuzzy operators and fuzziness measures induced by fuzzy operators, *Fuzzy Sets and Systems,* **8** : 149–163.

[32] J. Dombi (1982b), Basic concepts for a theory of evaluation : the aggregative operator, *European Journal of Operational Research,* **20** : 221-233.

[33] D. Dubois and H. Prade (1980), *Fuzzy Sets and Systems : Theory and Applications*, (Academic Press, New York).

[34] D. Dubois and H. Prade (1982), A class of fuzzy measures based on triangular norms – A general framework for the combination of uncertain information, *Int. J. of General Systems,* **8** : 43–61.

[35] D. Dubois and H. Prade (1983), Ranking fuzzy numbers in the setting of possibility theory, *Inform. Sci.,* **30** : 183-224.

[36] D. Dubois and H. Prade (1984a), A theorem on implication functions defined from triangular norms, *Stochastica,* **VIII** : 267–279.

[37] D. Dubois and H. Prade (1984b), Criteria aggregation and ranking of alternatives in the framework of fuzzy set theory, in: H.-J. Zimmermann, L.A. Zadeh and B. Gaines (Eds.), *Fuzzy Sets and Decision Making*, (TIMS Studies in Management Sciences, vol. 20) : 209-240.

[38] D. Dubois and H. Prade (1984c), Fuzzy set-theoretic differences and inclusions and their use in the analysis of fuzzy equations, *Control and Cybernetics,* **13** : 129–145.

[39] D. Dubois and H. Prade (1985), A review of fuzzy set aggregation connectives, *Inform. Sci.,* **36** : 85–121.

[40] D. Dubois and H. Prade (1986), Weighted minimum and maximum in fuzzy set theory, *Inform. Sci.,* **39** : 205-210.

[41] D. Dubois and H. Prade (1990), Aggregation of possibility measures, in: J. Kacprzyk and M. Fedrizzi (Eds.), *Multiperson Decision Making Using Fuzzy Sets and Possibility Theory* (Kluwer, Dordrecht) : 55-63.

[42] D. Dubois and H. Prade (1991a), On the ranking of ill-known values in possibility theory, *Fuzzy Sets and Systems*, **43** : 311-317.

[43] D. Dubois and H. Prade (1991b), Fuzzy sets in approximate reasoning, Part 1: Inference with possibility distributions, *Fuzzy Sets and Systems,* **40** : 143–202.

[44] D. Dubois and H. Prade (1991c), Putting rough sets and fuzzy sets together, in: R. Slowinski (Ed.), *Intelligent Decision Support* (Kluwer, Dordrecht) : 203–231.

[45] D. Dubois, J.C. Fodor, H. Prade and M. Roubens (1994), Aggregation of decomposable measures with application to utility theory, submitted.

[46] J.J. Dujmovic (1974), Weighted conjunctive and disjunctive means and their application in system evaluation, *Univ. Beograd. Publ. Elektrotechn. Fak.* : 147-158.

[47] J.J. Dujmovic (1975), Extended continuous logic and the theory of complex criteria, *Univ. Beograd. Publ. Elektrotechn. Fak.* : 197-216.

[48] H. Dyckhoff (1985), Basic concepts for a theory of evaluation : Hierarchical aggregation via autodistributive connectives in fuzzy set theory, *European Journal of Operational Research,* **10** : 221-233.

[49] W. M. Faucett (1955), Compact semigroups irreducibly connected between two idempotents, *Proc. Amer. Math. Soc.,* **6** : 741–747.

[50] P.C. Fishburn (1970a), Intransitive indifference with unequal indifference intervals, *Journal of Mathematical Psychology,* **7** : 114-149.

[51] P.C. Fishburn (1970b), *Utility Theory for Decision-making*, (J. Wiley, New-York).

[52] P.C. Fishburn (1972), *The Theory of Social Choice,* (Princeton University Press, Princeton).

[53] P. C. Fishburn (1973), Binary choice probabilities: On the varieties of stochastic transitivity, *Journal of Math. Psychology,* **10** : 327–352.

[54] P.C. Fishburn (1977a), Multicriteria choice functions based on binary relations, *Operations Research,* **25** : 989-1012.

[55] P.C. Fishburn (1977b), Condorcet social choice functions, *SIAM J. Appl. Math.*, **33** : 469–489.

[56] J.C. Fodor (1988), t-norms and generalized copulas, *BUSEFAL,* **36** : 4–7.

[57] J.C. Fodor (1991a), Aggregation of decomposable measures, in : Abstracts of the 13th Linz Seminar on Fuzzy Set Theory : 28-31.

[58] J.C. Fodor (1991b), Aggregation of preferences – an axiomatic approach with applications, *Annales Univ. Sci. Budapest, Sectio Computatorica,* **12** : 86-92.

[59] J.C. Fodor (1991c), A remark on constructing t-norms, *Fuzzy Sets and Systems,* **41** : 195–199.

[60] J.C. Fodor (1991d), On fuzzy implication operators, *Fuzzy Sets and Systems,* **42** : 293–300.

[61] J.C. Fodor (1991e), Strict preference relations based on weak t-norms, *Fuzzy Sets and Systems,* **43** : 327–336.

[62] J.C. Fodor (1992a), Traces of fuzzy binary relations, *Fuzzy Sets and Systems,* **50** : 331–341.

[63] J.C. Fodor (1992b), An axiomatic approach to fuzzy preference modelling, *Fuzzy Sets and Systems,* **52** : 47–52.

[64] J.C. Fodor (1993a), Fuzzy connectives via matrix logic, *Fuzzy Sets and Systems,* **56** : 67–77.

[65] J.C. Fodor (1993b), A new look at fuzzy connectives, *Fuzzy Sets and Systems,* **57** : 141–148.

[66] J.C. Fodor (1993c), On contrapositive symmetry of implications in fuzzy logic, *Proceedings of the First European Congress on Fuzzy and Intelligent Technologies, Aachen* (Verlag der Augustinus Buchhandlung Aachen) : 1342–1348.

[67] J.C. Fodor and T. Keresztfalvi (1994), Characterization of the Hamacher family of t-norms, *Fuzzy Sets and Systems* (to appear).

[68] J.C. Fodor and S. Ovchinnikov, On aggregation of T-transitive fuzzy binary relations, *Fuzzy Sets and Systems* (to appear).

[69] J.C. Fodor and M. Roubens (1991a), Fuzzy preference modelling — an overview, *Annales Univ. Sci. Budapest., Sectio Computatorica,* **XII** : 93–100.

[70] J.C. Fodor and M. Roubens (1991b), Aggregation of choice functions, unpublished.

[71] J.C. Fodor and M. Roubens (1992), Aggregation of strict preference relations in MCDM procedures, in: V. Novák et al. (Eds.), *Fuzzy Approach to Reasoning and Decision-Making*, (Academia, Praha) : 163–171.

[72] J.C. Fodor and M. Roubens (1993), Preference modelling and aggregation procedures with valued binary relations, in : B. Lowen and M. Roubens (Eds.), *Fuzzy Logic : State of the Art*, (Kluwer Academic Press, Dordrecht) : 29-38

[73] J.C. Fodor and M. Roubens (1994a), Valued preference structures, *European Journal of Operational Research*, to appear.

[74] J.C. Fodor and M. Roubens (1994b), Structure of transitive valued binary relations, *Prépublication n° 94.001*, (Institut de Mathématique, Université de Liège, Belgium), to be published in IEEE Transactions on Fuzzy Systems.

[75] J.C. Fodor and M. Roubens (1994c), On meaningfulness of means, *Prépublication n° 94.002*, (Institut de Mathématique, Université de Liège, Belgium).

[76] M. J. Frank (1979), On the simultaneous associativity of $F(x,y)$ and $x + y - F(x,y)$, *Aeq. Math.,* **19** : 194–226.

[77] L. W. Fung and K. S. Fu (1975), An axiomatic approach to rational decision-making in a fuzzy environment, in: L.A. Zadeh, K.S. Fu, K. Tanaka and M. Shimura (Eds.), *Fuzzy Sets and their Applications to Cognitive and Decision Processes* (Academic Press, New York–San Francisco–London) : 227–256.

[78] J.A. Goguen (1967), L-fuzzy sets, *J. Math. Anal. Appl.,* **18** : 145–174.

[79] J. A. Goguen (1969), The logic of inexact concepts, *Synthese,* **19** : 325–373.

[80] M. Grabish (1992), Applications des mesures floues et intégrales floues, *Actes des Deuxièmes Journées Nationales sur les Applications des Ensembles Flous*, (Nîmes) : 139-147.

[81] M. Grabish (1993), On the use of fuzzy integral as a fuzzy connective, *Proceedings of the Second IEEE International Conference on Fuzzy Systems*, San Francisco : 213-218.

[82] M. Grabisch, T. Murofushi and M. Sugeno (1992), Fuzzy measure of fuzzy events defined by fuzzy integrals, *Fuzzy Sets and Systems*, **50** : 293–313.

[83] M. Grabish and M. Sugeno (1992), Multi-attribute classification using fuzzy integral, *Proceedings of the IEEE International Conference on Fuzzy Systems*, San Diego : 47-54.

[84] P.R. Halmos (1950), *Measure Theory,* (Van Nostrand, New York).

[85] H. Hamacher (1978), *Über logische Aggrationen nichtbinär explizierter Entscheidungskriterien; Ein axiomatischer Beitrag zur normativen Entscheidungstheorie,* (Rita G. Fischer Verlag, Frankfurt).

[86] H. Hashimoto (1983a), Canonical form of a transitive fuzzy matrix, *Fuzzy Sets and Systems,* **11** : 157–162.

[87] H. Hashimoto (1983b), Convergence of powers of a fuzzy transitive matrix, *Fuzzy Sets and Systems,* **9** : 153-160.

[88] U. Höhle (1982), Integration with respect to fuzzy measures, *Proc. IFAC Symposium on Theory and Application of Digital Control*, New Delhi : 35-37.

[89] U. Höhle (1988), Quotients with respect to similarity relations, *Fuzzy Sets and Systems*, **27** : 31–44.

[90] U. Höhle and L. N. Stout (1991), Foundations of fuzzy sets, *Fuzzy Sets and Systems*, **40** : 257–296.

[91] C.-L. Huang and M.-J. Liu (1987), *Group Decision Making under Multiple Criteria* (Lectures Notes in Economics and Mathematical Systems 281, Springer-Verlag, Berlin).

[92] Y. Kamen and S. Ovchinnikov (1993), Meaningful means and ordered sets, Working paper.

[93] A. Kandel and W.J. Byatt (1978), Fuzzy sets, fuzzy algebra, and fuzzy statistics, Proceedings of IEEE 66 : 1619-1639.

[94] R. Keeney and H. Raiffa (1976), *Decisions with Multiple Objectives; Preferences and Value Trade-offs*, (J. Wiley, New York).

[95] J.G. Kemeny (1959), Mathematics without numbers, *Daedalus,* **88** : 577-591.

[96] J.G. Kemeny and J.C. Snell (1962), *Mathematical Models in the Social Sciences*, (Blaisdell, New York).

[97] S. Kim (1988), On the possible scientific laws, *Technical Report RRR-22-88*, (Rutgers Center for OR, Rutgers University, New Brunswick, N.J).

[98] T. Kitagawa (1934), On some class of weighted means, *Proceedings of the Physico-Mathematical Society of Japan*, **16** : 117–126.

[99] L. Kitainik (1993), *Fuzzy Decision Procedures with Binary Relations*, (Kluwer, Dordrecht).

[100] L. Kitainik, S. Orlovski and M. Roubens (1993), Expert Assistant FICCAS : Fuzzy Information, Cluster Choice and Knowledge Acquisition System, *Fuzzy Sets and Systems,* **58** : 105–118.

[101] E. P. Klement (1981a), An axiomatic theory of operations on fuzzy sets, (Institut für Mathematik, Johannes Kepler Universität Linz, Institutsbericht 18).

[102] E. P. Klement (1981b), Operations on fuzzy sets and fuzzy numbers related to triangular norms, in: *Proceedings of the Eleventh International Symposium on Multiple-valued Logic*, (Oklahoma) : 218–225.

[103] W. Kolodziejczyk (1987), Canonical form of a strongly transitive fuzzy matrix, *Fuzzy Sets and Systems,* **22** : 297–302.

[104] A.N. Kolmogoroff (1930), Sur la notion de la moyenne, *Accad. Naz. Lincei Mem. Cl. Sci. Fis. Mat. Natur. Sez.,* **12** : 388-391.

[105] H. L. Larsen and R. R. Yager (1990), Efficient computation of transitive closure, *Fuzzy Sets and Systems,* **38** : 81–90.

[106] B. Leclerc (1984), Efficient and binary consensus functions on transitively valued relations, *Mathematical Social Sciences*, **8** : 45-61.

[107] A. Levenglick and H.P. Young (1978), A consistent extension of Condorcet election principle, *SIAM J. Applied Math.,* **35** : 285-360.

[108] S. Li (1990), The simplest method of ascending value to find fuzzy transitive closure, *Fuzzy Sets and Systems,* **38** : 91–96.

[109] C. H. Ling (1965), Representation of associative functions, *Publ. Math. Debrecen,* **12** : 189–212.

[110] R. Lowen (1978), On fuzzy complements, *Inform. Sci.,* **14** : 107–113.

[111] J. Łukasiewicz (1970), *Selected Works* (Reidel, Amsterdam).

[112] R. D. Luce (1956), Semiorders and a theory of utility discrimination, *Econometrica,* **24** : 178–191.

[113] R. D. Luce (1958), A probabilistic theory of utility, *Econometrica,* **26** : 193–224.

[114] R.D. Luce (1964), A generalization of a theorem of dimensional analysis, *J. Math. Psycho.,* **1** : 278–284.

[115] F. Marcotorchino and P. Michaud (1979), *Optimisation en Analyse Ordinale des Données*, (Masson, Paris).

[116] J.L. Marichal (1992), Caractérisation de méthodes de rangement basées sur des relations binaires valuées, Mémoire our l'obtention du grade de licencié en sciences mathématiques, Université de Liège.

[117] J.L. Marichal and M. Roubens (1993), Characterization of some stable aggregation functions, in : *Proc. Intern. Conf. on Industrial Engineering and Production Management*, (Mons, Belgium) : 187–196.

[118] K. Menger (1951), Probabilistic theories of relations, *Proc. Nat. Acad. Sci. (Math.),* **37** : 178–180.

[119] K. Menger (1979), *Selected Papers in Logic and Foundations, Didactics, Economics,* (Reidel, Dordrecht).

[120] N.R. Miller (1980), A new solution set for tournaments and majority voting : further graph-theoretical approaches to the theory of voting, *Amer. J. Polit. Sci.*, **24** : 68–96.

[121] M. Miyakoshi and M. Shimbo (1985), Solutions of composite fuzzy relational operations with triangular norms, *Fuzzy Sets and Systems*, **16** : 53–63.

[122] B. Monjardet (1984), Probabilistic consistency, homogeneous families of relations and linear $\wedge$-relations, in: E. De Greef, J. van Buggenhaut (Eds.), *Trends in Mathematical Psychology*, (Elsevier Science Publishers, Amsterdam) : 271-281.

[123] B. Monjardet (1988), A generalization of probabilistic consistency: linearity conditions for valued preference relations, in: J. Kacprzyk and M. Roubens (Eds), *Non-Conventional Preference Relations in Decision Making*, (Springer-Verlag, Berlin) : 36–53.

[124] F.J. Montero and J. Tejada (1986), Some problems on the definition of fuzzy preference relations, *Fuzzy Sets and Systems*, **20** : 45-53.

[125] F.J. Montero and J. Tejada (1988), A necessary and sufficient condition for the existence of Orlovski's choice set, *Fuzzy Sets and Systems*, **26** : 121–126.

[126] P. S. Mostert and A. L. Shields (1957), On the structure of semigroups on a compact manifold with boundary, *Ann. of Math.,* **65** : 117–143.

[127] T. Murofushi and M. Sugeno (1991), Fuzzy t-conorm integrals with respect to fuzzy measures: generalization of Sugeno integral and Choquet integral, *Fuzzy Sets and Systems*, **42** : 51–57.

[128] M. Nagumo (1930), Über eine Klasse der Mittelwerte, *Japanese Journal of Mathematics*, **6** : 71-79.

[129] K. Nakamura (1986), Preference relations on a set of fuzzy utilities, *Fuzzy Sets and Systems*, **20** : 147-162.

[130] S.A. Orlovski (1978), Decision-making with a fuzzy preference relation, *Fuzzy Sets and Systems,* **1** : 155-167.

[131] S.A. Orlovski (1983), Problems of Decision-Making with Fuzzy Information, Working Paper, (International Institute for Applied Systems Analysis, Laxenburg, Austria).

[132] D.K. Osborne (1970), Further extensions of a theorem of dimensional analysis, *J. Math. Psycho.,* **7** : 236–242.

[133] S. Ovchinnikov (1980), Involutions in fuzzy set theory, *Stochastica,* **IV** : 227–231.

[134] S. Ovchinnikov (1981), Structure of fuzzy binary relations, *Fuzzy Sets and Systems,* **6** : 169–195.

[135] S. Ovchinnikov (1983), General negations in fuzzy set theory, *J. Math. Anal. Appl.,* **92** : 234–239.

[136] S. Ovchinnikov (1984), Representations of transitive fuzzy relations, in: H.J. Skala, S. Termini and E. Trillas (Eds.), *Aspects of Vagueness,* (D. Reidel Publishing Company, Dordrecht) : 105-118.

[137] S. Ovchinnikov (1986), On the transitivity property, *Fuzzy Sets and Systems,* **20** : 241–243.

[138] S. Ovchinnikov (1991), Similarity relations, fuzzy partitions, and fuzzy orderings, *Fuzzy Sets and Systems,* **40** : 107–126.

[139] S. Ovchinnikov (1992a), On aggregation of max-product transitive valued binary relations, submitted.

[140] S. Ovchinnikov (1992b), Aggregating transitive fuzzy preference relations, submitted.

[141] S. Ovchinnikov and T. Riera (1982), On fuzzy classifications, in : R. Yager (Ed.), *Fuzzy Sets and Possibility Theory : Recent Developments,* (Pergamon Press, New York) : 120–132.

[142] S. Ovchinnikov and M. Roubens (1991), On strict preference relations, *Fuzzy Sets and Systems,* **43** : 319–326.

[143] S. Ovchinnikov and M. Roubens (1992), On fuzzy strict preference, indifference and incomparability relations, *Fuzzy Sets and Systems,* **47** : 313-318 and **49** : 15-20.

[144] A. B. Paalman – De Miranda (1964), *Topological Semigroups,* (Mathematisch Centrum, Amsterdam).

[145] P. Perny (1992), Modélisation, agrégation et exploitation des préférences floues dans une problématique de rangement, Thèse pour l'obtention du titre de Docteur en méthodes scientifiques de gestion, Ph.D. Thesis, Paris-Dauphine University.

[146] P. Perny and B. Roy (1992), The use of fuzzy outranking relations in preference modelling, *Fuzzy Sets and Systems,* **49** : 33–53.

[147] M. Pirlot (1993), A characterization of "min" as a procedure for exploiting valued preference relations and related results, Working Paper, (Faculté Polytechnique de Mons, Belgium).

[148] N. Rescher (1969), *Many-valued Logic,* (McGraw-Hill, New York).

[149] F.S. Roberts (1971), Homogeneous families of semiorders and the theory of probabilistic consistency, *J. Math. Psycho.,* **8** : 248–263.

[150] F.S. Roberts (1976), *Discrete Mathematical Models*, (Prentice-Hall, New Jersey).

[151] F.S. Roberts (1979), *Measurement Theory with Applications to Decision-making, Utility and the Social Sciences,* (Addison-Wesley Pub., Reading).

[152] M. Roubens (1989), Some properties of choice functions based on valued binary relations, *European Journal of Operational Research,* **40** : 309-321.

[153] M. Roubens (1990), Inequality constraints between fuzzy members and their use in mathematical programming, in R. Slowinski and I. Teghem Jr. (Eds.), *Stochastic versus Fuzzy Approaches to Multiobjective Mathematical Programming under Uncertainty*, (Kluwer, Dordrecht) : 321-330.

[154] M. Roubens and Ph. Vincke (1985), *Preference Modelling,* (Springer-Verlag, Berlin).

[155] M. Roubens and Ph. Vincke (1988), Fuzzy possibility graphs and their application to ranking fuzzy numbers, in: J. Kacprzyk and M. Roubens (Eds.), *Non-Conventional Preference Relations in Decision Making,* (Springer-Verlag, Berlin) : 119-128.

[156] B. Roy (1968), Classement et choix en présence des points de vue multiples (la méthode ELECTRE), *Revue Française d'Informatique et de Recherche Opérationnelle,* **8** : 57-75.

[157] B. Roy (1978), Un algorithme de classements fondé sur une représentation floue des préférences en présence de critères multiples, *Cahiers du Centre d'Etudes de Recherche Opérationnelle,* Bruxelles, vol. **20** : 2-24.

[158] B. Roy and Ph. Vincke (1984), Relational systems of preference with one or more pseudo-criteria : some new concepts and results, *Management Science,* **30** : 1323-1335.

[159] B. Roy and P. Vincke (1987), Pseudo-orders : definition, properties and numerical representation, *Mathematical Social Sciences,* **14** : 263-274.

[160] B. Scheweizer and A. Sklar (1983), *Probabilistic Metric Spaces,* (North-Holland, Amsterdam).

[161] D.S. Scott and P. Suppes (1958), Foundation aspects of theories of measurement, *J. Symb. Logic*, **23** : 113-128.

[162] A.K. Sen (1970), *Collective Choice and Social Welfare,* (Holden-Day, Inc. San Francisco).

[163] W. Silvert (1979), Symmetric summation: A class of operations on fuzzy sets, *IEEE Trans. Systems, Man Cybernet.,* **9** : 667–659.

[164] J.M. Skalka, D. Bouyssou and D. Vallée (1992), ELECTRE III et IV : Aspects méthodologiques et guide d'utilisation, Documents du LAMSADE, Université de Paris Dauphine, **25** (4ème édition).

[165] P. Slater (1961), Inconsistencies in a schedule of paired comparisons, *Biometrika,* **48** : 303-312.

[166] P. Smets and P. Magrez (1987), Implication in fuzzy logic, *Int. J. of Approximate Reasoning,* **1** : 327–347.

[167] S.S. Stevens (1946), On the Theory of Scales of Measurement, *Science*, **103** : 677–680.

[168] M. Sugeno (1975), *Theory of fuzzy integrals and its applications*, Ph.D.Thesis, Tokyo Institute of Technology, Tokyo.

[169] M. Sugeno (1977), Fuzzy measures and fuzzy initegrals: a survey, in: M.M. Gupta, G.N. Saridis and B.R. Gaines (Eds.), *Fuzzy Automata and Decision Processes*, (North-Holland, Amsterdam) : 89-102.

[170] M. Sugeno and T. Murofushi (1987), Choquet integral as an integral form for a general class of fuzzy measures, *Proceedings of the 2nd IFSA Congress*, Tokyo : 408-411.

[171] P. Suppes (1959), Measurement, empirical meaningfulness and three-valued logic, in : C.W. Churchman and P. Ratoosh (Eds.), *Measurement : Definitions and Theories*, (Wiley, New York) : 129–143.

[172] M.G. Thomason (1977), Convergence of power of a fuzzy matrix, *Math. Anal. Appl.,* **57** : 476-480.

[173] R.H. Tong and P.P. Bonissone (1980), A Linguistic Approach to Decision making with Fuzzy Sets, *IEEE Transactions on Systems, Man and Cybernetics*, **10** : 716-723.

[174] E. Trillas (1979), Sobre funciones de negación en la teoría de conjuntos difusos, *Stochastica,* **III** : 47–60.

[175] E. Trillas and L. Valverde (1981), On some functionally expressable implications for fuzzy set theory, in: Proc. of the 3rd Intern. Seminar on Fuzzy Set Theory (Linz, Austria) : 173–190.

[176] E. Trillas and L. Valverde (1985), On implication and indistinguishability in the setting of fuzzy logic, in: J. Kacprzyk and R.R Yager (Eds.), *Management Decision Support Systems using Fuzzy Sets and Possibility Theory,* (Verlag TÜV Rheinland, Köln) : 198–212.

[177] L. Valverde (1985), On the structure of F-indistinguishability operators, *Fuzzy Sets and Systems,* **17** : 313–328.

[178] Ph. Vincke (1988), P-Q-I-preferences structures, in: J. Kacprzyk and M. Roubens (Eds.), *Non-conventional Preference Relations in Decision Makinig,* (Springer-Verlag, Berlin).

[179] Ph. Vincke (1991), Exploitation d'une relation non valuée dans une problématique de rangement complet, Documents du LAMSADE, Université de Paris Dauphine, **62**.

[180] Ph. Vincke (1992a), *Multicriteria Decision Aid*, (J. Wiley, New York).

[181] Ph. Vincke (1992b), Exploitation of a crisp relation in a ranking problem, submitted.

[182] S. Weber (1983), A general concept of fuzzy connectives, negations and implications based on t-norms and t-conorms, *Fuzzy Sets and Systems,* **11** : 115–134.

[183] S. Weber (1984), $\perp$-decomposable measures and integrals for archimedean t-conorms $\perp$, *J. Math. Anal. Appl.*, **101** : 114–138.

[184] Wu Wangming (1992), Representation of Archimedean implications, (unpublished).

[185] R.R. Yager (1977), Multiple objective decision making using fuzzy sets, *Intern. J. of Man-Machines Studies,* **9** : 375-382.

[186] R. R. Yager (1980), On the measure of fuzziness and negation II. Lattices, *Information and Control,* **44** : 236–260.

[187] R.R. Yager (1988), On ordered weighted averaging aggregation operators in multicriteria decisionmaking, *IEEE Trans. on Systems, Man and Cybernetics*, **18** : 183–190.

[188] R. R. Yager (1991), Connectives and quantifiers in fuzzy sets, *Fuzzy Sets and Systems,* **40** : 39–75.

[189] R. T. Yeh (1973), Toward an algebraic theory of fuzzy relational systems, *Proc. Int. Congr. Cybern. Namur* : 205–223.

[190] R.T. Yeh and S.Y. Bang (1975), Fuzzy relations, fuzzy graphs, and their applications in clustering analysis, in : L.A. Zadeh, K.S. Fu, K. Tanaka and M. Shimura (Eds.), *Fuzzy Sets and Their Applications to Cognitive and Decision Processes,* (Academic Press, New York) : 125–149.

[191] L. Zadeh (1965), Fuzzy sets, *Information and Control,* **8** : 338–353.

[192] L. A. Zadeh (1971), Similarity relations and fuzzy orderings, *Inf. Sci.,* **3** : 177–200.

[193] L.A. Zadeh (1978), Fuzzy sets as a basis for a theory of possibility, *Fuzzy Sets and Systems*, **1** : 3–28.

[194] L. Zadeh (1983), A computational approach to fuzzy quantifiers in natural language, *Comput. Math. Appl.,* **9** : 149–184.

[195] S. Zahariev (1987), An approach to group choice with fuzzy preference relations, *Fuzzy Sets and Systems,* **22**: 203-213.

[196] C. Zahn (1971), Graph theoretical methods for detecting and describing gestalt clusters, *IEEE Trans. Comput.*, **20** : 66–86.

[197] H.-J. Zimmermann and P. Zysno (1983), Decisions and evaluations by hierarchical aggregation of information, *Fuzzy Sets and Systems,* **10** : 243-260.

Index

THEORY AND DECISION LIBRARY

SERIES D: SYSTEM THEORY, KNOWLEDGE ENGINEERING AND PROBLEM SOLVING

1. E.R. Caianiello and M.A. Aizerman (eds.): *Topics in the General Theory of Structures.* 1987 ISBN 90-277-2451-2
2. M.E. Carvallo (ed.): *Nature, Cognition and System I.* Current Systems-Scientific Research on Natural and Cognitive Systems. With a Foreword by G.J. Klir. 1988 ISBN 90-277-2740-6
3. A. Di Nola, S. Sessa, W. Pedrycz and E. Sanchez: *Fuzzy Relation Equations and Their Applications to Knowledge Engineering.* With a Foreword by L.A. Zadeh. 1989 ISBN 0-7923-0307-5
4. S. Miyamoto: *Fuzzy Sets in Information Retrieval and Cluster Analysis.* 1990 ISBN 0-7923-0721-6
5. W.H. Janko, M. Roubens and H.-J. Zimmermann (eds.): *Progress in Fuzzy Sets and Systems.* 1990 ISBN 0-7923-0730-5
6. R. Slowinski and J. Teghem (eds.): *Stochastic versus Fuzzy Approaches to Multiobjective Mathematical Programming under Uncertainty.* 1990 ISBN 0-7923-0887-5
7. P.L. Dann, S.H. Irvine and J.M. Collis (eds.): *Advances in Computer-Based Human Assessment.* 1991 ISBN 0-7923-1071-3
8. V. Novák, J. Ramík, M. Mareš, M. Černý and J. Nekola (eds.): *Fuzzy Approach to Reasoning and Decision-Making.* 1992 ISBN 0-7923-1358-5
9. Z. Pawlak: *Rough Sets.* Theoretical Aspects of Reasoning about Data. 1991 ISBN 0-7923-1472-7
10. M.E. Carvallo (ed.): *Nature, Cognition and System II.* Current Systems-Scientific Research on Natural and Cognitive Systems. Vol. 2: On Complementarity and Beyond. 1992 ISBN 0-7923-1788-2
11. R. Słowiński (ed.): *Intelligent Decision Support.* Handbook of Applications and Advances of the Rough Sets Theory. 1992 ISBN 0-7923-1923-0
12. R. Lowen and M. Roubens (eds.): *Fuzzy Logic.* State of the Art. 1993 ISBN 0-7923-2324-6
13. L. Kitaimik: *Fuzzy Decision Procedures with Binary Relations.* Toward a Unified Theory. 1993 ISBN 0-7923-2367-X
14. J. Fodor and M. Roubens: *Fuzzy Preference Modelling and Multicriteria Decision Support.* 1994 ISBN 0-7923-3116-8

KLUWER ACADEMIC PUBLISHERS – DORDRECHT / BOSTON / LONDON

MIX
Papier aus verantwortungsvollen Quellen
Paper from responsible sources
FSC® C105338

If you have any concerns about our products,
you can contact us on
ProductSafety@springernature.com

In case Publisher is established outside the EU,
the EU authorized representative is:
Springer Nature Customer Service Center GmbH
Europaplatz 3, 69115 Heidelberg, Germany

Printed by Libri Plureos GmbH
in Hamburg, Germany